Showing resistance

Manchester University Press

To buy or to find out more about the books currently available in this series, please go to: https://manchesteruniversitypress.co.uk/series/studies-in-design-and-material-culture/

STUDIES IN DESIGN & MATERIAL CULTURE

Showing resistance

Propaganda and Modernist exhibitions in Britain, 1933–53

Harriet Atkinson

Manchester University Press

Published by Manchester University Press
Oxford Road, Manchester, M13 9PL
www.manchesteruniversitypress.co.uk

British Library Cataloguing-in-Publication Data
A catalogue record for this book is available from the British Library

ISBN 978 1 5261 5741 6 hardback

First published 2024

Typeset
by Cheshire Typesetting Ltd, Cuddington, Cheshire

For John and Jane

Contents

Illustrations

Every reasonable attempt has been made to obtain permission to reproduce copyright images. If any proper acknowledgement has not been made, copyright holders are invited to contact the author via Manchester University Press.

Preface

'There are no impartial observers. The battlefront is everywhere', announced American actor Paul Robeson at a major political meeting held in London in 1937, beseeching the audience to engage with the fight against fascism.[1] Robeson's impassioned speech underlines the major theme and impetus of this book, describing, as it does, the spatial and imaginative expansion of culture implied in an omnipresent battlefront. This is the imagination that nurtured exhibitions as propaganda across a multiplicity of sites and spaces, for the two decades that are the focus of this book.

This book has been in formation for over a decade. Its focus on exhibitions as propaganda started to develop in the aftermath of the global financial crash of 2008 and gained further momentum after the election of Donald Trump as President of the United States of America in 2016, when democratic governments increasingly came to be accused of transmitting 'fake news', 'alternative facts' and editing news content to manipulate its meanings.[2] In response, activist movements were in search of new ways to visualise their intersecting causes (from women's rights to Black Lives Matter, climate change and anti-capitalism).[3] All of this created a renewed interest in histories and forms of protest, which became clearer from my conversations with students, friends and colleagues. This book responds to this interest in visual, material and design cultures of the mid-twentieth century, considering how and why artists and designers made public political interventions through exhibitions from the 1930s to the 1950s, and giving a long view of the uses of exhibitions as propaganda within democracies.

While, during the last decade, protest cultures were manifest in the spectacle of processions, placards and posters reported regularly on the news, tragedies like London's Grenfell Tower fire in 2017 (when more than seventy people died in a tower block blaze, avoidable had there been sufficient government investment) provoked an ad hoc exhibition to be thrown up under London's major West Way road.[4] This display memorialised the tragedy and, more importantly, informed the passing public of the injustices meted out by a Conservative local council whose penny-pinching approach to social housing provision had had catastrophic consequences. Such spontaneous displays have become regular responses to public tragedies, such as 9/11 and 7/7.[5]

My doctoral study and the book that came of it focused on government engagements with British design in the immediate postwar period, through the multiple events of the Festival of Britain, held across Britain in 1951. This was an interest sparked by working in close proximity at the Department for Culture, Media and Sport with the vexed Millennium Dome project of 2000.[6] My recognition, when writing that earlier book, that the forms and formations of the Festival of Britain were anticipated in exhibitions from the 1930s, made manifest in British government work during the Second World War and in the emerging welfare state, became ever clearer through analysis, yet there were no published accounts of the period or of this type of exhibition. This book acts as the prequel to my Festival work. But instead of considering formations of nationhood with and through exhibitions or how institutions and official bodies used exhibitions to amplify their hegemonic positions, this book also considers exhibitions as an urgent form of communications media that gave a voice and a platform to groups on the margins.

The focus of this book was crystallised when I heard Professor Fred Turner introducing his idea of 'the democratic surround' at California College of the Arts in 2015.[7] During his talk, about artists and intellectuals in postwar America who had developed new models of media and collaboration in response to the rise of fascist and communist politics, I recognised a parallel and, as yet, untold story about the creation of multi-image environments, shaped and influenced by refugee artists in Britain.

Ever the slow academic, this book has been imagined, researched and written over a long period punctuated by childbirth, shattering elections, Brexit grief, in snatched moments on commuter trains, in cafés and in and out of home-schooling my children whilst in lockdown from March 2020. The Coronavirus pandemic gave the subject of my research a strong resonance when exhibition-going, an activity I had always taken for granted, became impossible as museums and galleries were closed, causing me intense reflection on the forms, purposes and meanings of exhibitions past, present and future. The archives and libraries providing the meat of this book were closed, meaning I had to find new ways of researching. With none of the sources or spaces I was used to working in available and without the benefit of regular conversation with friends and colleagues, I had to re-learn how to write as a process of solitary thinking. Philosopher

Walter Benjamin's words on the three steps of writing – 'Work on good prose has three steps: a musical stage when it is composed, an architectonic one when it is built, and a textile one when it is woven' – were endlessly reassuring, with their understanding of slow and painstaking work.[8] Given all that has happened since the spark of the idea for this book was lit, it seems astonishing that it is in the world at last.

Notes

1 *Paul Robeson Speaks* (Secaucus, NJ: Citadel, 1978).
2 Brian McNair, *Fake News: Falsehood, Fabrication and Fantasy in Journalism* (Abingdon: Routledge, 2021).
3 My recent podcast series *Graphic Interventions* (2021) illuminated this: https://open.spotify.com/show/2slSCoQJ22bVjATCYP0sqj. It includes interviews with artists and designers who have made recent political interventions through their work including Paris 68 Redux, Ed Hall, Conversations from Calais, See Red Women's Workshop, Protest Stencil, OOMK and Spelling Mistakes Cost Lives.
4 'Grenfell Tower: What Happened', *BBC News*, 29 October 2019: www.bbc.co.uk/news/uk-40301289 (accessed 13 May 2023).
5 Memorials for 9/11 are discussed in Margaret Olin's *Touching Photographs* (Chicago: University of Chicago, 2011) and 7/7 discussed in Gillian Rose's *Doing Family Photography: The Domestic, the Public and the Politics of Sentiment* (Abingdon: Routledge, 2010).
6 Harriet Atkinson, *The Festival of Britain: A Land and Its People* (London: I.B. Tauris, 2012).
7 Fred Turner was keynote speaker at the Design History Society's annual conference, '"How we live, and How we might live": Design and the Spirit of Critical Utopianism', 11–13 September 2015 at California College of the Arts, San Francisco, California. He was discussing his book *The Democratic Surround: Multimedia & American Liberalism from World War II to the Psychedelic Sixties* (Chicago and London: University of Chicago Press, 2013).
8 Walter Benjamin, *One-Way Street and Other Writings* (London: Harcourt Brace Jovanovich, 1978), p. 61; first published 1925–26.

Acknowledgements

In the long process of making this book, I have accrued many debts of gratitude. Undoubtedly the largest I owe to Jeremy Aynsley: my PhD supervisor, mentor and colleague for twenty years, firstly at the Royal College of Art and, latterly, at the University of Brighton. The completion of this manuscript coincides with Jeremy's retirement and I want to take this opportunity to acknowledge his enormous academic generosity to me, as well as the many others who have been fortunate enough to work with him.

This book has grown through countless conversations with friends and colleagues at the University of Brighton's School of Humanities and Social Science; Centre for Design History; Centre for Memory, Narrative and Histories; and Centre for Spatial, Environmental and Cultural Politics; and my magnificent students at the University of Brighton across History of Art and Design and Graphic Design and Illustration. I am so grateful to so many at my university for their capacity to nurture research despite the increasingly hostile financial and regulatory climate in which post-'92 universities like mine function. I want to mention some key people in research who have helped me: Ingrid Pugh, Stuart Hedley, Jo Allen, Lauren Greenslate, Andrew Church, Andrea McKoy, Rupert Brown, Emmy Sale, Chris Matthews, Laura Shockley and the Brighton Bid Support and Post-Award teams.

Several generous people read parts of this manuscript while in progress: Jeremy Aynsley, Monica Bohm-Duchen, Sue Breakell, Cheryl Buckley, Elizabeth Darling, Henry Irving, Zeina Maasri, Annebella Pollen and Claire Wintle. To all of them I owe a huge debt of thanks for improving it immeasurably (although, it goes without saying, all remaining errors and shortcomings are my own responsibility). I am indebted to the research and insights of many people and have enjoyed discussing this project, and the wider research interests in which the work sits, in many places over the years, from conference and seminar rooms to pubs, cafés, libraries and kitchens, over Zoom and by email, with Leah Armstrong, Nicola Ashmore, Richard Atkinson, Jenny Black, Valeria Carullo, Verity Clarkson, Becky Conekin, David Crowley, Clare Cumberlidge, Barry Curtis, Liz Farrelly, Andy Friend, Lisa Godson, Robert Harland, the late Clifford Hatts, Zoe Hendon, Max Henrion, Josie Kane, Jessica Kelly, Lucy Kimbell, Yunah Lee, Grace Lees-Maffei, Gill Levin, Sarah Lichtman, Jo Littler, Aasiya Lodhi, Anna Lovell, Jessica Lovell, Sam Lyons, Anne Massey, David Matless, Tania Messell, Lynda Morris, Sorcha O'Brien, Peter Peri, Megha Rajguru, Will Rea, Irit Rogoff, Michael Rosen, Cat Rossi, Linda Sandino, Adrian Shaughnessy, Jana Scholze, Daniel Snowman, Ceri Smith, Dora Souza Dias, Penny Sparke, Deborah Sugg Ryan, Annie Sutherland, Lou Taylor, Claudia Treacher, Martha Turland, Michael Tymkiw, Vanessa Vanden Berghe, Gillian Whiteley, Lesley Whitworth and Jonathan Woodham.

At Manchester University Press (MUP), effusive thanks to my insightful and patient editors Emma Brennan and Alun Richards; to the editors of the Design and Material Culture series, originally Christopher Breward and James Ryan, subsequently Elizabeth Currie, Sally-Anne Huxtable, Livia Lazzaro Rezende and Wessie Ling. Series founder Paul Greenhalgh's 1988 book *Ephemeral Vistas*, which made linkages across exhibitions over several decades, has had a talismanic quality whenever I have questioned my own wisdom in setting out to unite a disparate series of exhibitions in one account. I can't express warmly enough my gratitude to MUP's brilliant, anonymous peer reviewers, whose patient and generous advice at the book proposal and draft stages was utterly invaluable in shaping and improving the book. Warm thanks, too, to Ellie Howard for excellent picture research.

I have been helped and advised by archivists, curators and librarians at New York Public Library; National Archives Washington; National Archives London; Design Archives at University of Brighton (in addition to those already mentioned, also by Sirpa Kutilainen and Jen Grasso); Victoria & Albert Museum's Archives of Art and Design; Victoria & Albert Museum's RIBA collection; Caterina Tiezzi at the London Transport Museum; RIBA library and reading rooms; Caitlin Condell at the Cooper Hewitt Museum; Ian Carter and John Delaney at the Imperial War Museum; Whitechapel Gallery archives; Tate Archives; Henry Moore Institute; Peter Suschitzky; Kurt Kaindl and Stefanie Pirker from FOTOHOF, Salzburg; Churchill Archives; Museum of Domestic Architecture. I have been sustained through regular visits to the British Library (BL) – which is amongst my favourite places in the world – and am grateful to the stimulating conversation of my brilliant fellow BL worker Rachel Wrangham, to the BL's consistently friendly staff, including expert baristas at Origin coffee, whose chat and excellent coffee punctuated long and sometimes lonely research days, and to the librarians who smiled sympathetically whilst issuing another stack of books.

I have presented fragments of this material at various events over recent years: 'Reconfiguring Relations: Britain and the Bauhaus', Tate Britain, October 2019; 'Lessons to Learn? Past Design Experiences and Contemporary Design Practices', ICDHS conference Zagreb, October 2020; 'The Writer as Psychological Warrior: Intellectuals, Propaganda, and Modern Conflict', Durham University, July 2021; 'Media Building: Architecture, Communications and the Built Environment from Fleet Street to Facebook', Salford University, July 2021; 'Lloyd Corporation *Today's gift is tomorrow's commodity*', Brighton CCA, March 2022; 'Exhibitions, New Nations and the Human Factor, 1873–1939', Institut national d'histoire de l'art, Paris, April 2022; 'Art Exhibitions as Intersections in Post War Europe', Södertörn University, Stockholm, May 2022; and 'Exhibitions as Interiors?', Modern Interiors Research Centre at Kingston University, February 2023. Related elements have been published as 'Exhibitions as Political "Demonstrations": Artists International Association's For Liberty Exhibition, London 1943', *ICDHS Conference Proceedings*, 2020 and 'Visual storytelling in the Ministry of Information's wartime exhibitions' in Beatriz Lopez, James Smith and Guy Ward (eds), *British Writing, Propaganda, and Cultural Diplomacy in the Second World War and Beyond* (London: Bloomsbury Academic, 2024). I am indebted to the University of Brighton for awarding the Rising Stars grant in 2017–18 that allowed me to develop the seed of an idea that led to this book and am deeply grateful for the Arts and Humanities Research Council Leadership Fellowship Project Reference: AH/S001883/1, during which I wrote this book. As part of that project, I was lucky to have an excellent and insightful project advisory group comprising Professor Jeremy Aynsley (Chair, University of Brighton), Dr Frank Gray (Director, Screen Archives South East), Professor Annebella Pollen (University of Brighton), Carla Mitchell (Development Director, Four Corners film), Dr Rafal Niemojewski (Director, Biennial Foundation) and Sue Breakell (Archive Leader, University of Brighton Design Archives). As part of my AHRC Fellowship I was fortunate to collaborate with Four Corners London on making the short documentary film *Art on the Streets*, which draws its ideas from the research in this book, working with Jane Dibblin, Elena Reimeryte, Giacomo Baraldi, Nicholas D. Celano, Carla Mitchell, Hugh Hartford and Kate Bilbow.

Finally, big love to all my family and, in particular, V, J, M & S.

Abbreviations

AASTA	Association of Architects Surveyors and Technical Assistants
ABCA	Army Bureau of Current Affairs
AIA	Artists International Association
ARC	Artists' Refugee Committee
ASA	Atomic Scientists Association
ATO	Architects' and Technicians' Organisation
BBC	British Broadcasting Corporation
BIAE	British Institute of Adult Education
BIIA	British Institute of Industrial Arts
BUF	British Union of Fascists
CAI	Council for Art and Industry
CEA	Cinematograph Exhibitors' Association
CIAM	Congrès Internationaux d'Architecture Moderne
COI	Central Office of Information UK
CPRE	Council for Preservation of Rural England
DIA	Design and Industries Association
D.O.O.D.	*De Olympiade Onder Dictatuur*
EDA	Electrical Distributors' Association
EMB	Empire Marketing Board
FGLC	Free German League of Culture
GPO	General Post Office
IAFE	International African Friends of Ethiopia
ICA	Institute of Contemporary Arts, London
ILP	Independent Labour Party
IWM	Imperial War Museum
KdF	Kraft durch Freude
LCC	London County Council
LOC	Library of Congress, Washington

LPTB	London Passenger Transport Board
LTM	London Transport Museum
LWP	London Women's Parliament
MARS	Modern Architectural Research Group
MODA	Museum of Domestic Design & Architecture, Middlesex
MOI	Ministry of Information UK
MOMA	Museum of Modern Art, New York
NCCL	National Council for Civil Liberties
NJCSR	National Joint Committee for Spanish Relief
NPU	Nigerian Progress Union
OKB	Oskar-Kokoschka-Bund
OWI	Office of War Information US
PPU	Peace Pledge Union
RIBA	Royal Institute of British Architects, London
SIA	Society of Industrial Artists
TDA	Timber Development Association
TGA	Tate Gallery Archives
TNA	The National Archives, London
UERL	Underground Electric Railways Company of London Limited
UoBDA	Design Archives, University of Brighton
UNESCO	United Nations Educational, Scientific and Cultural Organisation
USAD	Union of Students of African Descent
USNA	United States National Archives
V&A	Victoria and Albert Museum
WDC	Warwick Digital Collections
WMRC	Warwick Modern Records Centre

Timeline of exhibitions

1933

- *Anti-War Exhibition*
- *Design in Modern Life*

1934

- *Exhibition of German-Jewish Artists' Work: Sculpture-Painting-Architecture*
- *Flying over the Empire*
- *New Homes for Old*
- *The Social Scene*
- *Unit One Exhibition*

1935

- *Artists Against Fascism and War*
- *Cambridge Exhibition on Fascism & War*
- *Empire's Airway*
- *New Homes for Old*
- *Noise Abatement Exhibition*

1936

- *Artists Aid Spain*
- *Do Not Pass by the Special Areas*
- *Drawings of Felicia Browne*
- *Exhibition on Working-Class Housing by the Architects and Technicians Organisation*
- *International Surrealist Exhibition*
- *New Homes for Old*
- *Peeps at the Colonial Empire*
- *Post Office Exhibition*
- *The Empire's Airway*
- *Timber through the Ages*
- *World Communications*

1937

- *Empire Tea*
- *Exhibition for Peace, Democracy and Cultural Development*
- *Exposition Internationale des Arts et des Techniques dans la Vie Moderne*
- *News Chronicle Schools Exhibition*
- *Peace Exhibition*
- *Spain: The Child and the War*
- *Spanish Exhibition*
- *The Highway Code: An Exhibition*
- *Women's Anti-War Exhibition*

1938

- *Clear Smoke from the Air*
- *Exhibition of the Elements of Modern Architecture*
- *Food for Fitness*
- *Guernica*
- *New Homes for Old*
- *Twentieth Century German Art*
- *Workers' Empire Exhibition*

1939

- *Art for the People*
- *First Group Exhibition of German, Austrian, Czechoslovakian Painters and Sculptors*
- *Guernica*
- *One Man's War against Hitler: Exhibition of John Heartfield*

1940

- 'Five to Ten' Exhibition of Children's War Pictures
- London Pride

1941

- Fire Guard
- Life Line
- Poison Gas
- Refugee Artists and Their British Friends

1942

- Aid to Russia
- Allies inside Germany
- Artists Aid Russia
- Battle for Fuel
- Comrades in Arms (A Picture of Russia at War)
- Dig for Victory
- Eastern Front
- Make Do and Mend
- Off the Ration
- The War as Seen by Children
- Twenty-Five Years of Progress: National Exhibition of Soviet Life in War and Peace

1943

- America Marches with the United Nations
- Artists Aid China Exhibition
- Artists Aid Jewry
- Back the Attack
- Cinema Exhibition
- Colonial Life
- For Liberty: Paintings on War, Peace and Freedom
- John Olsson, the Story of an Average American
- RAF Exhibition of Bomb Damage on Germany
- Red Army Week Exhibition
- The Army Exhibition: The Equipment of a Division
- The Evil We Fight
- The Nature of the Enemy
- The Two Mrs Britains
- The War as Seen by Children
- 25 Years of the USSR and the Red Army

1944

- Kurt Schwitters
- LCC Plan for London, Health Centres
- Planning Your Kitchen
- The Resistance Exhibition
- Traffic
- Young America

1945

- La Guerre et la Paix (War and Peace)
- Pacific '45
- The War in Wax
- Victory Over Japan

1946

- Atomic Energy and Uranium
- Germany Under Control

1947

- Atom Train: The Travelling Exhibition on Atomic Energy
- Builder and the State
- Hospital Domestic Aids
- The Daily Express Atomic Age Exhibition
- The Miner Comes to Town

1948

- Britain Goes Ahead
- How Goes Britain
- Raising the Standard
- The Health of the People
- The Nation and the Child

1949

- Colonial Exhibition

1951

- Festival of Britain

1952

- The Mirror and the Square

Introduction: exhibitions as 'propaganda in three dimensions'

Bombsites, shop windows and ticket halls

For the two decades from 1933, exhibitions were mounted in station ticket halls and factory workers' canteens, in the windows of high street grocery stores, evacuated department stores and on newly bulldozered bombsites. Their themes were diverse – from the nature of freedom to the culinary possibilities of the potato – and used for myriad functions. They communicated urgent, persuasive messages and practical information, intended to change people's personal behaviours. They signalled international alignments and solidarities, and acted as fundraising vehicles for important causes, inspiring social change. They gave voice to the voiceless: empowering working-class people living in poor housing conditions, recently arrived refugees, those suffering demeaning employment conditions, women taking up the struggle for equality and people exasperated by the British government's failure and inaction in the face of the rising fascist threat. This is the first extended study of such persuasive exhibitions, mounted from the interwar period to the early Cold War in Britain. It spotlights a twenty-year period – 1933 to 1953 – when artists and designers developed a form of exhibitions suited to communicating ideas and ideals, a form then taken up by politicians and bureaucrats as a means for direct political intervention.

1933 is my chosen starting point, principally because it was the date when Hitler came to power in Germany, supported by a powerful cultural propaganda machinery, including a preponderance of impactful propaganda exhibitions. 1933 is a logical starting point for this study, being the formation date of three significant artists' and architects' groups in Britain: Unit One, the Artists International Association (AIA) and the Modern Architectural Research Group (MARS), as I discuss in subsequent chapters. Each of these groups embraced exhibitions as key forms of representation for their ideas and ideals. 1953 is this book's end-point as it marked the waning of the British government's very frequent use of exhibitions to communicate policy and progress to home audiences. It was the year that the AIA – a group central to this study – dispensed with its clause requiring members to be aligned with its Leftist political aims.

This book analyses exhibitions as propaganda across many sites and scales in Britain in the years from 1933 to 1953.[1] Each chapter draws on a series of examples that allow me to present the different paradigms through which exhibitions were conceived in these years: as vehicles for projection, promotion and publicity; as activism; as manifestos; as demonstrations; as counter-arguments; as weapons of war; as solidarities; and as welfare. It looks at how exhibitions described both at the time and since as 'propaganda', focused towards communicating partial, persuasive messages, were produced by small activist groups, commercial organisations and companies, and by the British government alike.

Common elements were transmogrified across these contexts, all linked by their intention to build consent around particular ideas, issues and experiences, despite their strikingly different social and political impetuses. The term 'propaganda' is often associated now with extremist political messages, because of its use by various totalitarian regimes, but in the 1930s and 1940s the term was used to signal acts of persuasion and information-sharing considered socially beneficial and benign – a way of maintaining a peaceable and unified society. Sociologist Jacques Ellul's 1965 analysis of how propaganda operates in different political contexts underpins this study of British propaganda. Ellul contrasts the 'propaganda of agitation', as used most aggressively and conspicuously within authoritarian states, and the 'propaganda of integration', described as 'the propaganda of developed nations' and a 'propaganda of conformity', calling for 'total adherence to a society's truths and behavioural patterns'. It is 'the propaganda of integration' and conformity that is the focus of this book, rather than the 'propaganda of agitation' (although agitation inevitably shaped the wider context).[2]

In carrying persuasive messages, the exhibitions in this book can be described as 'propaganda exhibitions', created to persuade people, although the propagandists were not always in positions of power or authority. The people pivotal to this study of exhibitions beyond galleries were artists, architects, designers, scriptwriters, business managers and bureaucrats (note that curators were not amongst them). At times there was divergence, sometimes straight-out contradiction between the ideas of these makers of 'political' exhibitions – that is to say, exhibitions made by those engaging with explicit political agendas, who took particular political positions and aligned themselves with one side in struggles. They were almost overwhelmingly either on the far Left (aligned with the Communist

Party of Great Britain or the Independent Labour Party) or more centrist (in the case of exhibitions organised by the Ministry of Information and then the Central Office of Information). This did not stop them from borrowing forms across ideological lines, given that the political imperative, not stylistic orthodoxy, was the overriding concern.

I have searched hard for evidence that the organised Right in Britain, particularly Oswald Mosley and the British Union of Fascists (BUF), used exhibitions to represent their political vision but have found none accounted for in the literature, such as *Action*, the newspaper of the BUF published from 1936 to 1940.[3] The BUF did, however, use all manner of other means of communication including *Action*, monographs, pamphlets, fortnightly bulletins, radio broadcasts, uniforms, insignia and banners, picnics, youth camps, marches and rallies at major venues (both in London at the Royal Albert Hall, Earls Court and beyond). I can only conclude that the BUF's focus away from using visual means through which to raise their cause was related to the limited number of artists aligned with the British fascists who were available to take on this work. Cyril Connolly's comment of 1938 comes to mind here: 'We are having to choose between democracy and fascism, and fascism is the enemy of art. It is not a question of relative freedom; there are no artists in Fascist countries'.[4] Connolly's formulation was wishful thinking, as we know from much excellent scholarship referenced across this book, which explores the flowering of art and exhibitions supported by fascist regimes in Italy, Germany and Spain.

While explicitly articulated political positions are the overt focus of this book, Tony Bennett's well-known discussion of 'the exhibitionary complex', in which he casts all exhibitions as inherently political, 'vehicles for inscribing and broadcasting the messages of power', is an important formulation for considering the panoptic visual control and power dynamics operating in all exhibitions. Bennett – channelling Foucault – provides the gateway to another orthodoxy that this book takes up and extends: that exhibitions are not – and never were – neutral.[5] Instead, they are the potent context in which ideologies take shape as modern myth, thereby affecting wider culture.[6] Exhibitions, as 'acts of exposure', are inherently persuasive, as cultural theorist Mieke Bal suggests, even when they are not being developed as propaganda. This makes political exhibitions double-layered propaganda: being implicitly persuasive, expository forms that, at times, take up explicitly political arguments.[7]

Terms used in this book

I use many overlapping, interlinked terms in this book to describe these exhibitions, which I will expand on over the course of this and subsequent chapters, but introduce briefly here. The one term that covers all of the many manifestations of this form across this period and contexts is *propaganda exhibitions*, describing exhibitions used as persuasive devices across contexts and political traditions. Other terms I use include *exhibitions as communications* to assert exhibitions' emergence alongside other forms of early twentieth-century media and their connection with the 'communications paradigm' through which this form emerged.[8] *Exhibitions as demonstrations* takes up a phrase used by activist artists in the 1930s to describe exhibitions used for protest, evoking the idea of exhibitions as active, performative, provocative and participative forms for manifesting solidarities.[9] *Manifesto exhibitions* describes exhibitions mounted in Britain in the 1930s to present the ideas of avant-garde artists' groups. *Political exhibitions* describes exhibitions that explicitly addressed political issues, often taking anti-fascist or anti-imperialist positions. *Modernist exhibitions* describes exhibitions either made by Modernist designers or which adopted forms, tropes or ideas associated with Modernism.[10] *Didactic exhibitions* takes up photography historian Olivier Lugon's phrase to describe exhibitions that were spatially and textually constructed to teach their audiences.[11] *Factographic exhibitions* takes up the term 'factography', which art historian Benjamin Buchloh used to describe an art tradition concerned with rendering aspects of reality visible without interference or mediation, and suggests that it can be used to describe the qualities of exhibitions in this book.[12] *Useful exhibitions* is a play on Charles R. Acland and Haidee Wasson's phrase 'useful cinema', to describe films more involved with functionality than beauty, which I use to describe exhibitions that were put to 'work', with functionality the guiding concern.[13] *Documentary exhibitions* describes the overlap between exhibitions and other forms, such as photography and film, which evolved within the British documentary tradition. *Information exhibitions* is a phrase used to describe exhibitions that functioned to communicate practical, everyday advice to the public. To consider any of the exhibitions in this book merely as 'information' is somehow to suggest their neutrality, rather than situating them within the ideological and economic complexities of the period, so this is not a phrase that I find useful, except in distinguishing exhibitions as communication from exhibitions of 'original' or unique objects (such as art and artefacts). I distinguish the forms of exhibition above from *displays*, *trade fairs* or *commercial exhibitions*, which were focused on showing and selling things with a profit-making motive, and use *exhibits* to signal small sub-sections of wider exhibitions.

My overall intention is to show that the exhibitions featured in this book were more than platforms or envelopes for presenting objects or images; instead, they were a calculated and didactic means of communication for war and peacetime, a form through which to teach conformity and adherence to appropriate truths and behavioural patterns and an affective form intended to provoke emotional responses.

The remit of this book

These exhibitions were marked by a complex visual, textual and spatial hybridity; they were a way of representing ideas and a collective endeavour through which to meet, build relationships and share ideas about life, work and beliefs in modern Britain. For marginalised people in Britain on the eve of the Second World War, the process of making exhibitions provided opportunities to build social and cultural capital. This was particularly true for commercial artists and designers who came to exhibition making with limited money or institutionalised capital, given that in the highly class-ridden British art school system their work or their training in technical colleges or apprenticeships was considered of low status (as discussed in Chapter 1).[14] For those who had arrived in Britain during the years immediately preceding the Second World War following training elsewhere, there was work to be done to achieve recognition by joining emerging professional organisations, forming collaborations in practice, building up bonds of friendship and experiencing conviviality in the midst of trauma. Exhibition making provided the points of contact through which such relationships could be built.[15] It allowed artists, designers and architects, some of whom were marginalised after arriving in Britain with limited financial means and few contacts, to meet, share concerns and make a collective, public response. They built shared solidarities that connected them with the world beyond. Exhibition networks acted as connective tissue within communities, as crucial stepping-stones within careers and as a micro model of formation for the developing creative professions. Some artists' networks, such as the AIA, provided a crucial platform and voice to women who, while marginalised in more established artistic circles, became centrally important. Sculptor Betty Rea – 'dynamo' of the AIA, as Misha Black later described her role as AIA Secretary[16] – was one such woman, declaring in 1935, 'It is time the artists began to think what kind of future they want and what they can do to get it', recognising their agency to create a direction.[17] Exhibition making was one route by which artists could think through the future and give form to it.

For the politically engaged artists and designers central to this account, exhibitions operated as nodes of ideological resistance for political dissenters and subordinated groups, offering shelter and collective activities for newly arrived people seeking refuge from the Nazis, as well as for people on the Left actively seeking to create and to defend anti-fascist and anti-imperialist social spaces (as discussed in Chapters 3, 4 and 5). Exhibitions were versatile enough both to document and to make manifest invisible values and truths. Occasionally exhibition making operated as pretext, the basis on which refugee artists and architects were allowed to enter Britain when their work was due to be exhibited (as discussed in Chapter 3).[18] They were an acceptable reason for activist-makers to come together in public spaces, at a time when public gatherings were contentious and curtailed by legislation (as discussed in Chapter 4).

They offered a channel for eloquent interventions into contemporary discourse, enabling activists to protest the British government's policy of non-intervention during the Spanish Civil War, for instance (as discussed in Chapter 3). In the context of the Second World War, this particular form of exhibition-as-communication was repurposed for a different ideological context by the British government, becoming part of their armoury of propaganda (as discussed in Chapter 5), and in the postwar period exhibitions became naturalised once again as one amongst many forms of communication in the service of the embryonic welfare state (as discussed in Chapter 7).

This book features many artists and designers over the two decades of its focus. The key exhibition makers central to this book had diverse origins, united by an internationalist, democratising collective vision, mostly living and working in exile within a limited geography, in Britain's capital city, London. This intersects with recent histories of the lives and careers of artists, designers and architects, who settled in London in this period from Central and Eastern Europe, including artists Naum Gabo, László Moholy-Nagy, John Heartfield, Ludwig Meidner and Oskar Kokoschka and architects Serge Chermayeff, Walter Gropius and Ernö Goldfinger, living for a short period of the war close together, within a small area of north London.[19] We can consider the work of these artists and architects, existing in exile in London, in relation to art historian T. J. Demos's proposition of 'modernity-as-exile', which he describes as 'defined by the dislocating ravages and alienating effects of capitalism and nationalism as much as by the psychic disequilibrium of traumatic *unheimlichkeit*, as it is comprehended by Marxist and Freudian thought'. The particular mobilisation of exhibitions as forms by this group was inextricably related to their vilification in home contexts, displacement from home and alienation from previous contacts. Regardless of the geographic displacement of many of this book's subjects, Demos suggests Modernism's epoch is defined more generally as one of 'transcendental homelessness', making migration the ongoing counternarrative to nationalisms and other overarching narratives.[20]

Much of London, during the period of this book, was drab and run-down, its restaurants noted for their poor food and much of the city's population living in deteriorating Victorian housing stock.[21] In this context, exhibitions-in-progress acted as nodes and spaces of convergence, contact and transnational encounter for incoming artists, rendering the practices, designers and spaces of production cosmopolitan. In thinking about British exhibitionary practices of this period as cosmopolitan, I look to design historian Zeina Maasri's discussion of the intersection of visual culture, design and politics in Beirut from the late 1950s to the mid-1970s in her recent book *Cosmopolitan Radicalism*. Maasri, in focusing on the city of Beirut rather than on the nation of Lebanon, adopts a non-essentialist understanding of place that takes into account, using geographer Doreen Massey's phrase, a 'global sense of the local'; a place formed by networks of social relations.[22] In doing so Maasri seeks 'to trouble any putative binary between the "West" and the "non-West"'.

Taking up this idea, in a different time and place, I show the complexities of discussing the work of designers who were based, often briefly, in London during these years. Through the lens of exhibitionary cultures, this book traces the roots of change in Britain from the 1930s, building on the work of historians like Marc Matera and Priyamvada Gopal, who discuss the work and impact of anti-colonial campaigners centred in London from the early twentieth century.[23] The role of refugee artists and designers in giving shape to British exhibitionary cultures in these decades is a fulcrum of this study, as is the key role of anti-colonial activists in identifying exhibitions as spaces of counter-argument. Literary theorist Rebecca L. Walkowitz's reflections on negotiations of distance and proximity, interconnections of the personal and international, in relation to the concept of 'critical cosmopolitanism' are useful when considering the geographies covered by this book.[24] The initial focus of this book is 'British' exhibitions only in so far as the events happened in and from Britain, but many of the events central to this book described or connected with other geographies, through diaspora communities and the transnational networks of artists and designers who worked on them, or through the influences, ideas and ideologies gleaned through education and political activities that shaped their visual and material forms and imaginaries. But, as the book shows, the geographies of the exhibitions analysed here changed during the twenty years the book covers, from predominantly being mounted in sites in London before the Second World War to being dispersed across the United Kingdom and beyond, once they became instruments of the wartime British government and the postwar welfare state. The psychic geographies of the artists and designers who mounted the exhibitions were transmogrifying, as this book shows, from internationalist interests connecting them with Russia, China, the popular front in Spain and inter-Imperialist battles to more solidly nationalist visions, as they were drawn further into working and performing in the national interest. In this sense, 'Britain' is the most accurate description of this book's geography, not 'London', and this is reflected in the book's title, but many of its endeavours were pursued in the interests of transnational connections and solidarities.

There has been significant historical focus on migration to Britain from the late 1940s, centring the major cultural shift brought about by migration after the Second World War.[25] This is reinforced by accounts like V. S. Naipaul's autobiographical novel *The Enigma of Arrival* where he recalls: 'Because in 1950 in London I was at the beginning of that great movement of peoples that was to take place in the second half of the twentieth century ... Cities like London were to change. They were to cease being more or less national cities; they were to become cities of the world'.[26] This book locates the start of such changes within the consciousness shown through exhibitions from the early 1930s.

Central actors in this account, in addition to those cited above, are artist and gallerist Roland Penrose; critic Herbert Read; painters Nan Youngman and Betty Rea; photographer Edith Tudor-Hart; and designers Misha Black, F. H. K. Henrion, Richard Levin, Hans Schleger and Milner Gray. In a sense, this book acts as prosopography, a collective account, of makers who lived at close quarters; with exhibition making as the significant point of contact, a collective means of expression and acceptable subterfuge for expressing challenging opinions. Their interests and beliefs were amplified through membership of a collection of overlapping and interlocking international artists', designers' and architects' organisations, including the British-based Artists International Association (AIA) (introduced in Chapter 4) and the Free German League of Culture (FGLC) (introduced in Chapter 6), as well as through alignment with the International Brigades. These were not bounded by national preoccupations but connected with transnational concerns coalescing around anti-fascist and anti-colonial causes.

This account of Modernist exhibitions centres on the work and imagination of refugees who took up residence temporarily in Britain during these years. It offers exhibitions as a form of provisional 'home-from-home' and a vehicle for critiquing society. It differs from accounts that separate Britain's avant-garde from radical politics or commerce in this period. Instead, it expands the focus on Modernist cultures in Britain to include commercial art and culture, practices of packaging design and advertising, shop window displays and trade fairs, and cultures of lecturing, writing and publishing, linking these with activist political work, which often ran alongside each other concurrently in careers. Rather than seeing these activities as marginal and subservient to more highly prized creative practices, this book foregrounds and centres such creative labour, understanding exhibition making as connected with – and integral to – art, design and architectural practice.

It shows the wider cultures of British Modernism as developed in and through meetings, articles, speeches, posters, leaflets and magazines, as much as through works of art, design, architecture and literature. It takes the focus beyond the formal qualities of Modernism and traces how Leftist radicals in the 1930s, activated by their identification with the International Brigades during the Spanish Civil War and displaced by rising authoritarian states in their home countries, increasingly became identified with and worked for the British establishment, helping create the visual and material representation of the early welfare state, with exhibition making central to each of these moments. In this sense, the geographical imaginary changes through the twenty years of this book, from local and international in the interwar period to national and nationalist from the Second World War until the end of the book in 1953, when the Cold War was unsettling political beliefs and certainties.

Continuing to address these histories has been given added impetus by the recent resurgence in Britain of chauvinist nationalism, with its associated anti-immigration rhetoric in the wake of the Brexit vote, as expressed through regular alarmist headlines about the 'migrant crisis' in right-wing newspapers like the *Daily Mail*. In that sense, this book contributes to literatures providing a longer view on refugees in Britain (or people whose context has, at times, been de-politicised by being described as 'émigrés'). It intends to highlight the refugee experience in Britain, showing how designers who had arrived under duress in Britain in this period became pivotal to the nation's visual, material and architectural cultures despite what was, at times, harsh and inhospitable treatment.[27] The particular form of exhibitions that are the focus here were very emphatically the product of a cosmopolitan imaginary, having more in common with designs evolving from the 1920s onwards in Germany, Russia, Italy and France, as this book shows.[28]

Exhibitions as the 'materialisation of persuasion'

By the early 1950s in Britain, the end of the period of this book, it was well established that exhibitions existed within a complex system of communication modes, entangled with information, publicity and public relations.[29] Exhibition makers, with their deep understanding of principles of persuasion and of commercial advertising, were part of the 'invisible government' shaping and manipulating society, as US proponent of public relations Edward Bernays had described the work of publicists in the late 1920s.[30] By 1949, influential British designer Misha Black (introduced more fully in Chapter 1) was alive to his participation in this 'invisible government'. He described the hybrid exhibitions he created as 'propaganda in three dimensions'.[31] '[T]he essential function of a propaganda exhibition', Black wrote, 'is to implant or sustain a general idea in the mind of the visitor … which may later affect his actions'. Black did not distinguish between exhibitions for commercial or official contexts, believing they existed on a continuum. His experience of working with exhibitions as publicity and propaganda had been honed through earlier commissions for advertising and commercial stands. All were, he thought, elements in 'the materialisation of persuasion'.[32] This evocative phrase of Black's was the title of the AHRC Fellowship from which this book came. 'Materialisation' is a useful term for discussing this kind of exhibition because of its lack of formal specificity, reflecting the multiple forms that exhibitions took during these years: reflecting function, rather than stylistic orthodoxy.

Exhibitions were a key element in Britain's evolving public relations culture, which is the focus of Chapter 2. Government-funded bodies shaped public relations culture in Britain, including the Empire Marketing Board (EMB), the General Post Office (GPO), the Underground Electric Railways Company of London (UERL) and its successor the London Passenger Transport Board (LPTB), all of which used exhibitions to promote their work and ethos. Publicity and propaganda exhibitions emerged in Britain while the science of persuasion was being honed in the profession of publicity and public relations.[33]

A major conundrum for all involved with the business of publicity in Britain from the 1930s was how to create impactful propaganda, appropriate to a social democracy, that mirrored the undoubted strengths and qualities of the 'agitation' being created by Nazi Germany and fascist Italy. Editor of BBC *Listener* magazine Richard S. Lambert acknowledged the problem in his 1938 book *Propaganda*. Whilst naming exhibitions as part of the wider media of propaganda, Lambert acknowledged that artists and designers were in a bind, attracting suspicion if considered to be producing propaganda.[34] The conundrum as to how to deal with what had become regarded as deceitful mechanisms used by totalitarian regimes was acknowledged by Misha Black. 'Propaganda in a democracy', Black wrote, 'must be based on the principle of persuasion, consent and participation'.[35] Essentially, an exhibition's job was 'the manufacturing of consent', to take up commentator Noam Chomsky's description of the basic institutional structures and relationships within which the US mass media operates.[36] Assessing the success or impact of this material was a long-running interest, though often largely inconclusive.

Modernist exhibitions and exhibitions of Modernism?

This book takes up the idea of exhibition design as a significant strand in Modernism, a practice in its own right. It builds on studies of exhibitions' political agency, the most notable being Mary Anne Staniszewski's 1988 *The Power of Display*, which draws on case study exhibitions from New York's Museum of Modern Art (MOMA) to make the case for understanding installation design as an 'aesthetic medium and historical category', to be regarded as significant in its own right, rather than merely providing context for showing works of art and design.[37] Innovating in forms of installation was certainly not the guiding principle in all exhibitions in this book, however. Indeed, often the installation was a marginal concern, while the context within which the exhibition was imagined and created, the ideas it espoused or single elements of its design warranted its inclusion here.

Considering the agency of exhibitions as Modernist practice in and of itself, as this book does, has an established lineage. Art critic Brian O'Doherty's seminal 1976 essay on the gallery space as 'white cube' broke with the previous interest in Modernist activities *within* the exhibition space, shifting the focus so that 'context' became 'content'. O'Doherty wrote:

> The history of modernism is intimately framed by [the gallery] space; or rather the history of modern art can be correlated with changes in that space and in the way we see it. We have now reached a point where we see not the art but the space first.[38]

Art historian Rosalind Krauss took up this re-focus on the agency of exhibition spaces with her suggestion that the 'space of exhibition' was, 'in fact what we know as the history of modernism … within this space it is constituted as a representation of the plane of exhibition, the surface of the museum, the capacity of the gallery to constitute the objects it selects for inclusion as art'.[39] For Krauss, writing in the early 1980s, art works had themselves become accounts of the exhibition spaces that they were part of, which she characterised as 'exhibitionality'. Both O'Doherty and Krauss focused on the qualities of exhibitions as part of the complex ecology of the contemporary art world, a self-referential means in and of themselves, to be conceptualised as Modernist objects rather than merely a series of Modernist moving parts. Given the sites and spaces privileged in this account – from stations to bomb sites – their work begs the question: what is exhibition without the gallery?[40]

Factographic exhibitions

Modernist exhibitions created in Britain outside galleries are the major focus of this book. The exhibitions and activities central here were ensembles created from images and didactic texts, used to communicate political positions. They were more in the tradition of 'factography' (concerned with rendering aspects of reality visible without interference or mediation) than of 'faktura' (concerned with the condition of the coloured surface), to take up art historian Benjamin Buchloh's distinction.[41] Frequently assembled from picture library images held, for example, by the wartime Ministry of Information (MOI), the exhibitions' images were severed from original meanings or contexts; the viewers' connection with a 'true' context made irrelevant.[42]

Creating exhibitions in community or public spaces was a guiding principle within the authoritarian regimes, by the period of this book. In the Soviet Union, museums had been created in industrial plants, an idea K. I. Vorobyvov discussed in his 1931 essay 'Museums in Industrial Enterprises'.[43] Such exhibitions, highlighting local popular traditions, were realised in a handful of large factories. In Nazi Germany a programme of art exhibitions, the *Fabrikausstellungen* ('factory exhibitions'), were staged inside factories between 1934 and 1942. These were modelled on the Italian fascist leisure organisation Opera Nazionale Dopolavoro's model for Kraft durch Freude (Strength through Joy) (KdF) and their factory exhibition concept, as well as the impetus of the Freunde der bildenden Künste (Friends of Fine Art).[44] The quasi-display, quasi-performance of Leftist agitational propaganda or 'agitprop', as it became known, has parallels in the AIA's exhibitionary activities of the 1930s, as I discuss in Chapter 4. Agitprop used the street and mass action, reinventing the revolution as carnival. The Berlin-based troupe Rote Raketen (Red Rockets) was one such group, using a mix of theatre and cabaret to develop class-consciousness in audiences during the late 1920s.[45]

Far from being dependent on institutional spaces, most exhibitions in this book were held outside purpose-built galleries, with the official expectations that often circumscribe them, as subsequent chapters show. While some were mounted inside public interiors, others were held in the open air, becoming part of a mutating mosaic of cityscapes in the process of being remade after the devastation of war.[46] In being held beyond established purpose-built spaces, they were developed with the democratising instinct to take art to where the general public could see it, each focused towards spotlighting a set of political ideas happening beyond the exhibition space, sometimes locally (the need for improved housing stock), often far from home (campaigns in the Spanish Civil War, anti-fascists working underground inside Germany during the Nazi regime or the enslavement of colonised peoples across the world). They operated as three-dimensional manifestos – combining image, text and space – closer to political tracts than to works of art. In locating the appropriate exhibitionary form for communicating such political ideas, artists and designers looked within and beyond Britain.

Common tropes and rhetorical devices for presentation of Modernist exhibitions that occurred and recurred in British exhibitions from the 1930s, as I show over the course of this book, included creating a comparison between past, present and future to argue for a particular way forward; using a problem/solution paradigm whereby visitors were shown the problem then the proposed solution; addressing visitors by speaking direct to them, in the present tense; appealing direct to people by helping them identify with an individual who acted as a proxy for explaining a whole profession; using one family or individual as a kind of avatar to follow through the narrative of a whole exhibition; and using narratives to show progression through scenarios (including sequences of objects or images reinforced spatially through ramps and multi-layered spaces).

Recovering exhibitionary traces

Researching the exhibitions in this book has been slow and challenging: many were held outside archiving institutions, created as immediate responses to current affairs at moments of profound crisis. Many comprised photographic 'exhibition prints' chosen for their interchangeability, fleetingness, reproducibility and disposability.[47] They were displayed for a short period, to highlight an urgent, but now long surpassed, issue and often only very piecemeal visual material remains. Mostly they were not subject to the care and concern shaping the curation of original works of art, to be kept and conserved with pristine traces of provenance. Instead, they were constructed from hastily assembled combinations of reproducible elements: photographs, text, props and architectural elements. Many were principally focused on the 'argument' or 'story', rather than the means of compelling it.

Their limited archival trace might include as little as a one-line recollection in an artist's memoir, a passing mention in a contemporary trade journal or a single, grainy press photograph, making the kind of sustained formal, visual analysis of exhibitions of the kind that is so admirable in some art historical accounts of exhibitions almost impossible. This absence has pushed me to be more reliant than I might like on a range of contemporaneous display and trade literature – magazines like *Display* and *Shelf Life* – as evidence, meaning that the available appraisals of the exhibitions in my account are necessarily weighted towards questions of comparative technique and promotion of the trade. I have spent much time examining pamphlets and flyers, often the only remaining fragments, bringing to mind poet W. H. Auden's line from his 1937 poem 'Spain': 'To-day the expending of powers/On the flat ephemeral pamphlet and the boring meeting', with pamphlets thankfully less ephemeral and mostly less boring than the poet imagined. Wherever possible, I have tried to counter-balance this with other types of literature, in what was already a long and painstaking research process, made more battle-like during the archive and library closures of the pandemic, the backdrop to the main writing period of this book.

Although many of the protagonists in this book were exercised about how 'successful' the exhibitions they were mounting were, being at pains to show through strong visitor numbers or press attention how efficacious they had been, in many ways this kind of success is peripheral to this book, which is much more focused on exhibitions as the materialisation of certain cultural and political debates. A few British exhibitions have achieved mythical status, given how many accounts exist, how many times they provide a historical scaffold as the start or end-point; in Britain, these include the 1851 Great Exhibition, the 1946 Britain Can Make It exhibition and the 1951 Festival of Britain. Their dominance is often enabled by extensive and well-kept archives, amplified by the pivotal point in history that they have come to act as shorthand for. But it is important not to think that the profusion of archival evidence and historical accounts makes these mythical exhibitions somehow of wholly greater significance. Being partially archived makes the exhibitions that are the subject of this book of no lesser significance; it is simply a reflection of their contribution as immediate, functional and short-lived and mounted by groups who did not have particular regard for legacy. The challenge of writing this book has thus been to act as archaeologist, piecing together fragments of material about events largely lost from memory and omitted from history.[48]

On many occasions, the only remaining trace has been a written account in the trade press or periodicals, with visual evidence limited or non-existent. Due to the scarcity of evidence about some of the exhibitions, this book acts in places as a record and reconstruction of these events and I use ekphrasis – the vivid written description of visual and material encounters – as a method for evoking and linking fragmentary details. This is particularly true in my discussion of the *Cambridge Anti-War Exhibitions* (in Chapter 4), for example, which brought together many of the most talented exhibition designers of their day but for which tantalisingly little visual evidence remains. In his essay 'Ekphrasis and the Other', W. J. T. Mitchell discusses the possibilities and limitations of this strategy.[49] He acknowledges that in a literal sense ekphrasis is impossible: writers can never give their readers sight. But by practising presenting 'otherness', this limitation can be overcome. For Mitchell, ekphrasis is a key to difference within language, focusing the interarticulation of perceptual, semiotic and social contradictions within verbal representation.[50]

Researching this book has reinforced my understanding of the complexity and multivalence of exhibitions as objects.[51] As starting point and culmination, their study intersects the disciplinary boundaries of cultural politics, visual culture, art, architectural and design histories. When seen in retrospect, their meanings often appear unfixed and difficult to pin down, their forms 'promiscuous', as art historian Michael Tymkiw notes, being adopted and rejected between successive events, across wide geographies and political divisions. This promiscuity, in the case of the exhibitions in this book, came from the porousness created by mass media, which allowed images of certain prominent exhibitions to be shared across ideological divides. Indeed, as the archival research for this book has elucidated, visual examples were exchanged fluidly between Germany and the US and the US and the UK, with designers in Britain 'borrowing' from many contexts, including those on the extreme political Left and Right beyond Britain.[52]

The form of these exhibitions is a major focus of this book but often, during the research process, the relationships that were enabled through an exhibition's process of making came to preoccupy me as much as its forms and materiality. The question of how, where and by whom creative labour is carried out is an enduring theme of this book and an increasingly central focus of exhibition histories.[53] The centrality to exhibitions' formation of many makers, including designers, artists, writers, organisers and others, means the kind of labour involved in bringing these objects to life crosses creative and disciplinary boundaries. How they were commissioned and financed, through government ministries and voluntary engagements, is also my focus.

Vision in motion

The exhibitions in this book were a complex amalgam of the plethora of associations and influences their designers absorbed through print media, visits, training and beyond. This makes tracing the origins of forms difficult. But the ideas of those associated with the German Staatliches Bauhaus, the school of design, architecture and applied arts generally referred to as the Bauhaus, had a major impact on the evolution and form of many of the exhibitions discussed in this book, as is clear through a formal comparison as well as through the accounts of the exhibition designers themselves and the critical appraisal in the contemporary press, as this book will show. Perhaps the most profound way in which these British exhibitionary practices shared Bauhaus ideals was through their taking up of what design historian Justus Nieland describes as the 'sweeping communications paradigm', whereby exhibitions were forced to act as a communications medium, part of an extended ecology of intermedial production.[54] The work of Bauhaus members was known in Britain in the 1920s and early 1930s through writings, lectures, periodicals, books, window displays and exhibitions. Exhibitions by Bauhaus faculty such as Herbert Bayer, László Moholy-Nagy and Walter Gropius's much publicised *Ausstellungsstand der Baugewerkschaften* (*Exhibition of the Building Workers Unions*), held in Berlin in 1931, were known to British audiences through the pages of magazines like *Commercial Art and Industry*, *Display*, *Architectural Review* and the *DIA Quarterly*, each of which carried regular reports on developments in Germany.

Hungarian-born László Moholy-Nagy's ideas were known in Britain through their translation from German into English. His earliest book, *Malerei-Photographie-Film* (*Painting-Photography-Film*) of 1925 considered the merging of media across types, introducing the idea of the 'typophoto', which he claimed as the 'visually most exact rendering of communication': the merging of typography, 'communication composed in type', and photography, 'the visual presentation of what can be optically apprehended'. Going further, he saw that photographs might replace words in the form of 'phototext', the potential for 'poly-cinema' by experimenting with sequential projections and 'photograms' (a photographic image made without a camera by placing objects directly onto the surface of a light-sensitive material) allowing for experimentation with the space–time continuum.[55] In *Von Material zu Architektur* of 1928, published in English in 1932 as *The New Vision*, Moholy-Nagy explained his experiments with kinetics, light and space.[56] Finally, in *Vision in Motion*, published posthumously in English in 1947, Moholy-Nagy gave a more general view of his ideas of the interrelatedness of art and life, including paying homage to artists whose ideas had influenced him, such as Dada artist Kurt Schwitters' use of political collage, photomontage and preoccupation with typography.[57] An early Moholy-Nagy experiment with art and communication was his act of 'painting' three pictures by telephone, achieved by dictating the painting to the foreman of a sign factory, using a colour chart and a piece of graph paper.

Moholy-Nagy's multi-faceted interests were shared by the creators of British exhibitions discussed in this book in terms of their conception, their experimentation with diverse materials (celluloid, photographs, aluminium, plexiglass, gallalith and more) and, on a more prosaic level, in modelling specific formal qualities (such as peep-holes and port-holes). Ultimately, Moholy-Nagy's interest in interrelatedness, 'integration' and 'assemblage' led him to experiment with stretching the bounds of communication; to combine visual, material, spatial and textual means; to experiment with materials; to investigate immersive and experimental modes of cinematic spectatorship and 'mobile perception'; and bolstered his interest in light, space, time and colour. These experiments informed the culture of Modernist exhibition making that is the subject of this book.[58]

From 1930, Herbert Bayer theorised the impact of exhibition design as working with film and cinematic perception, 'a new discipline … an apex of all media and powers of communication and collective efforts and effects', he wrote, which shaped the wider field in which exhibitions were being conceptualised, even if they did not literally deploy film as part of their exhibitionary repertoire.[59] Bayer's idea of the extended vision – set out in his well-known 'Diagram of Field of Vision' – was an interpretation from Moholy-Nagy's writings, informed by his regard for the work of exhibition designers like El Lissitzky. Bayer's work transposed Moholy-Nagy's visual ideas into three dimensions, making exhibitions into forms of *Gesamtkunstwerk* (total work of art).

Ex-Bauhaus faculty Gropius, Moholy-Nagy and Bayer all eventually settled in the US, where they continued to create multi-image, multi-source media environments suited to conveying the present and future of a social democracy, which art historian Fred Turner calls 'surrounds'.[60] Their work, Turner argues, continued the application of ideas originating in 1930s multi-screen displays and immersive theatre. These same makers were influential in the British context. Through living in exile in London in the 1930s, architect Walter Gropius (living in London from 1934 to 1937) and artist László Moholy-Nagy (living in London from 1935 to 1937) had a direct impact on the form of exhibitions in Britain. In London, Gropius's work and ideas were shown in exhibitions and lectures and through espousal by prominent advocates such as Herbert Read, whose 1934 book *Art and Industry* explained Gropius's educational programme to a British audience.[61] While in Britain, Moholy-Nagy worked across a range of commercial projects and was directly involved in creating exhibitions that are the focus of this book, most notably planning the 1938 MARS Group exhibition (as discussed in Chapter 2).[62] Although Herbert Bayer fled from Germany straight to the US, his work was experienced directly by British audiences in 1943, when a modified version of his 1942 exhibition *Road to Victory* for MOMA toured Britain under the revised title *America Marches with the United Nations* (as discussed in Chapter 7).

Designers in Britain, such as Misha Black, shared Moholy-Nagy's and Bayer's interests in modes of perception, in the integration of elements in design (visual, textual, spatial) and in the continuities between exhibitions and moving images, which this book explores across many contexts. Showing his awareness of exhibition's particular appeal to its audiences and its comparative role as media, Black wrote in 1949 that 'the exhibition takes equal place with the film in completely encompassing the spectator and allowing only those distractions which are deliberately planned to accentuate the effect', an idea developed in his lectures and writings over several years.[63] As subsequent chapters show, other influential visual thinkers spent significant time in Britain during this period, including Kurt Schwitters (from 1940 to 1948), Naum Gabo (from 1936 to 1946) and John Heartfield (from 1938 to 1950), shaping and influencing visual presentation.

Catalogues, as portable elements of an exhibition, were another focus for viewers' dispersed negotiation between text, image, object and space, allowing for an amplification of ideas within the exhibition space itself and a crossover of these ideas from the public space of the exhibition into the sphere of someone's private possession. The importance of this element of the designed exhibition entity was borne out in Herbert Bayer's meticulous printed catalogues for exhibitions in 1929, 1930 and 1933. In his catalogue for the German section of the Society of Applied Arts at the Grand Palais, Paris of 1930 (designed by Gropius, Bayer and Moholy-Nagy), Bayer developed his ideas about 'extended vision', while the exhibition catalogue accompanying the Berlin version of *Die Camera* (*The Camera*) exhibition of 1933, designed by Bayer, consistently followed the design principles of the Bauhaus (despite the exhibition's projection of Nazi ideology through sections including a visual 'history of the Nazi movement' in sixteen photomurals).[64] In the interstices between text and image were diagrammatic and pictogram languages, enabling exhibition designers to communicate complex technical information to public audiences at a glance.[65] In Germany, pictograms and diagrams were used from the 1920s, for example in the 1929 *Gas und Wasser* (*Gas and Water*) exhibition designed by Schmidt and Gropius.[66] In Britain, the best-known visual system of this period was Isotype, an information language pioneered by Otto and Marie Neurath in the 1920s, which was used in many exhibitions designed and mounted during the 1930s and the wartime (as discussed in Chapters 4 and 7).

Documentary exhibitions

The history of documentary exhibitions in Britain is inextricably linked to the development of the documentary film movement, with both occupying a place in the extended network of communication forms in the 1930s and 1940s.[67] This was not least because several influential designers, such as Moholy-Nagy, worked across film, exhibitions and other commissions while living in Britain, including for the GPO (as discussed in Chapter 2).[68] The content and focus of exhibitions and documentary films had much in common, sharing a deep functional preoccupation with explaining how everyday life in Britain worked, in both war and peacetime. GPO films like *The Horsey Mail* (1938), *Night Mail* (1936) and *North Sea* (1938) explained how everyday feats were underpinned by social and technological infrastructure, as a way of creating public understanding and gaining support. This was a central endeavour of GPO exhibitions (as Chapter 2 shows).[69] The GPO Film Unit became the Crown Film Unit in 1940, carrying many of these preoccupations and overlaps into the war effort.

Beyond the GPO, films and exhibitions shared characteristics in common within the documentary movement and, after the outbreak of war, in their common abilities to educate and entertain.[70] The continuities between British cultures of exhibition and of cinema in this period are striking: exhibitions were often mounted in cinemas (as discussed in Chapter 5) and films were taken outside cinemas, to be shown in other public places.[71] It is no mere coincidence that those who ran cinemas in interwar and wartime Britain were called 'exhibitors' and represented in the 1930s and 1940s by the Cinematograph Exhibitors' Association (CEA). British (film) exhibitors recognised and traded upon the linkage between their films and the environments in which they operated.[72] The crossover between media was evident in British trade exhibitions, such as the series of Radiolympia exhibitions running through the 1930s and 1940s (after their launch in 1922), which ran in parallel with radio exhibitions such as the New York Radio World's Fair, held from 1922, and Berlin *Funkausstellung*, held from 1924.[73] All of these put on display novel sound and communication technologies, setting out to help audiences understand the wider cultural and social implications of these devices and ultimately enthusing them towards adopting them.

Many of the events in this book can be described as 'documentary exhibitions', given their fit within the well-known definition of 'documentary' as 'the creative treatment of actuality'[74] and their shared conjunction of photography and text informed by a strong Left social consciousness and inflected with ideology (as will be clear in Chapter 2). These exhibitions emerged at the same historical juncture as better-known manifestations of the British documentary tradition and sat close to documentary film, photography and writing (particularly in its shared wartime iteration within the Ministry of Information).[75] Indeed, writer George Orwell, whose *Road to Wigan Pier* of 1936 combined descriptive narrative with photographs and is considered a classic of documentary realist writing, was one of many writers employed to write scripts for exhibitions mounted by the MOI.

John Grierson was central to the establishment of the British documentary film movement, active in the interwar period (at the EMB 1926–33 and then the GPO Film Unit 1933–40) when, as Zoe Druick and Deane Williams note, social liberalism attempted to negotiate a third way between planned economies and free markets, and during the command economies of wartime. Grierson recognised realist documentary as 'a troubled and difficult art' but as early as 1933 said he looked upon 'the cinema as a pulpit and use[d] it as a propagandist'.[76] As with documentary films, several of the exhibitions discussed here raise questions of authenticity, appearing to show named members of the public going about their everyday lives but creating ambiguity as to whether these were in fact 'real' or staged.[77]

Exhibitions and the politics of spectatorship

This book is about spectatorship, in particular how exhibitions, in and from Britain, were designed to arouse the emotions, interests, passions and pride of audiences. This led them at times to question the status quo or, at other times, to accept it. This focus relates not only to the optical and material dynamics of exhibitions but also to the textual and discursive rhetoric found in and around these events (in wall panels, exhibited slogans, catalogues, opening speeches, lecture programmes, magazine articles, radio broadcasts, advertisements and administrative papers) and to the context and relationships within which these entities were made. In this sense, this book focuses on the entire 'apparatus', to use Michel Foucault's term describing the enveloping administrative mechanisms and knowledge structures, rather than merely the finished, exhibited product.[78]

The politics of spectatorship is a central focus of many recent studies of twentieth-century exhibitions, which address questions of perception and attention in exhibition spaces, dominant scopic regimes and how exhibitions enabled individual spectators to engage with ideas of collective culture and responsibility, reflecting the recurring focus of Modernist visual culture studies on the relationship between images and the individual.[79] Writing in 1936, philosopher Walter Benjamin famously discussed the experience of 'simultaneous collective reception', which was possible within architecture and film but not through paintings.[80] While in his seminal 1984 essay 'From Faktura to Factography', art historian Benjamin Buchloh describes how montage in exhibitions was initially presented with a simultaneity of opposing views, rapidly changing angles and unmediated transitions from part to whole, which had embodied the relationship between individuality and collectivity as one that was constantly to be redefined. He traces how this was displaced by unified spatial perspectives, often a bird's-eye view.[81]

More recently, art historian Jonathan Crary, in his study of how perception and attention were transformed from the late nineteenth century alongside the emergence of new technological forms of spectacle, display, projection, attraction and recording, proposes that Western modernity has demanded that individuals disengage from 'a broader field of attraction', to isolate 'a reduced number of stimuli', in order to 'pay attention'. 'Modern distraction', Crary writes, 'can only be understood through its reciprocal relation to the rise of attentive norms and practices'. 'Visuality', Crary argues, 'should not be over-emphasised lest it be cut off from richer and more historically determined notions of "embodiment", in which an embodied subject is both the location of operations of power and the potential for resistance'. Crary continues, 'Spectacular culture is not founded on the necessity of making a subject see, but rather on strategies in which individuals are isolated, separated, and inhabit time as disempowered'.[82] Often such ideas, as in Crary's work, are presented as binaries: concentration versus distraction; visuality versus embodiment.

While exhibitions of the Right have often been understood as spectacle, exhibitions of the Left have commonly been understood within the paradigm of 'carnival', to use the phrase of philosopher Mikhail Bakhtin, as in characterised by a mocking or satirical challenge to authority and the traditional social hierarchy. The 'carnival', Bakhtin writes, is 'not a spectacle seen by the people; they live in it, and everyone participates because its very idea embraces all the people'. Emphasising its performative and playful character, Bakhtin states, 'while carnival lasts, there is no other life outside it'.[83] Such binaries of Left (as carnival) and Right (as spectacle) have been disrupted more recently. Art historian Michael Tymkiw, in his 2018 study of Nazi exhibition practices, departs from the idea of visitors stunned into passivity by monumentalised imagery, instead arguing that Nazi exhibitions, particularly those mounted during the mid to late 1930s, encouraged 'engaged spectatorship', beckoning visitors to become involved in forms of social and political change upon leaving 'the highly constructed environment of an exhibition space'.[84] The Nazis used exhibitions to encourage 'an empathetic mode of spectatorship', zigzagging walls articulating the link between visitors and the idealised workers depicted in images.[85] Subject matter was reinforced through formal qualities, for example in the virulently anti-Semitic exhibition *Der ewige Jude* (*The Eternal Jew*) where plunging walls created anxiety and 'fragmentation' to signify the chaos, brokenness and disunity of the Other, as well as 'dissonance' to elicit aversion among spectators.[86]

Exhibitions, displays, demonstrations

In the twenty-first century, the word 'exhibition' most usually connotes a gallery-based show containing art or artefacts, while those ensembles without the auratic pull of original objects mounted to convey information, perhaps in commercial or public space, we might call 'displays'. Central to this account is another term not immediately associated with contemporary exhibitions but which was coined in the 1930s to describe a particular strand of activist exhibition, that is 'demonstration'. A group central to this book, the anti-fascist Artists International Association, used 'demonstration' to describe their exhibitions mounted from 1933, evoking the performative idea of exhibitions as active, provocative and participative, useful vehicles for propaganda and persuasion.

As I discuss in Chapter 4, from the outset the AIA envisaged fighting fascism through exhibitions, identifying themselves as a radical exhibiting society. In an account of the AIA's origins they recalled: 'At first [exhibitions] were used primarily [by the AIA] as **demonstrations**'.[87] It is informative to trace the etymology of 'demonstration', which comes from the Latin 'to point out', to 'make aware in a clear and public way', which led in the mid-nineteenth century to a connection with public protest.[88] By the 1930s, the decade the AIA was founded, 'demonstration' was in common parlance to indicate political protest, often used to describe working people's 'demonstrations' of anger about poor conditions they were experiencing during the Depression. Indeed, Hunger Marches and other forms of demonstration were regularly the subjects of AIA members' artworks. Taking up this language of politics and indicating their vision of audiences being animated by contact with culture, exhibitions as 'demonstrations' were key to the AIA's politically engaged vision, akin to active acts of protest, about making manifest and visible issues that were abstract, invisible or not seen and noticed.[89]

Britain was not the only place where this language of exhibition as demonstration was being used in the 1930s nor was the idea purely the preserve of the Left. Historian Jeffrey T. Schnapp quotes fascist journalist Margherita Sarfatti writing in Italy in 1933, describing *Mostra della*

Rivoluzione Fascista, the major exhibitionary celebration of the ten-year anniversary of Mussolini's March on Rome as a demonstration, which led to his coming to power, as follows:

> that which opened in Rome is not simply *the exhibition* but something greater; it is the demonstration of the Fascist Revolution. And here I employ the verb 'to demonstrate' in its literal and figurative, its mathematical and physical meanings. The show makes the Revolution plain, palpable, and intelligible, while at the same time providing proof, a definitive proof of the experiment's success by means of figures and calculations.[90]

Sarfatti's use of an exhibition as 'demonstration' was in parallel with the AIA's, despite her radically different politics.

By using this language, the AIA moved the focus from overly privileging the visual, instead revealing exhibition as bringing ideas closer to the spectator, by being focused on human interactions (in the case of people 'demonstrating' processes) and escaping the confines of the gallery and the museum, to liminal spaces suspended between the sites of the everyday – station ticket halls, shop windows, bombsites, factory canteens – spaces encountered during a daily journey to work or a lunch break. In doing so, the AIA adopted the language of working-class labour and the factory floor to create a sense of participatory and collective 'doing', rather than top-down 'instruction', giving a suggestion of the workings being revealed, so that others could engage, with both the process and the outcome.

Art, design and architectural historians have largely focused in on exhibitions either as 'surface', as visual spectacle, as sets of assembled objects, or else analysed their formal, spatial qualities. The way in which they provoke movement, a kinetic process, the dynamics of navigation through exhibition spaces, has been more elusive.[91] This is why 'demonstration', associated with movement and interaction (either in the form of demonstrating a process or of demonstration as protest), is useful in suggesting the necessity of movement in order to engage with this subject matter. Exhibitions as demonstrations demand engagement in their hybridity: as image, material, text, space and sequence, as this book shows.

The structure of this book

The exhibitions that form the main focus of this book range across two decades from 1933 to 1953, from tiny and fleeting to major, well-documented events. What they have in common is their engagement with contemporary political themes and issues, their use as propaganda and their role as nodes for artists and designers in interwar, wartime and postwar Britain. The current chapter, 'Introduction: exhibitions as "propaganda in three dimensions"', has introduced the major contexts and themes of the book. Chapter 1, 'Banishing chaos, vulgarity and mediocrity: training as an exhibition designer', sets out the precedents and the contexts for the particular form of exhibitions, exploring the history and evolution of using exhibitions for propaganda and to promote trade and industry in Britain before 1933. The rest of the book presents a series of interconnected arguments about how exhibitions operated as propaganda in Britain from 1933 to 1953. Each chapter centres on a few exhibitions that speak to a particular propagandistic paradigm or proposition. Designers and groups recur across different times and places. Chapter 2, 'Exhibitions as projection, promotion, policy and activism in three dimensions', considers the way that the government, authorities and trade bodies used exhibitions for promotion, projection and for sharing policy during the 1930s, with Charing Cross Station a particularly popular site. Chapter 3, 'Exhibitions as manifestos', considers how exhibitions took the form of manifestos in 1930s Britain. Chapter 4, 'Exhibitions as demonstrations', considers how exhibitions allowed activists to appropriate public space in order to share their political positions. Chapter 5, 'Counter-exhibitions', considers how exhibitions allowed activists to raise visible counter-arguments. Chapter 6, 'Exhibitions as solidarities', discusses how organisations beyond central government used exhibitions to share solidarities, in particular for fundraising for Russia; while refugee organisations such as the Free German League of Culture used exhibitions as a meeting point, a form of visibility and a creative outlet. Chapter 7, 'Exhibitions as weapons of war', shows how exhibitions became wartime propaganda in the context of the Ministry of Information, where exhibitions as 'propaganda in three dimensions' were adopted to build patriotic wartime citizens and to communicate urgent practical information to audiences at home and abroad. Chapter 8, 'Exhibitions as welfare', shows how exhibitions were used as a source of public information and propaganda for the embryonic welfare state in the postwar period from 1945 to 1953, charting how exhibitions were used as a source of public information within the embryonic Cold War, becoming institutionalised as part of the communications media of the welfare state. Exhibitions formed the focus for an entangled group of designers, working across many contexts from major commercial clients to activist political groups, making them at different times sites of consumption, sites of conformity and sites of resistance.

Notes

1 Important work on the history of propaganda in twentieth-century Britain includes Jo Fox, *Film Propaganda in Britain and Nazi Germany: World War II* (Oxford: Berg, 2006); James Chapman, *The British at War: Cinema, State and Propaganda, 1939–45* (London: I.B. Tauris, 2011); David Welch, *Persuading the People: British Propaganda in World War II* (London: British Library, 2016).

2 Jacques Ellul, *Propaganda: The Formation of Men's Attitudes* (New York: Vintage, 1973), p. 70; originally published in 1965.
3 Held in the collection of the British Library.
4 Cyril Connolly, *Enemies of Promise* (London: Routledge Kegan Paul, 1938), p. 2.
5 Tony Bennett, 'The Exhibitionary Complex', *New Formations*, vol. 4, Spring 1988. Museums Are Not Neutral is a global movement challenging museums to be spaces of positive change for our times. See, for example, La Tanya S. Autry's 'Changing the Things I Cannot Accept: Museums are Not Neutral', *Artstuffmatters* Blog, 15 October 2017: https://artstuffmatters.wordpress.com/museums-are-not-neutral/ (accessed 18 January 2021).
6 In his influential essay on myth, philosopher and critic Roland Barthes explained how the 1958 *Family of Man* exhibition, mounted at MOMA in a collaboration between photographer Edward Steichen and designer Herbert Bayer, functioned as myth; *Mythologies* (London: Jonathan Cape, 1972), pp. 117–18, 155–6.
7 Mieke Bal, *Double Exposures: The Subject of Cultural Analysis* (London: Routledge, 1996), pp. 2, 5.
8 Justus Nieland, *Happiness by Design: Modernism and Media in the Eames Era* (Minneapolis, MN: University of Minnesota Press, 2020), pp. 8–10.
9 British Library, *AIA Bulletin*, No. 81, January 1944.
10 Mary Anne Staniszewski, *The Power of Display: A History of Exhibition Installations at the Museum of Modern Art* (Cambridge, MA: MIT Press, 1998).
11 'Didactic exhibitions' is a phrase used by Olivier Lugon in 'Dynamic Paths of Thought: Exhibition Design, Photography and Circulation in the Work of Herbert Bayer', in Francois Albera and Maria Tortajada (eds), *Cinema Beyond Film: Media Epistemology in the Modern Era* (Amsterdam: Amsterdam University Press, 2010), pp. 117–44.
12 Benjamin H. D. Buchloh, 'From Faktura to Factography', *October*, no. 30, Autumn 1984, pp. 82–119.
13 To echo Charles R. Acland and Haidee Wasson's phrase 'useful cinema'; *Useful Cinema* (Durham, NC: Duke University Press, 2012).
14 Pierre Bourdieu's concept of cultural capital is important here and, in particular, his analysis of the way *social capital* is an aggregate 'of the actual or potential resources which are linked to possession of a durable network of more or less institutionalised relationships of mutual acquaintance and recognition – or in other words, to membership of a group'. Bourdieu, 'Forms of Capital', in J. E. Richardson (ed.), *Handbook of Theory of Research for the Sociology of Education* (Westport, CT: Greenwood Press, 1986).
15 This is something I discuss in '"Lines of Becoming" Misha Black and Entanglements through Exhibition Design', *Journal of Design History*, vol. 34, issue 1, March 2021, pp. 37–53.
16 Henry Moore Institute, Betty Rea collection, *Betty Rea 1904–1965: Sculpture*, June 1965.
17 Katy Deepwell, 'Anti-fascist Activities amongst Women Artists in the 1930s in Britain', in Concha Lomba Serrano and Alberto Castán Chocarro (eds), *Las mujeres en el sistema artístico (1804–1939)/ Women in the Art Scene (1804–1939)* (Madrid: Universidad de Madrid, 2022). Betty Rea, Foreword to *5 on Revolutionary Art* (London: Wishart, 1935), p. 7.
18 As shown in the work of the Artists' Refugee Committee, discussed in Chapter 3.
19 A phenomenon discussed in previous studies, for example by Monica Bohm-Duchen, *Insiders Outsiders: Refugees from Nazi Europe and Their Contribution to British Visual Culture* (London: Lund Humphries, 2019); Daniel Snowman, *The Hitler Émigrés: The Cultural Impact on Britain of Refugees from Nazism* (Vintage Digital, 2010); and David Cesarani and Tony Kushner, *The Internment of Aliens in Twentieth Century Britain* Routledge, 1993).

20 T. J. Demos, *The Migrant Image: The Art and Politics of Documentary During Global Crisis* (Durham, NC: Duke University Press, 2013), pp. 1–2.

21 The smoggy, downbeat atmospheres and moods of 1940s and 1950s Britain are the focus of Lynda Nead's book *The Tiger in the Smoke: Art and Culture in Post-War Britain* (New Haven, CT and London: Yale University Press, 2017).

22 Zeina Maasri, *Cosmopolitan Radicalism: The Visual Politics of Beirut's Global Sixties* (Cambridge: Cambridge University Press, 2020), p. 5, quoting Doreen Massey, 'A Global Sense of Place', *Marxism Today*, June 1991, pp. 24–9.

23 Marc Matera, *Black London: The Imperial Metropolis and Decolonization in the Twentieth Century* (Berkeley, CA: University of California Press, 2015) and Priyamvada Gopal, *Insurgent Empire: Anticolonial Resistance and British Dissent* (London: Verso, 2019) both discuss anti-colonial campaigners based in London from the early twentieth century and address their use of exhibitions as amongst the anti-colonial tools at their disposal.

24 Rebecca L. Walkowitz, *Cosmopolitan Style: Modernism Beyond the Nation* (New York: Columbia University Press, 2006), p. 4.

25 Wendy Webster, *Mixing It: Diversity in World War Two Britain* (Oxford: Oxford University Press, 2018) focuses on the period after 1940. See also Courtney J. Martin, 'Exiles, Émigrés and Cosmopolitans: London's Postwar Art World', in Okwui Enwezor, Katy Siegel and Ulrich Wilmes (eds), *Postwar: Art Between the Pacific and the Atlantic* (Munich: Prestel, 2016).

26 V. S. Naipaul, *The Enigma of Arrival* (New York: Alfred A. Knopf, 1987), pp. 141–2.

27 This is also the focus of the Insiders/Outsiders Festival directed by art historian Monica Bohm-Duchen from 2019, accompanied by the book *Insiders Outsiders*.

28 It is possible that small exhibitions formed a focus for the efforts during the First World War in Britain, although the evidence for these is elusive, but exhibitions with designers involved started to take shape after the First World War (see, for example, the account of Milner Gray in 'Exhibitions In or Out?', *Art and Industry*, October 1952, p. 110).

29 As asserted by Misha Black in his 1949 essay 'Propaganda in Three Dimensions', in A. Blake (ed.), *The Black Papers on Design: Selected Writings of the Late Sir Misha Black* (Oxford: Pergamon Press, 1983).

30 Edward L. Bernays, *Propaganda: The Public Mind in the Making* (Port Washington, NY: Kennikat Press, 1928), p. 9. The 'invisible government' Bernays described was essentially the work of cultural hegemony, as described by Antonio Gramsci in *Selections from the Prison Notebooks* (London: Lawrence & Wishart, 1971) and Raymond Williams in *Culture and Society* (London: Chatto & Windus, 1958). British MP Arthur Ponsonby raised some parallel ideas the same year in *Falsehood in Wartime* (London: Allen & Unwin, 1928). For earlier developments, Charles Higham, early champion of publicity in Britain, recognised that commercial advertising techniques would be useful in disseminating information or propaganda to the public, giving it the legitimacy and respectability to advertising that it had previously lacked, as Jim Aulich argues in *War Posters: Weapons of Mass Communication* (London: Thames & Hudson, 2011), p. 8. See Charles Higham, *Looking Forward: Mass Education through Publicity* (London: Nisbet, 1920).

31 Black, 'Propaganda in Three Dimensions', pp. 119–29.

32 Misha Black (ed.), *Exhibition Design* (London: Architectural Press: 1950), p. 22. Jacques Rancière discusses the way that advertising and Modernist formalism shared goals and principles in 'The Surface of Design' in the *Future of the Image* (London: Verso, 2007), pp. 91–108.

33 This is the subject of Scott Anthony's study *Public Relations and the Making of Modern Britain* (Manchester: Manchester University Press, 2012). Stephen Tallents was central to these organisations, as Anthony shows. See also Mark Wollaeger,

Modernism, Media and Propaganda: British Narrative from 1900 to 1945 (Princeton, NJ: Princeton University Press, 2006).

34 Richard S. Lambert, *Propaganda* (London: Thomas Nelson & Sons, 1938), pp. 118–19.

35 Black, 'Propaganda in Three Dimensions', p. 128. Reporting during the war in *Architectural Review* magazine on the MOI's extensive exhibitions programme, writer G. S. Kallmann cautioned that producers of propaganda exhibitions such as these needed to understand mass psychology'; G. S. Kallmann, 'The Wartime Exhibition', *Architectural Review*, October 1943, p. 105.

36 Noam Chomsky, *Manufacturing Consent: The Political Economy of the Mass Media* (New York: Pantheon, 1988).

37 Staniszewski, *The Power of Display*.

38 Brian O'Doherty, first published as 'Inside the White Cube', *Artforum*, vol. 14, no. 7, March 1976, p. 24; later as *Inside the White Cube* (San Francisco: Lapis, 1986), p. 14.

39 Rosalind Krauss, 'Photography's Discursive Spaces: Landscape/ View', *Art Journal*, vol. 42, no. 4, *The Crisis in the Discipline*, Winter 1982, pp. 311–19.

40 Extending this theme of Modernism's antagonistic relationship with cultural institutions, Andreas Huyssen discusses 'the battle against the museum' as 'an enduring trope of modernist culture' in 'Escape from Amnesia: The Museum as Mass Medium', in *Twilight Memories: Marking Time in a Culture of Amnesia* (New York and London: Routledge, 1995).

41 Buchloh, 'From Faktura to Factography'.

42 For discussion of the evolution of stock photography see Paul Frosh, *The Image Factory: Consumer Culture, Photography and the Visual Content Industry* (Oxford and New York: Berg, 2003).

43 K. I. Vorobyyov, 'Museums in Industrial Enterprises', translated by Caroline Rees in Arseny Zhilyaev (ed.), *Avant-Garde Museology* (Minneapolis, MN: University of Minnesota Press, 2015), pp. 443–52.

44 Michael Tymkiw, *Nazi Exhibition Design and Modernism* (Minneapolis, MN: University of Minnesota Press, 2018), pp. 74–5.

45 Toby Clark, *Art and Propaganda in the Twentieth Century: The Political Image in Mass Culture* (London: Weidenfeld & Nicolson, 1997), p. 26.

46 Erkki Huhtamo points out that visual media culture does not exist solely in interior spaces in 'Messages on the Wall: An Archaeology of Public Media Displays', in *Urban Screens Reader* (Amsterdam: Institute of Network Cultures, 2009), pp. 15–28.

47 Olivier Lugon distinguishes this use of photography from more permanent 'collection prints', in 'Photography and Exhibition in Germany around 1930', in *Object: Photo. Modern Photographs 1909–1949: The Thomas Walther Collection at the Museum of Modern Art* (New York: Museum of Modern Art, 2014) and 'The Ubiquitous Exhibition: Magazines, Museums and the Reproducible Exhibition after World War II', in Thierry Gervais (ed.), *The 'Public' Life of Photographs* (London: MIT Press, 2016), p. 123.

48 Michael Shanks, *Experiencing the Past: On the Character of Archaeology* (London: Routledge, 1992).

49 W. J. T. Mitchell, 'Ekphrasis and the Other', Chapter 5 of *Picture Theory: Essays on Verbal and Visual Representation* (Chicago and London: University of Chicago Press, 1994), pp. 151–81.

50 Mitchell, 'Ekphrasis and the Other', p. 180.

51 Key archival collections for this study include National Archives London, British Library, Imperial War Museum, UoBDA, RIBA, V&A, Mass Observation, London Transport Museum, Transport for London, Tate, Tyne & Wear archives, Wellcome Collection, National Archives Washington, Library of Congress, New York Public Library and private collections.

52 Tymkiw uses this word to describe the easy borrowing of form; *Nazi Exhibition Design*, p. 17. Tymkiw discusses Eiermann's study trip to the US in spring of 1936, which gave him plenty of material for future German Nazi shows (p. 152). Amongst the official records of US wartime propaganda exhibitions at Library of Congress, I found images of Nazi exhibitions, as well as UK propaganda exhibitions.

53 The production of exhibitions is the focus of Kate Guy, Hajra Williams and Claire Wintle's edited book *Histories of Exhibition Design in the Museum: Makers, Process and Practice* (London: Routledge, 2023) and Harriet Atkinson, Verity Clarkson and Sarah Lichtman (eds), *Exhibitions Beyond Boundaries* (London: Bloomsbury Academic, 2022).

54 Nieland, *Happiness by Design*, p. 8.

55 László Moholy-Nagy, *Painting Photography Film* (London: Lund Humphries, 1986), pp. 38, 40.

56 László Moholy-Nagy, *Von Material zu Architektur* (*The New Vision*) was published in English in 1932 and shared information about Bauhaus methods, in particular the merging of theory and practice in design.

57 László Moholy-Nagy, *Vision in Motion* (Chicago: Hillison and Etten, 1947) was published posthumously. Sibyl Moholy-Nagy, *Experiment in Totality* (New York: Harper and Brothers, 1950), pp. 24–5, 31.

58 For an account of forms of attention and observation after the Second World War, see Orit Halpern, *Beautiful Data: A History of Vision and Reason since 1945* (Durham, NC and London: Duke University Press, 2014).

59 Herbert Bayer published his 'Diagram of Field of Vision' within the catalogue of the German Werkbund section of *Exposition de la Société des Artistes Décorateurs*, Paris, 1930, published as the 'Diagram of 360° Field of Vision' in 1935. Bayer later wrote 'Aspects of Design of Exhibitions and Museums', *Curator: The Museum Journal*, vol. 14, no. 3, 1961, pp. 257–88.

60 Turner, *The Democratic Surround.*

61 A Gropius exhibition held in Britain in 1934 was opened by Raymond Unwin and accompanied by a lecture Gropius gave to the DIA in May 1934 on 'The Future of the Town' at London School of Hygiene and Tropical Medicine, *RIBA Journal*, 19 May 1934, p. 690. RIBA Ove Arup papers ArO/2/12/1. Herbert Read had espoused the ideas and ideals of Walter Gropius in his 1934 book *Art and Industry: The Principles of Industrial Design* (London: Faber & Faber, 1934), which laid out Gropius's educational programme for English-speaking audiences.

62 As Valeria Carullo shows in *Moholy-Nagy in Britain, 1935–7* (London: Lund Humphries, 2019), Moholy-Nagy's displays for Simpson's and other shops made it into the pages of contemporary trade magazines. See also Achim Borchardt-Hume (ed.), *Albers and Moholy-Nagy: From the Bauhaus to the New World* (London: Yale University Press, 2006).

63 Black, 'Propaganda in Three Dimensions'. See also Black, *Exhibition Design*.

64 Ulrich Pohlmann, '"Not Autonomous Art but a Political Weapon": Photography Exhibitions as a Means for Aestheticising Politics and Economy in National Socialism', in *Public Photographic Spaces: Exhibitions of Propaganda from Pressa to the Family of Man, 1928–55* (Barcelona: MACBA, 2009), pp. 275–98.

65 Bayer referred to this kind of presentation in his essay of 1961. Halpern, *Beautiful Data*.

66 Tymkiw illustrates this in *Nazi Exhibition Design*, p. 51 and at Fig. 1.16 on p. 52.

67 Film historian Justus Nieland traces the history of designers experimenting with film as a medium back to the interwar avant-gardes, especially as they intersected with the techno-utopian agendas of the Bauhaus and the more 'quotidian, instrumental practices of industrial and educational film and other modes of … "useful cinema"' (Nieland, *Happiness by Design*, pp. 7–8).

68 Moholy-Nagy made films while in Britain including a documentary about lobster fishermen in Littlehampton, as discussed in Scott Anthony and James G. Mansell, *The Projection of Britain: A History of the GPO Film Unit* (London: Palgrave Macmillan, 2011), p. 15 and Carullo, *Moholy-Nagy in Britain*.
69 Scott Anthony in Anthony and Mansell, *The Projection of Britain*, p. 11. See Vinzenz Hediger and Patrick Vonderau (eds), *Films That Work: Industrial Film and the Productivity of Media* (Amsterdam: Amsterdam University Press, 2014).
70 See Richard Farmer, *Cinemas and Cinemagoing in Wartime Britain, 1939–45: The Utility Dream Palace* (Manchester: Manchester University Press, 2016), p. 14.
71 Hollie Price, *Picturing Home: Domestic Life and Modernity in 1940s British Film* (Manchester: Manchester University Press, 2021).
72 Farmer, *Cinemas and Cinemagoing*.
73 Anne-Katrin Weber, *Television before TV: New Media and Exhibition Culture in Europe and the USA, 1928–1939* (Amsterdam: Amsterdam University Press, 2022).
74 Forsyth Hardy, *John Grierson: On Documentary* (London: Collins, 1946).
75 John Taylor, 'Picturing the Past', *Ten.8*, no. 11, 1983, pp. 15–31 discusses George Orwell's *The Road to Wigan Pier* (London: Left Book Club, 1937).
76 Zoe Druick and Deane Williams (eds), *The Grierson Effect: Tracing Documentary's International Movement* (London: Palgrave Macmillan, 2014), p. 2.
77 Brian Winston, 'To Play the Part That Was in Fact His/ Her Own', in Druick and Williams, *The Grierson Effect*, p. 43.
78 As discussed by Francois Albera and Maria Tortajada in their 'Introduction to an Epistemology of Viewing and Listening Dispositives', in *Cinema Beyond Film* (Amsterdam: Amsterdam University Press, 2010), pp. 10–11.
79 Buchloh, 'From Faktura to Factography'; Charlotte Klonk, *Spaces of Experience: Art Gallery Interiors from 1800 to 2000* (New Haven, CT: Yale University Press, 2009).
80 Walter Benjamin, 'The Work of Art in the Age of Mechanical Reproduction', in *Illuminations* (London: Bodley Head, 2015), p. 228.
81 Buchloh, 'From Faktura to Factography'.
82 Jonathan Crary, *Suspensions of Perception: Attention, Spectacle and Modern Culture* (Cambridge, MA: MIT Press, 1999), pp. 1–3.
83 Mikhail Bakhtin, *Rabelais and His World* (Bloomington, IN: Indiana University Press, 1984), p. 7.
84 Tymkiw, *Nazi Exhibition Design*, p. 269.
85 Tymkiw, *Nazi Exhibition Design*, p. 187.
86 Tymkiw, *Nazi Exhibition Design*, pp. 178, 183.
87 British Library, *AIA Bulletin*, no. 81, January 1944 (original emphasis).
88 Glynnis Chantrell (ed.), *The Oxford Dictionary of Word Histories* (Oxford: Oxford University Press, 2002).
89 Exhibitions therefore operated as montage in the way that art historian Matthew Teitelbaum has discussed it, as seeking 'not merely to represent the real but also to extend the idea of the real to something not yet seen'; *Montage and Modern Life 1919–1942* (Cambridge, MA and London: MIT Press, 1992), p. 8.
90 Jeffrey T. Schnapp, 'Fascism's Museum in Motion', *Journal of Architectural Education*, vol. 45, no. 2, February 1992, p. 88. See also Jeffrey T. Schnapp, 'Epic Demonstrations: Fascist Modernity and the 1932 Exhibition of the Fascist Revolution', in Richard J. Golsan (ed.), *Fascism, Aesthetics, and Culture* (Hanover, NH: University Press of New England, 1992), pp. 1–37.
91 This is the focus of Oliver Lugon's seminal essay 'Dynamic Paths of Thought', pp. 117–44. There is a strand of contemporary museum visitor studies that maps and engages with how visitors move through exhibition spaces; see for example George E. Hein, *Learning in the Museum* (London: Routledge, 1998) and Helen Rees Leahy, *Museum Bodies: The Politics and Practices of Visiting and Viewing* (London: Routledge, 2016).

1

Banishing 'chaos, vulgarity and mediocrity': training as an exhibition designer

The major focus of this book is the transformation of perceptions of the possibilities of exhibitions as communication in Britain from the 1930s. While exhibitions had long been understood as the focus for acts of diplomacy and for revitalising international trade, as this book will show, it was not until the Second World War that exhibitions' form and content became used more systematically in Britain for expressing political positions and opinions, in the context of the reinstated Ministry of Information (MOI). The pretext for these changes was the recurring accusation that the British government and its agencies were unable to make exhibitions work in Britain's favour, despite a national identification with the inception of the international exhibitions tradition in 1851. British attempts to put this failure right, through training and by initiating new organisations to drive up design standards, is the central focus of this chapter.

Implicit in these developments was the seriousness with which exhibitions were regarded by British authorities as potent vehicles for carrying significant public messages. By the early twentieth century, exhibitions were established channels for enacting British diplomatic relations, with the British Pavilion opening at the Venice Biennale in 1909 as one of the Biennale's earliest national buildings.[1] The British government's Board of Trade invested heavily in exhibitions to drive up standards and sales in manufacturing. This chapter traces the varying routes through which exhibition designers learnt skills for the job: in training, apprenticeships, and art and architecture schools in Britain, including through the arrival in London of the German Reimann School, offering specific courses on exhibition and display design.

Although focused on attracting audiences to enjoy a day out, early twentieth-century commercial exhibitions were potent platforms for unsanctioned political interventions. The *Daily Mail* reported a 'suffragette invasion' at the first Ideal Home Exhibition of 1908, recounting the moment when a party of women's rights activists took the opportunity to speak to the predominantly female exhibition crowd about the rights denied to women who made homes.[2] Although the displays did not themselves serve to reinforce these political messages, the Ideal Home Exhibition served as a powerfully evocative platform from which to speak of women's discontent with the status quo and to share a vision of emancipation, given that the home was one of the few domains in which female decision-making was influential, through purchasing power rather than political might. Historian Zoe Thomas discusses various early twentieth century exhibitions that did, however, reinforce political messages of the British suffrage movement. These included a grand exhibition of the National Union of Women's Suffrage Societies staged at London's Olympia in spring 1914, which taught visitors about suffrage and put women's needs centre-stage in displays, as well as linking women with organisations that could offer support, such as The Women's Taxpayers' Agency and Legal Advice Bureau.[3]

Driving up standards: the British Industries Fairs

The conviction that trade fairs would boost the national economy, whilst displaying a country's industrial prowess, drove their evolution from the mid-nineteenth century. Through the early twentieth century, British officials continued to consider trade exhibitions as potent vehicles for commercial displays and promotion, despite Britain being regularly considered to produce inferior products by comparison with international competitors. The drive to improve Britain's exports was the pretext for several initiatives, including the long-lived British Industries Fair (BIF). Inaugurated by the British Board of Trade soon after the outbreak of the First World War in 1915, BIFs were intended to exhibit samples of British goods, to inspire home manufacturers to emulate them, for the benefit of British industry.[4] But as they discovered, the act of displaying goods could be counter-productive, risking making their deficiencies all the more apparent. Reporting from Shepherd's Bush BIF in 1927, *Commercial Art* noted the 'weaknesses conspicuous in our industrial output' compared with the 'facile ingenuity' of continental designers, attributing this to the structural separation between industry and 'first-rate talents' in British art schools.[5]

Despite the Board of Trade's best efforts, by the mid-1930s critics still had little positive to say about BIF and the state of British manufacturing. Visiting BIF 1934, *Shelf Appeal* singled out two Modernist 'bright spots'. One was commercial artist Edward McKnight Kauffer's stand for the GPO, with its simple black and white motif and tubular steel furniture. (Kauffer is introduced more fully in Chapter 2.) The other bright spot was Richard Levin's stand for the BBC, incorporating huge photomontages into the curved, streamlined structure (Figure 1.1). Levin, who had started his

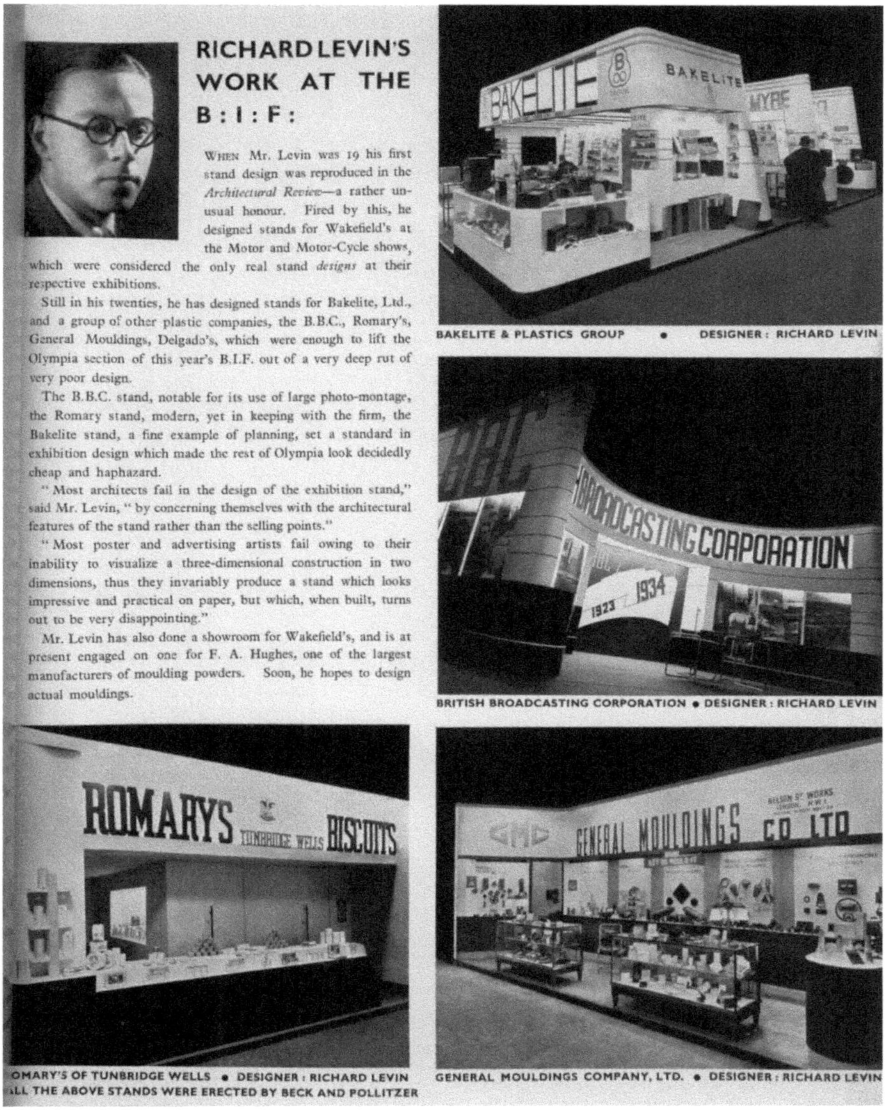

RICHARD LEVIN'S WORK AT THE B : I : F :

WHEN Mr. Levin was 19 his first stand design was reproduced in the *Architectural Review*—a rather unusual honour. Fired by this, he designed stands for Wakefield's at the Motor and Motor-Cycle shows, which were considered the only real stand *designs* at their respective exhibitions.

Still in his twenties, he has designed stands for Bakelite, Ltd., and a group of other plastic companies, the B.B.C., Romary's, General Mouldings, Delgado's, which were enough to lift the Olympia section of this year's B.I.F. out of a very deep rut of very poor design.

The B.B.C. stand, notable for its use of large photo-montage, the Romary stand, modern, yet in keeping with the firm, the Bakelite stand, a fine example of planning, set a standard in exhibition design which made the rest of Olympia look decidedly cheap and haphazard.

"Most architects fail in the design of the exhibition stand," said Mr. Levin, "by concerning themselves with the architectural features of the stand rather than the selling points."

"Most poster and advertising artists fail owing to their inability to visualize a three-dimensional construction in two dimensions, thus they invariably produce a stand which looks impressive and practical on paper, but which, when built, turns out to be very disappointing."

Mr. Levin has also done a showroom for Wakefield's, and is at present engaged on one for F. A. Hughes, one of the largest manufacturers of moulding powders. Soon, he hopes to design actual mouldings.

BAKELITE & PLASTICS GROUP • DESIGNER : RICHARD LEVIN

BRITISH BROADCASTING CORPORATION • DESIGNER : RICHARD LEVIN

OMARY'S OF TUNBRIDGE WELLS • DESIGNER : RICHARD LEVIN

LL THE ABOVE STANDS WERE ERECTED BY BECK AND POLLITZER

GENERAL MOULDINGS COMPANY, LTD. • DESIGNER : RICHARD LEVIN

1.1 Review of Richard Levin's exhibition stands at the British Industries Fair, including his work for the BBC, which was noted by *Shelf Appeal* of March 1934 for incorporating 'large photo-montage'. Uncredited photographer. 'Richard Levin's Work at the BIF', *Shelf Appeal* (March 1934), p. 321 © Gillian Levin. All rights reserved and permission to use the figure must be obtained from the copyright holder.

career in stage design with Gaumont-British, the largest British cinema chain of the time, was a prolific exhibition designer through the 1930s. He worked for a period with Arundell Ltd, alongside Misha Black, making stands for companies, as well as designing regularly for the BBC home and international service.[6] Levin was born in Britain in 1910, his Russian father (from what is now Latvia) and Dutch mother had arrived in Britain in 1908. Levin went on to work on exhibitions across many of the contexts in this book, including at the MOI and overseeing a major element of the 1951 Festival of Britain.[7] He worked regularly alongside exhibition designer Misha Black whose own parents had moved to Britain from Russia (now Azerbaijan) in 1912. Many of the designers that are central to this book shared Levin and Black's Jewish heritage.

Levin was one of the most distinguished exhibition designers of the interwar period. Speaking about his work at BIF 1934 to *Shelf Appeal*, he explained that most exhibition stands failed by being overly focused on architectural features, rather than on 'selling points', while poster and advertising artists' stands failed as they were unable to visualise three-dimensional constructions.[8] Levin's view was increasingly being recognised by those who agreed that a stand's formal qualities should be secondary to its message.

Shelf Appeal's critics continued to decry BIF's 'chaos and vulgarity', as visitors to BIF 1935 paid inflated charges to enter an overly vast space filled with a chaotic series of 'new firms' showing 'gadgets' and 'knick-knacks'. Richard Levin's laminated Bakelite armchair for Bakelite Ltd was another rare high point in an event otherwise dismissed by the design press as 'easily the worst from every standpoint'.[9] Editor and designer Noel Carrington, writing in *News Chronicle* after visiting BIF 1935, declared it 'utterly obsolete', 'tedious, ill-arranged and flyblown as a provincial church bazaar', vast and seemingly endless.[10] Clearly BIF's exhibition strategies were failing, as criticisms continued, with *Shelf Appeal* describing BIF 1936 as 'a monument to mediocrity' and noting that enlarged photographs or 'photomurals' were for the first time to the fore in displays, with the hope that the general standard of exhibition stands was slowly improving, thus showing the magazine's criteria for 'good design' to be closely aligned with manifestations of international Modernism.[11]

Instilling 'a new spirit': the Design and Industries Association and British Institute of Industrial Arts

For those puzzling over how to improve British exhibitions and the standards of goods in them, two interlinked issues recurred from the 1910s: the question of what constituted 'good design' and the nature of 'appropriate' national style. How to improve the quality of design in industrial products without losing sight of a perceived character of 'Englishness' was a guiding consideration, particularly when Modernism was considered an international import. From its founding in 1915, the same year as BIF, the Design and Industries Association (DIA) aimed 'to instill a new spirit of design into British industry', using exhibitions for promotion.[12]

The DIA was formed following the visit of a group of British makers to the German Werkbund (German League of Works) exhibition held in Cologne in 1914. The first major exhibition of the Werkbund, an association of artists, architects, designers and industrialists, it set out to demonstrate to visitors what had been achieved through German government support for design.[13] German design in general, and German exhibition design in particular, were long-running points of comparison and competition for Britain. The DIA was a response to the perceived successes of the German Werkbund in stimulating better-quality industrial design and enhancing the relationships between manufacturers, consumers and the economy while trying to correct the heavily criticised, lacklustre Arts and Crafts triennial exhibition of 1912.[14] Responding directly to the 1914 Werkbund exhibition, and demonstrating exhibitions' capacity to inform and inspire across different contexts, the DIA organised an exhibition at Goldsmiths' Hall, London in March 1915, foregrounding examples of 'well-designed' German and Austrian commercial products and acting as a recruiting ground for the DIA.[15]

Reporting on the Leipzig Fair in spring 1926, DIA member and businessman Harry Peach noted how impressed he was by Germans employing 'the best brains they can to help them, whether for shop window display or exhibitions' and allowing architects 'a much bigger place in commercial design than with us'. German printing was the area that most impressed Peach, with posters, advertising and packaging particularly 'gay', 'bright' and 'full of character' and posters with 'character and liveliness', 'more individuality' and 'brighter colours'.[16] Standards appeared to be improving in Britain in 1927 when the DIA arranged a British section at Leipzig that magazine *Commercial Art* noted as taking a 'creditable place', although offering no information as to its contents or installation.[17] The DIA continued to use exhibitions as propaganda for their work, both at home and abroad: to represent their ideas about 'good design', the need for reform of manufacturing practices and to share their censorious vision of acceptable form.[18] The DIA's efforts were largely ignored by the trade, who refused to be preached to, and it never grew to be a substantial organisation, despite continuing efforts through exhibitions.[19]

The short-lived British Institute of Industrial Arts (BIIA), founded in 1920 with Treasury support, was another government-supported exhibiting effort, whose aims at raising standards of design in industry and improving public taste overlapped with the DIA's.[20] The BIIA had a permanent Exhibition Gallery in Knightsbridge from which to showcase exemplars of industrial art and handicrafts including metalwork, ceramics and glassware; however, modes of display and installation were conventional. Three temporary displays of modern products held at the V&A included *Industrial Art for the Slender Purse* of 1929, focused on cost-saving during the economic slump. The BIIA lasted until 1933, having had limited impact.[21]

The faltering progress of advertising design was another concern for Britain, with British eyes on Germany.[22] *Commercial Art*, reporting on *The Advertising Exhibition* held at Olympia in 1927, noted the healthy state of German advertising: 'until recently, industrial exhibitions in [Britain] have suffered by comparison with similar enterprises abroad', their exhibits displayed in a 'haphazard' way.[23] The *Regent Exhibition of Advertising of To-day and To-morrow*, overseen by architect Joseph Emberton and held at Dorland Hall on Lower Regent Street in 1934, included a section on political advertising, showing how fascist and communist politics were propagated and inviting people to contribute work.[24]

The amplification and dissemination of exhibitions in and from Britain in the 1930s was aided by the proliferation of trade journals, which had long reported on Britain's mixed efforts towards exhibiting on the world stage. *Architectural Review* magazine had a focus on emerging exhibition practices and was long established after its founding in 1896. The DIA had its own magazine, the *DIA Quarterly Journal*. *Display, the Official Magazine of the British Association of Display Men*, was launched in 1919. *Commercial Art: A Magazine of Printing & Advertising Progress* was produced in London from 1922 by the publishers of *The Studio*, conceived of as a trade journal for the British advertising industry. The *Daily Express* newspaper organised a National Display Competition annually from 1924 and in 1933 *Shelf Appeal* began, catering to the interests of commercial artists primarily working on display and packaging.

Britain eclipsed at international exhibitions

Britain's contributions at international expos in the 1920s were at best banal, nostalgic and eccentric. A notable example was the British contribution to *Der Internationalen Presse-Ausstellung* (*International Press Exhibition*), known as *Pressa*, held in Cologne in 1928. The British exhibits were a narrow selection, mainly from within the orbit of the DIA, including contemporary fine printing and book illustration, posters for London Transport and works by Crawfords advertising agency, accompanied by a potted palm and a Union Jack. *Das Berliner Tageblatt* newspaper dismissed Britain's effort as 'pious, aristocratic, historically reverent, at peace in its confidence'.[25] This contrasted with the newspaper's excitement about El Lissitzky's Soviet Pavilion, a montage environment that was wowing audiences. It used text to striking effect as a central element, with lettering in a range of bold, dominating sans serif fonts from tiny to human-height;

demanding attention and unsettling the viewer, following and reinforcing the structural elements. These were shown alongside powerful enlarged photographs on the horizontal and vertical axes over viewers' heads, producing rapid changes in rhythm and mood that disorientated and stimulated audiences. The 'L-E-N-I-N' display at *Pressa* used lettering to reinforce the sequential qualities of the exhibition, forcing visitors to move through the display in order to reveal its cryptic contents.

The gulf between the British and Soviet Pavilions' installations was stark: the British exhibits were wall-hung in a conventional space dominated by a Union Jack and a printing press, surrounded by wood engravings using historic typefaces. Meanwhile El Lissitzky led a Soviet team in the creation of a vast, chaotic but compelling photomural, 'The Task of the Press Is the Education of the Masses', which, *Das Berliner Tageblatt* enthused, displayed 'grandeur in its exposition of social conditions … Forward! In the struggle and into class consciousness'.[26] The Soviet contribution showed the potential of photography and of exhibition as a creative medium in its own right, a new discipline within the field of visual communication. Herbert Bayer later explained that Lissitzky's installation at *Pressa* had inspired him to take up exhibition design, although he saw Lissitzky's work as 'chaotic' while his own desire was for the imposition of artistic control.[27] Lissitzky's *Pressa* room was also admired in the British display press. Typographer Jan Tschichold, writing in London-based *Commercial Art* magazine, marvelled at his use of 'a new exhibition technique', which 'produced a new purely visual design of the exhibition space', bringing together many materials including glass, mirrors, celluloid, nickel, wood, lacquer, textiles and photographs, as well as experimenting with introducing technologies, 'continuous films, illuminated and intermittent letters' and 'rotating models'. He compared the exhibition space, in Lissitzky's hands, to a stage, 'on which the visitor himself seemed to be one of the players', a novel form of immersive experience for spectators.[28]

Britain was relatively slow to grapple with the arts of publicity and public relations. British civil servant Stephen Tallents, who led the short-lived Empire Marketing Board (EMB) from 1926 to 1933, which used exhibitions as a primary mode of promotion, described being profoundly moved by Germany's powerful contribution to the 1929 *International Exhibition* at Barcelona. Tallents saw Germany's ability, through Mies van der Rohe and Lilly Reich's German pavilion made of glass, steel and marble and fourteen other sections, to reinforce the 'sense of the industrial power of modern Germany' and to transmit the 'expression of a lonely, powerful and forward-looking spirit'. Through them, Tallents saw Germany harnessing exhibition as the medium through which to project itself as 'the industrial leader of Europe' and the place to find 'efficient, modern manufactured goods'. This visit galvanised Tallents to the view that British exhibitions must follow suit, to be used more effectively for purposes of communication and projection. 'England has always held herself aloof from the world's opinion', Tallents wrote. 'She can no longer afford that indifference ... she should set herself to throw a true and modern picture of her qualities on the screen of the world's mind'.[29] For Tallents, as for his contemporaries, there was a distinction between cultural propaganda promoting 'national aims and achievements' and political or economic propaganda. The former was acceptable and necessary within a democracy, while the latter was not.[30]

As well as putting the kind of 'good design' on show that had so impressed members of the DIA, while Britain was still far from having identified the power of photography in exhibitions, German Werkbund exhibitions demonstrated photography's communicative potential. *Film und Foto: Internationale Ausstellung des Deutschen Werkbundes* (*Film and Photography: International Exhibition of the German Werkbund*) was held in Stuttgart in May 1929 before touring internationally. The exhibition showed how photography might be used as a dynamic and persuasive tool for communication in contemporary society, with sections given over to individual nations, to techniques like photomontage and to photography's relationship with advertising. In his contribution to the show, Moholy-Nagy was able to show off the approach he had called 'Neue Sehen' (New Vision): the camera's ability to act as a kind of prosthesis, extending and improving upon what people perceive with their eyes. In *Film und Foto*, Moholy-Nagy illustrated the principles of 'New Vision' by means of a display that brought together varied photographs with different provenances (zoological, astronomical, botanical, aerial, medical, forensic, industrial, journalistic), with the logic that they could be 'read' beyond the purpose for which they were taken. The exhibition was accompanied by a full cinematic programme, which highlighted the visual parallels between the content and sequencing of photography and cinema, a connection that would not be made in British exhibitions until the mid-1930s.[31]

German exhibitions not only impressed the critics for their use of two-dimensional media, such as photography and film; they were noted for their innovation as spatial entities. Their designers embraced exhibitions' innately dualistic ability to prompt an embodied, sequential experience of moving through displays on the ground, whilst giving visitors an overview from differing spaces and levels.[32] In his design for the German section of the *Ausstellungsstand der Baugewerkschaften* (*Exhibition of the Building Workers Unions*) in Berlin 1931, Herbert Bayer, working with Marcel Breuer, Walter Gropius and László Moholy-Nagy, created opportunities for visitors to climb above ground level, looking down from bridges to change the interactions with spaces, to give them intimate experiences of individual sections and long views of themes. Gropius and Moholy-Nagy both spent time in Britain in the mid-1930s, shaping design and exhibition cultures (as discussed later in this chapter).

Bayer, who moved from Berlin to New York in 1938, believed exhibition design paralleled 'the psychology of advertising' in being calculated to change visitors' beliefs and behaviours, to get fully under their skin.[33] The parallels between circulating in an exhibition space and reading were apparent to him in his observation that 'an exhibition can be compared with a book insofar as the pages of the book are moved to pass by the reader's eye' and that 'the reading method of Western man is from left to right. The walking direction in exhibitions must, logically, be from left to right'.[34] Bayer continued to work with this concept of exhibition visitor-as-reader, capitalising on it in his design for the MOMA's *Bauhaus: 1919–1928* exhibition in 1938 and his 1942 MOMA *Road to Victory* exhibition. In the latter, narrative underlined and reinforced the meaning through inextricably linked sequences, a spatial structuring with curves, flows and the careful placing of information: 'planned circulation', as Bayer described it, as the key principle.[35] *Road to Victory* toured Britain in 1943 with the revised name *America Marches with the United Nations* (as discussed more fully in Chapter 7). British propaganda exhibitions' increasingly deft incorporation of text, narrative and sequence, which reinforced their potential as spatial arguments, is a major theme of this book.

The Gorell Committee and after

Acknowledging Britain's deficiency in contributions to international exhibitions, the Board of Trade formed the Gorell Committee in 1931 to investigate and advise on 'the desirability of forming in London a standing exhibition of articles of every-day use and good design of current manufacture, and of forming temporary exhibitions of the same kind' and 'the desirability of organizing local or travelling exhibitions of the same kind both at home and abroad'.[36] The resulting report of 1932 concluded that 'The Government and Local Education Authorities should vigorously promote the improvement of the art education of the country', focused on the role of 'exhibitions of industrial art' in improving 'the taste of designers, manufacturers, distributors and the general public' by displaying 'beautiful modern manufactured goods, due regard being paid to the purchasing power of the householder of moderate means'.[37]

The Council for Art and Industry (CAI), chaired by Frank Pick and funded by the Treasury, was founded to enact the recommendations of the Gorell Report. Civil servant Pick links many of the events analysed across this book as a key figure in British cultures of publicity, promotion and propaganda into the war period (until his early death in 1941). Pick, who identified exhibitions as a key promotions vehicle, had worked with the Underground Group from the start of his career, appointed as its Publicity

Officer in 1908 and rising through the ranks to become Managing Director in 1928. As a founding member of the DIA, Pick was invested in promoting 'good design' and after the formation of the London Passenger Transport Board (LPTB) in 1933, Pick acted as patron for Modernist designers who he commissioned to create publicity posters and exhibitions, as well as designing new forms of transport. At the CAI Pick established a Committee on Presentation and Display, tasked with addressing Britain's poor exhibitions record and arranging exhibitions of industrial products abroad. Sir Edward Crowe, Controller of the Department of Overseas Trade, amplified these endeavours, explaining in *Display* magazine the government's recognition of 'the full value of publicity, propaganda, and advertising, and particularly the value of display'.[38]

'The first significant exhibition of design', according to historian Kenneth Luckhurst, surveying exhibitions immediately after the Second World War, was the 1933 *British Industrial Art in Relation to the Home* exhibition.[39] This was mounted in response to Gorell and held at Dorland Hall in 1933, showing the public the 'best examples' of 'the new industrial art'.[40] Exhibition architect Oliver Hill worked with a committee including E. McKnight Kauffer, Wells Coates, Serge Chermayeff and Raymond McGrath to show a 'Minimum Flat', a 'Unit House' and a large stone wall incised by sculptor Eric Gill. Under the banner 'Designs for Living', the exhibition showed five living rooms, bedrooms, bathroom, kitchen and office, as well as a set of 'interiors in motion' for ocean liner, car, airliner, yacht and plane.[41] McGrath's illustrations for the design of a yacht cabin were shown and the Lawn Road flats designed by Wells Coates were built in 1933–34 after being shown as prototypes at the exhibition. Mounted in the same year that Wells Coates founded MARS (the Modern Architectural Research Group) and Berthold Lubetkin formed the Tecton Group, chiming with this Modernist zeal and self-mythification, the exhibition professed 'to give expression to a new type of civilisation'.[42]

Showing Gorell's perceived significance, the committee Report was reproduced as an appendix to Herbert Read's book *Art and Industry: The Principles of Industrial Design* in 1934. As an editor, poet, writer and critic, Read had an overview of British culture, which enabled him to become chief interlocutor for British Modernist art movements from the 1930s. In *Art and Industry*, designed by Herbert Bayer and published by Faber & Faber, Read addressed questions about how exhibitions could improve design education and foster 'good form', attributing the poor quality of manufactured goods to poor education. Read espoused the ideas and ideals of Walter Gropius, laying out Gropius's educational programme for an English-speaking audience. He noted the shortcomings of the technical schools system and how 'art' education had become separate from 'technical industrial education', art seen as distinct from the process of machine production. The Bauhaus, Read believed, offered a model where 'in every practical activity the artist is necessary, to give form to material', from regional plans to door-handles. All schools should be assimilated to the 'Fachschule' or else become factories.[43]

Despite recognising models of good practice, the standards of British exhibitions continued to be poor. The *British Art in Industry* exhibition of 1935 was another conspicuous failure. Held at Burlington House, it was intended to impress upon the British and foreign public the importance of beauty in the articles they purchased and to prove to manufacturers that British artists could supply 'original, attractive and technically suitable designs for the production of articles by mechanical means'.[44] Prominent artists were invited to contribute, with Graham Sutherland showing china and glass carrying strong graphic shapes, Vanessa Bell and Duncan Grant showing patterned furnishing fabrics and Paul Nash showing decorated glassware and printed chiffon.[45] Serge Chermayeff and others joined forces to lampoon the exhibition as 'fanciful, extravagant, nothing to do with design for mass-production'. It was, Chermayeff said, 'totally unfunctional, completely inefficient, ladylike, directed towards Mayfair gracious living', revealing the problematic sexist, class-ridden ambivalence at the heart of the 'good design' agenda, which could deteriorate very quickly into middle-class (male) do-gooders trying to impose their own taste on working-class customers.[46] Russian-born architect and industrial designer Chermayeff had moved to Britain as a child, training during the 1920s at schools in Germany, Austria, France and the Netherlands while working as a journalist and designer to support himself. On returning to Britain in the late 1920s, Chermayeff developed the Modern Design Studio at furniture designers Waring and Gillow, before launching his own practice as interior architect and taking on regular exhibition commissions.[47] The DIA, echoing Chermayeff's criticisms, denounced *British Art in Industry* as 'a grotesque failure', while Herbert Read believed the exhibition would do more harm than good unless it was demolished, lashing out at a 'modish' and 'Mayfairish' walnut and maple bed.[48]

Raymond Mortimer, writing in the *New Statesman*, bemoaned the standards it set as 'costly and depraved', while Pevsner admitted the exhibition was embarrassing: 'what such an exhibition might have been!' Redeeming features included a playful perfumery display by Misha Black, praised by *Shelf Appeal* for showing fifty objects 'without the least appearance of crowding' (Figure 1.2).[49] *Architectural Review*'s assistant editor J. M. Richards responded to the exhibition with a long piece about the character of modern industrial design entitled 'Towards a Rational Aesthetic' in which he emphasised that standardisation should not mean monotony, as it had done there.[50]

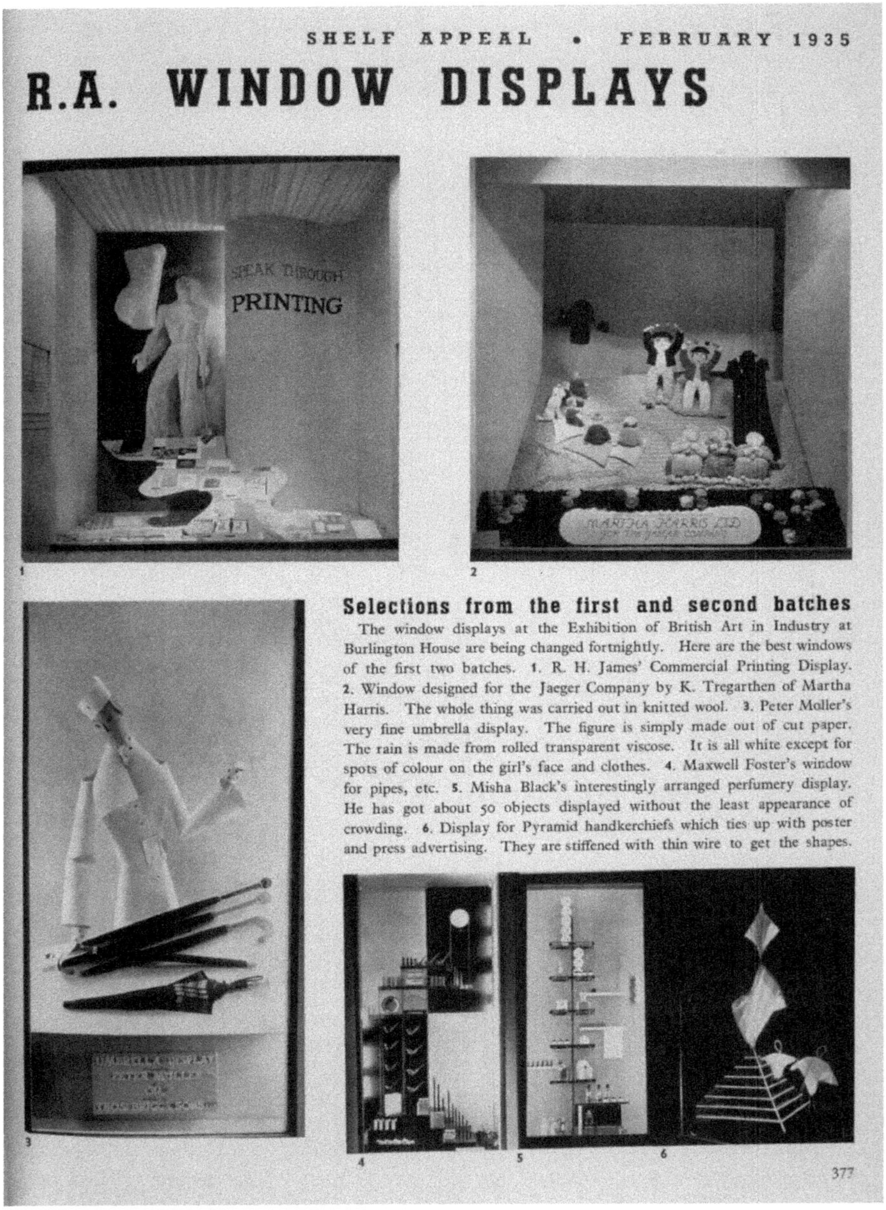

SHELF APPEAL • FEBRUARY 1935

R.A. WINDOW DISPLAYS

Selections from the first and second batches

The window displays at the Exhibition of British Art in Industry at Burlington House are being changed fortnightly. Here are the best windows of the first two batches. 1. R. H. James' Commercial Printing Display. 2. Window designed for the Jaeger Company by K. Tregarthen of Martha Harris. The whole thing was carried out in knitted wool. 3. Peter Moller's very fine umbrella display. The figure is simply made out of cut paper. The rain is made from rolled transparent viscose. It is all white except for spots of colour on the girl's face and clothes. 4. Maxwell Foster's window for pipes, etc. 5. Misha Black's interestingly arranged perfumery display. He has got about 50 objects displayed without the least appearance of crowding. 6. Display for Pyramid handkerchiefs which ties up with poster and press advertising. They are stiffened with thin wire to get the shapes.

377

1.2 Misha Black's 'interestingly arranged' perfumery display at the *British Art in Industry* exhibition, 1935. Image courtesy of the British Library.

Learning to be an exhibition designer

Herbert Read had proposed in his treatise *Art and Industry* that the standard of British design would only improve through education. Up to this point, exhibition design training had happened through apprenticeships and commissions for consultant design companies, which were increasingly central to the commercial ecology of shop window displays and trade fairs from the 1920s. The career trajectory of Misha Black, who is central to this book across all of its phases, is an important example of the haphazard nature of exhibition design training in early twentieth-century Britain. Black described becoming an exhibition designer 'by accident', starting work 'as an office boy in an exhibition firm' in 1927, at the age of 17, without formal art school training.[51] His first job was through London-based Arundell Display Ltd, managed by 'Lady' Page Wood,[52] which advertised itself in the 1920s as offering manufacturers, advertisers and retailers 'a complete window publicity service', with showrooms in Bush House for potential employers to see 'display novelties', innovative ways of treating display issues.[53] Black answered an advertisement in *The Times* inviting replies from 'anybody interested in learning shop-window dressing', which led to classes under Hans Kiesewetter, where he met and took on commissions alongside Lucy Rossetti, with whom he set up his first design practice, Studio Z.[54]

Other companies offering similar services included Wickham Limited and George Cuming Limited, which took on young apprentices such as Black to train up across several types of commission, including shop window displays, advertising in newspapers for people who were interested to work and learn semi-formally with authorities in the area.[55] Many companies put their stand designs to commercial display companies and consultants with specialist skills. Those regularly praised in the display press for commissioning interesting designers and incorporating new forms, typography and materials included Bakelite, the Electrical Distributors' Association (EDA) and the General Post Office (GPO).[56] The Gas Light and Coke Company, whose promotions work was often admired in the trade press, took on Misha Black as consultant designer for several years in the 1930s to design stands and other company branding elements.

Designing for exhibitions used transferable skills that could be developed in other commercial contexts including advertising poster and book cover design, packaging and shop window displays. Trade fairs such as BIF provided a lively training ground for commercial artists to hone their craft as designers and story-tellers. Dorothy Braddell's Crosse & Blackwell Marmalade stand for the Olympia *Food Exhibition* in 1925 showed her prowess as story-teller, incorporating a large model of the towers and spires of the four London churches mentioned in the 'Oranges and Lemons' nursery rhyme and a market scene, with exhibits on stalls with striped awnings playfully incorporating the firm's name. Braddell designed multiple stands for Shell-Mex Ltd, often with plate glass show windows behind which were positioned mechanical models of cars moving through a landscape.[57]

As well as being a route for commercially trained artists, exhibitions work offered an important training ground for young architects, especially given that the Depression had devastated architectural practice from the late 1920s, with building programmes slowing down or stopping entirely as a result.[58] Working on exhibitions filled the void, offering significant experience to recently qualified architects.[59] Some built their careers with companies that repeatedly commissioned them to design successive exhibition stands. Stands were often highly architectural, formed as miniature buildings, incorporating structural lettering, lighting and a furnished meeting area; becoming a chance to showcase new building forms and materials, novel forms of signage and contemporary furniture. Others were closer to the form of photographic exhibitions developed elsewhere, increasingly showcasing display trickery through the 1930s. Architect Rodney Thomas, who went on to found the Arcon architectural practice in 1943, worked with Ascot Gas Water Heater designing stands for nearly twenty years from 1936, which provided an excellent opportunity to experiment with building forms and materials.[60] Architect Joseph Emberton attributed the success of his whole architectural career to his first commission for an exhibition stand about artificial leather at Olympia in the early 1920s.[61]

By the mid-1930s, Modernist architects' contributions dominated the Building Exhibition, with manufacturers increasingly understanding the benefits of working with designers willing to experiment with forms in their stands.[62] The Venesta company pioneered the practice of commissioning stand designers direct: their 1930 stand for the Building Exhibition was designed by architects Le Corbusier and Charlotte Perriand;[63] they commissioned their 1932 stand from architect Wells Coates; their 1934 stand from new practice Tecton; and in 1936 they turned to reputable furniture designer R. D. Russell.[64] As I discuss in Chapters 7 and 8, architect Peter Moro and designer Robin Day, who collaborated on many exhibitions, testified to exhibitions' huge significance as training grounds within both their careers. Moro had been born in Heidelberg, training as an architect in Stuttgart, Berlin and Zurich before arriving in Britain in 1936, acquiring his Labour Permit initially as a specialist on spiral staircases, which allowed him to work with Berthold Lubetkin and the Tecton practice.[65]

Exhibition design was the least prestigious of the three-dimensional arts, as Misha Black acknowledged, describing exhibition designers as a 'small body of rather pathetic men'.[66] Yet designing exhibitions was an important source of income for artists and designers struggling to make ends meet. Artist Paul Nash – who worked variously as painter, printmaker, illustrator, photographer and commercial artist – recognised this, listing exhibitions as one of several ways of building artistic careers in an article of May 1933 for *The Journal of Careers*. In the article, Nash described how artists were admired but often not sufficiently remunerated for artistic work, discussing how difficult it was for artists to make a living by art alone and acknowledging the snobbery associated with admitting to taking on commercial design work, such as creating exhibitions.[67]

'A flutter in the dovecotes of display': Reimann in London

Standards in British exhibition design looked set to improve with the news that Berlin's 'famed' Reimann School was opening in London in January 1937. In anticipation, *Shelf Appeal* announced that soon there would be 'a flutter in the dovecotes of display' with the arrival of the School 'to design and sell display, as well as teach it', optimistic it could address British shortcomings in display.[68] Founded in Berlin in 1902 by Albert and Klara Reimann, the School had grown out of a regular weekend art school, quickly developing an international reputation.[69] Reimann staff knew how to bolster the School's reputation through publicity in the media, developing the in-house publication *Farbe und Form* and inviting prominent contributions. The School made sure that glowing reports on its work appeared regularly in the British trade press. Director Albert Reimann persuaded British readers of the merits of the School's training through an article in London-based *Commercial Art*, explaining how 'display men' were trained at 'the only higher technical school in Germany'.[70]

Rather than shying away from the commercial applications of training, Reimann embraced them, priding themselves on how excellently they equipped students for employment in the areas of design, mediating between production and consumption, courses including shop window display, stage exhibition and poster design, as well as packaging and commercial illustration. Reimann's window display courses under Georg Fischer were another draw, developed in tandem with department stores blossoming in Berlin.[71] The British trade press acknowledged the quality of German window displays as outshining British ones, something they attributed to poor training. An article in *Commercial Art*, written by an unnamed German designer, explained how German shop display training worked, highlighting the 'extraordinary mutual benefit' that could derive from a close relationship between the artist and merchant, and that 'commercial propaganda may be the means of raising the public taste to a higher level in matters of art'.[72]

The Reimanns' move to London was in direct response to increasing hostility from the Nazi regime. *Shelf Appeal* reported that the 'school of industrial art' in London would be led by Heinz Reimann, son of the founders, and run 'on a large scale as a profit-earning concern'. Its pointedly vocational focus was to be replicated in this new context: its intention for students to get 'good jobs on the strength of their training' paramount.[73] This was 'Design for Making a Living', a Reimann advert proudly declared, where 'training enables you to put your creative ability to practical use' (Figure 1.3).[74] The London School was set up to teach a mix of both amateurs and professionals already working in display, adapted to fit their timetables through teaching in both daytime and evening class sessions.

The relatively high fees meant students needed already to be earning. Fees meant, *Shelf Appeal* hoped in a spirit of encouraging its readers to put themselves forward to teach, that instructors would get better salaries than in art schools. Forty per cent of students were reported to come from outside Germany (including Scandinavia, Austria, France, England and the USA). That Reimann prepared students for 'real' work was reinforced by students not only using 'drawing boards' but working in studios 'to build displays' for stores and manufacturers that would, in themselves, bring income to the School and provide students with real experience. A staff of forty display instructors would be in charge, ten brought over from Berlin, as Heinz Reimann had been able to find few competent enough to teach in England, with students taught how to apply to display 'principles of dynamic, asymmetric balance, long applied to typography and layout'. *Shelf Appeal*, concluding its article about Reimann's London opening, stated, 'British display, judged by its conventions, its press, and on its own face value, is Britain's weakest selling point'.[75]

The London Reimann School and Studios opened on Regency Street, Westminster in January 1937, in an ex-warehouse converted into a 'modern school' for five hundred students, initially under the direction of ex-Architectural Association School Principal Howard Robertson. By the time of its opening its Principal was Canadian-born graphic designer Austin Cooper. The opening was marked with a sherry party, a tour of the new building and an exhibition sampling work of London teaching staff and Berlin-based students including a 'dramatisation of a Hoover', with 'ghost stairs' created from metal and stretched string, and an 'animated' window display for Ripolin paint using 'a black revolving roller' to suggest the can's contents were flowing down a slope. Cooper's opening speech referred to the exhibition: 'Remember it is the work of German students … But I doubt whether a higher standard of student work could be seen anywhere to-day', negotiating with sensitivity the possible perceived threat of the School as being too 'foreign' by reassuring that while the tried and tested system used in Berlin would be retained, the London School would present 'a tribute to British taste', by some unspecified means.[76] Design writer John Gloag spoke at the opening. 'There is', he said, 'an aching need for the proper training of people so that they work for commerce and industry without friction, without prejudice, without waste'. *Shelf Appeal*, reviewing the opening, reported

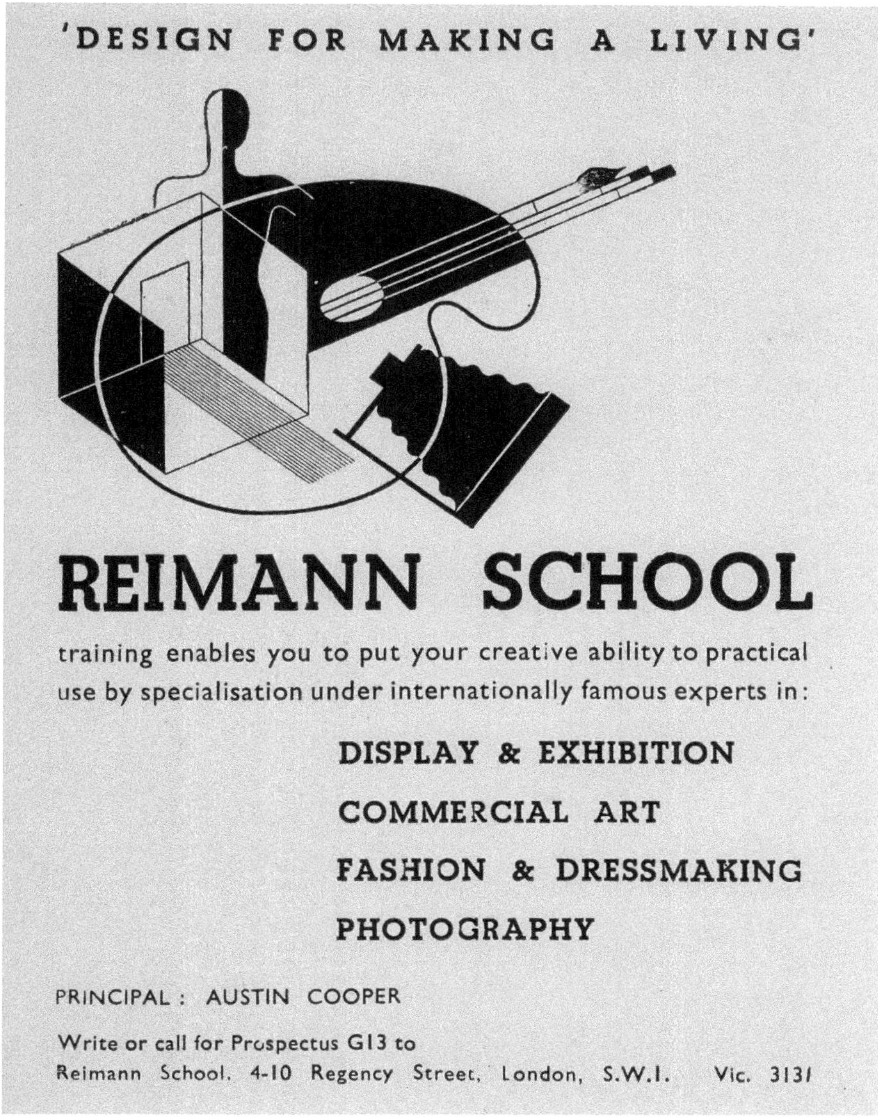

1.3 Advertisement for the Reimann School in London, 1937. Image courtesy of the British Library.

'display technique' vying with 'a spectacular list of instructors', while Frank Pick of the London Passenger Transport Board enthused about the inaugural exhibition as 'far above anything' he had ever seen – so good, in fact, that Pick thought it would make everything else in Britain look 'rubbish by contrast'.[77]

The London Reimann School's advisory council was a list of the great and the good of British commercial art.[78] The focus of specialisation was 'Display & Education', 'Commercial Art', 'Fashion & Dressmaking', 'Photography and Film' and 'Arts and Crafts (elementary and non-professional)', as trade advertisements made clear.[79] As *Shelf Appeal* noted again, the continuing paucity of British specialists in this area was reflected in the fact that only half of the forty London Reimann instructors was what they called 'English', among them illustrator Eric Fraser, industrial designer Milner Gray, typographer Robert Harling and graphic designer E. McKnight Kauffer. Only one of five display instructors was 'English', as such a talent 'did not exist in England'.[80] The lack of local instructors, *Shelf Appeal* explained, was because 'display is a three-dimensional job', 'construction and lighting and colour', 'not mere showcards and backgrounds as the average English art director sees it'. The magazine revealed an ongoing uncertainty over whether to embrace or gloss over the perceived foreignness of the School's work, influenced, as they described it, 'strongly by German practice'.[81] Richard Hamilton, later a celebrated Pop artist, was amongst the early tutors.[82] Natasha Kroll, former student at Reimann in Berlin and later an eminent display designer in Britain, was one of several designers who travelled from Berlin to join the London faculty (Figure 1.4).

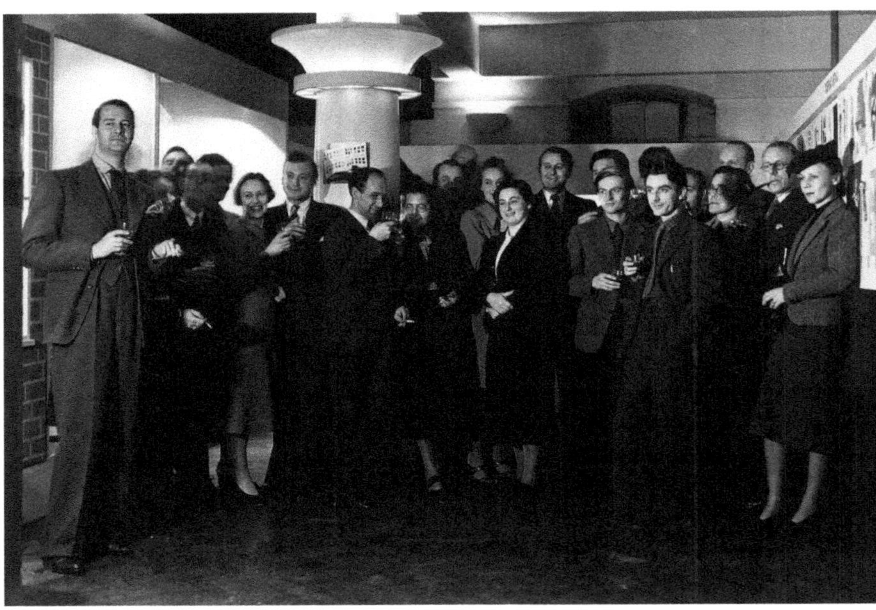

1.4 Faculty of the London Reimann School, with Natasha Kroll in the centre. Image courtesy of the University of Brighton Design Archives.

For some commentators the Reimann lived up to its promise: *Shelf Appeal* reported enthusiastically on progress, praising good quality work such as a raincoat display with string animating four coats worn by invisible people. It described how many department stores – the Lewis's chain, Swan & Edgar and Austin Reed – were sending staff to train at the School. Addressing potentially xenophobic critics, the magazine reported: 'If you ever had qualms that the school would introduce an alien note into British design', 'you will be reassured to learn that there is actually a bigger proportion of foreign students than foreign teachers', with, by this stage, only eight German of the forty teaching staff.[83] The head of exhibition and display planning was H. Loew, a student of Joost Schmidt and László Moholy-Nagy at the Bauhaus, where he had acted as Gropius's assistant in exhibition work.[84] Loew had joined the Reimann faculty after years as a freelance display and exhibition designer.[85]

An exhibition in autumn 1937 of the work of exhibition design students was an opportunity to show the departmental curriculum through 'an ingenious diagrammatic photo-mural showing the order of studies'. Much of its audience represented businesses 'anxious to engage trained students'.[86] By the end of the year Reimann taught shop window display, but not yet exhibition technique because of a shortage of expertise even on the Continent, despite the many propaganda exhibitions held in Germany.[87] Although there was much optimism for Reimann, it did not make a significant impact on the quality of British exhibitions and the training was not universally admired. Misha Black described his 'prejudice' against the school, considering their display work out of fashion, harking back to German display styles popular a decade earlier.[88]

Trying to push up standards

Designing impressive propaganda exhibitions demanded multiple skills – from understanding type and text layout, to using photography and film, to creating spaces as a setting to engage and envelope audiences – and these skills were still not widely shared and understood by British designers, even by the outbreak of the Second World War. This problem of how best to teach these skills associated with exhibitions work continued to puzzle all those involved. Art schools and technical colleges across Britain were often ill-equipped, their highly stratified and class-bound systems divided between teaching the higher status skills associated with fine art and architecture, on the one hand, and skills associated with practical and commercial applications on the other. This led designer Clifford Hatts, who started his career as exhibition designer after the Second World War, to observe the distinction up until the mid-1940s between the middle-class 'round boys', who pursued the more prestigious fine art and architecture training, and working-class 'flat boys', like himself, who had learnt commercial design skills through two-dimensional, graphic means at a technical college.[89] This stratified culture meant that nowhere were both the two- and three-dimensional skills associated with exhibition design being taught.

Brighton School of Art was one of several schools that ran courses associated with commercial display design in the mid-1930s. Brighton's 'Window and Shop Display' courses were entirely focused towards supporting commercial careers, benefiting shop workers and display assistants; providing, as they did, practical demonstrations, skills in window lighting, making suitable backgrounds, foregrounds and accessories and unit dressing.[90] The same was true elsewhere: at Leicester College of Arts and Crafts and Goldsmiths College, display courses were piecemeal and haphazard, focused towards commercial application.[91] In the process, any wider skills allowing students to develop a more conceptually nuanced sense of display as a way to communicate ideas, not just to sell products, were not being taught.

The failure of art and design schools to provide training for good exhibitions returned to the fore after Britain's critical failure at the Exposition Internationale des Arts et des Techniques dans la Vie Moderne held in Paris in 1937, where the British Pavilion was slated by critics. The British Pavilion, overseen by Frank Pick, working with lead architect Oliver Hill, contained a series of objects that were admired, including a painting by Edward Bawden and finely printed photographs of the English countryside. But they were shown beside sports displays about golf, tennis, polo, fishing and shooting game birds. The display's overall impact was eccentric and highly class-bound, leading bewildered *New Statesman and Nation* editor Kingsley Martin to ask: 'could this be England?'[92]

In the aftermath of this failure, the Board of Trade attempted once more to remedy the situation, setting up an enquiry on presentation and display, chaired by Pick, to ask all those involved with British exhibition design why they were 'not highly commended' and to investigate 'what had gone wrong'.[93] Those invited to give evidence to the panel give a fascinating insight into the profession of exhibition design just before the Second World War. Witnesses included architects (Serge Chermayeff, Joseph Emberton), art college principals, department store managers, the press, display trade bodies (the Association of Display Producers and Silk Screen Printers and the National Display Association), display firms (Beck and Pollitzer) and display designers (Misha Black, Grace Lovat Fraser). There was unanimous agreement from all witnesses on two points: first, that British art, architecture and design schools taught exhibition design badly, if at all, and second, that the pool of exhibition designers and design firms in Britain was extremely limited (less than a dozen).

An important focus of the hearings was on harnessing exhibitions' ability to act as communication. Several witnesses called in front of the Board of Trade committee were clear that this was a feature of foreign exhibitions but was not yet happening in Britain. Many designers described falling into exhibitions work by accident; however, architect Serge Chermayeff described his regular work on exhibitions as 'by intention'. Exhibitions were, he thought, an important form for explaining and showing modern architecture; 'we have to become our own propagandists', he said, likely referring to the exhibitions work of the Modern Architectural Research Group (MARS), which Chermayeff was involved with at the time (and which I discuss in more detail in Chapters 2 and 3). Chermayeff believed 'that if you put your morals, your principles, into the exhibit, with what has to be shown, you would educate the public and get them to accept those ideas by the way of the things displayed'. German exhibitions, he believed, were already managing to do this, explaining 'abstract facts' to the public, an approach he believed British government non-commercial exhibitions needed to adopt to show subjects like town planning, dealing with traffic, 'related to some particular feature in community life and not merely an efficient market place'.[94]

Exhibitions' shortcoming as communication in Britain – the major conclusion of the Board of Trade enquiry – was soon to be addressed head-on. But, as the next chapter will show, there were already some exceptions emerging during the 1930s, in the context of housing activism and the presentation of government policy through exhibitions. The strong economic imperatives towards the development of commercial exhibitions for promoting trade and industry in Britain continued during the first decades of the twentieth century, as did exhibitions focused on heading off poor quality goods and supporting their 'good design' agenda. British training in exhibition design was acknowledged as haphazard and piecemeal, with a renewed focus on improving mixed and deeply stratified art education from the mid-1930s. Authoritative art historian and critic Herbert Read supported this agenda. Like many contemporaries, Read saw importing the Bauhaus model as the answer to Britain's shortcomings. Early optimism about the potential impact on display design of the opening of the London Reimann School in 1937 turned out to be largely unfounded and the School's impact turned out to be more limited than hoped. The next chapter addresses the contexts in which exhibitions in Britain started to be seen as more effective as communications media.

Notes

1 Sophie Bowness and Clive Phillpot, *Britain at the Venice Biennale 1895–1995* (London: The British Council, 1995) and Annebella Pollen, *Art without Frontiers: The Story of the British Council, Visual Arts, and a Changing World* (London: British Council, 2023).

2 *Daily Mail*, 15 October 1908, p. 4 as quoted in Deborah Sugg Ryan, 'The Daily Mail Ideal Home Exhibition, 1908–39', in Maggie Andrews and Mary M. Talbot (eds), *All the World and Her Husband: Women in Twentieth-Century Consumer Culture* (London: Cassell, 2000).
3 Zoe Thomas, *Women Art Workers and the Arts and Crafts Movement* (Manchester: Manchester University Press, 2020), p. 80.
4 TNA BT54 Board of Trade: British Industries Fair: Minutes and Publications, https://discovery.nationalarchives.gov.uk/details/r/C3097. See also Kenneth Luckhurst, *The Story of Exhibitions* (London: Studio, 1951), p. 186. BIF ran until 1957 at rotating sites in London, Glasgow and Birmingham.
5 *Commercial Art*, 1927, p. 146.
6 Richard Levin interviewed on 3 September 1991 by Norman Swallow and Alan Lawson, https://historyproject.org.uk/interview/richard-levin (accessed 30 July 2023). Richard Levin's daughter Gill Levin explained more about his family history in an interview with the author over Zoom, 28 May 2023. See also Richard Levin obituary, *The Guardian*, 10 July 2000.
7 For further details see Atkinson, *The Festival of Britain*.
8 *Shelf Appeal*, March 1934, pp. 320–1.
9 *Shelf Appeal*, March 1935, pp. 422–5.
10 Noel Carrington writing in *News Chronicle*, 18 February 1935.
11 *Shelf Appeal*, March 1936, pp. 31, 36.
12 Fiona MacCarthy, *All Things Bright and Beautiful: Design in Britain, 1830 to Today* (London: Allen and Unwin, 1972), pp. 76–7.
13 This support had come in the wake of lobbying by Prussian civil servant Hermann Muthesius, posted to London between 1896 and 1904, who was inspired by design he had seen during this period. See Muthesius's three-volume study *Das Englische Haus* (*The English House*), 1904–5, and his writings for the German Werkbund from 1907, of which he was a founding member. Jeremy Aynsley, *Designing Modern Germany* (London: Reaktion, 2009), pp. 25, 62–7; Cheryl Buckley, *Designing Modern Britain* (London: Reaktion, 2007), pp. 49–54, 78–80.
14 Buckley, *Designing Modern Britain*, p. 49.
15 MacCarthy, *All Things Bright and Beautiful*, pp. 77–8.
16 Harry Peach, *Commercial Art*, April–May 1926, p. 89.
17 *Commercial Art*, 1927, p. 208.
18 Peter Rose, '"It Must Be Done Now": The Arts and Crafts Exhibition at Burlington House, 1916', *The Journal of the Decorative Arts Society 1850–the Present*, no. 17, 1993, pp. 3–12.
19 DIA exhibition *Household Things* held at Whitechapel Art Gallery in 1920; 'Exhibition of Household Things', *Manchester Guardian*, 26 August 1920, p. 6. DIA held exhibitions at Charing Cross Station in 1932, 1933, 1937 and 1938.
20 MacCarthy, *All Things Bright and Beautiful*, p. 82.
21 Jonathan M. Woodham, *A Dictionary of Modern Design* (Oxford: Oxford University Press, 2004). Other British groups using exhibitions to further their aims in the 1920s and 1930s included The Twentieth Century Group, founded in 1930 by architect Serge Chermayeff with Mansfield Forbes, Wells Coates, Howard Robertson and Jack Pritchard to promote exhibitions of contemporary design in relation to architecture and interior equipment; John Gold, *The Experience of Modernism* (London: Routledge, 1997), p. 118.
22 Poster designer Horace Taylor, writing in *Commercial Art*, argued that the problem of creating good advertising display was often, mistakenly, seen as an exhibition issue – a question of creating 'the ideal poster gallery' – but that 'the heterogeneous character of the hoardings is an essential part of the game', which should be 'motley in their appearance' and 'casually brought together'; 'The Ideal Hoarding', *Commercial Art*, June 1926, p. 102.

23 *Commercial Art*, 1927, p. 188.
24 *Shelf Appeal* reported on it in October 1934.
25 Quoted by Jeremy Aynsley in 'Pressa Cologne, 1928: Exhibitions and Publication Design in the Weimar Period', *Design Issues*, vol. 10, no. 3, Autumn 1994, p. 70.
26 Discussed by Aynsley in '"Pressa", Cologne, 1928. Exhibitions and Publication Design in the Weimar Period', in *Public Photographic Spaces*, pp. 99–100.
27 Staniszewski, *The Power of Display*, pp. 45, 48.
28 Jan Tschichold, 'Display that Has Dynamic Force: Exhibition Rooms by Lissitzky', *Modern Publicity Commercial Art Annual*, 1931.
29 Stephen Tallents, *The Projection of England* (London: Faber & Faber, 1932), pp. 34–5.
30 Philip M. Taylor, The *Projection of Britain: British Overseas Publicity and Propaganda 1919–1939* (Cambridge: Cambridge University Press, 1981) and Anthony, *Public Relations*.
31 Francesco Zanot, 'The Film und Foto Exhibition of 1929', in *Photoshow: Landmark Exhibitions that Defined the History of Photography* (London: Thames & Hudson, 2014), pp. 134–5. Andrés Mario Zervigón described the exhibition as an 'instructional photo essay', in 'The Peripatetic Viewer at Heartfield's Film und Foto Exhibition Room', *October*, no. 150, Fall 2014, pp. 27–48.
32 Lugon, 'Dynamic Paths of Thought', pp. 117–44.
33 Quoted by Alessia Tagliaventi in 'Photography at MoMA: Four Landmark Exhibitions', in *Photoshow: Landmark Exhibitions That Defined the History of Photography* (London: Thames & Hudson, 2014), pp. 170–1.
34 Bayer, 'Aspects of Design of Exhibitions and Museums', p. 276; Lugon, 'Dynamic Paths of Thought', pp. 117–44.
35 Staniszewski, *The Power of Display*; Bayer, 'Aspects of Design of Exhibitions and Museums', p. 260; Herbert Bayer, 'Fundamentals of Exhibition Design', *Production Manager*, vol. 6, no. 2, 1937.
36 Read, *Art and Industry*, appendix A, p. 134.
37 Published as *Art & Industry: Report of the Committee Appointed by the Board of Trade Under the Chairmanship of Lord Gorell on the Production and Exhibition of Articles of Good Design and Every-Day Use* (London: HMSO, 1932); Yasuko Suga, 'Modernism, Commercialism and Display Design in Britain', *Journal of Design History*, vol. 19, no. 2, 2006, p. 138. Read also reproduced The Gorell Report in Appendix A of *Art and Industry*, pp. 134–6.
38 'The Government and Display', *Display*, October 1934, p. 368.
39 Luckhurst, *The Story of Exhibitions*, p. 174.
40 MacCarthy, *All Things Bright and Beautiful*, pp. 104, 117–18.
41 *Shelf Appeal*, October 1934, p. 153.
42 MacCarthy, *All Things Bright and Beautiful*, p. 104.
43 Read, *Art and Industry*, pp. 6, 36, 39, 40, 131.
44 MacCarthy, *All Things Bright and Beautiful*, p. 106.
45 Susan Lambert, *Paul Nash as Designer* (London: Victoria & Albert Museum, 1975), p. 5.
46 *Shelf Appeal*, February 1935, p. 349.
47 Alan Powers, 'Obituary: Serge Chermayeff', *Independent*, 14 May 1996.
48 MacCarthy, *All Things Bright and Beautiful*, p. 107.
49 *Shelf Appeal*, February 1935, p. 377.
50 'Towards a Rational Aesthetic', *Architectural Review*, December 1935, pp. 211–20.
51 TNA BOT57/34 *Notes of a meeting held on Tuesday 1 February 1938, Committee on Presentation and Display*, evidence of Misha Black, p. 1.
52 'Lady' Page Wood's name appeared with scare quotes in the advertisement.
53 *Commercial Art*, vol. 3, no. 13, July 1927, p. XVI.

54 Kiesewetter had arrived in London from Germany the previous summer as Arundell Display Ltd's chief decorator; 'Two Continental Displays', *Display*, vol. 10, no. 4, July 1928, p. 167.
55 As Avril Blake notes in *Misha Black* (London: Design Council, 1984), p. 17. For more on shop window displays in Britain see the English publication in 1930 of Frederick Kiesler's *Contemporary Art Applied to the Store and Its Display* (London: Pitman, 1930); also Kerry Meakin, 'Women in British Window Display during the 1920s and 1930s', *History of Retailing and Consumption*, vol. 7, no. 2, 2021, pp. 115–36; and Emily Orr, *Designing the Department Store: Display and Retail at the Turn of the Twentieth Century* (London: Bloomsbury, 2019).
56 See, for example, *Shelf Appeal*, April 1934, p. 384.
57 *Commercial Art*, 1928, pp. 224–5.
58 As John Summerson noted in 'The MARS Group and the Thirties', in John Bold and Edward Chaney (eds), *English Architecture: Public and Private* (London: The Hambledon Press, 1993), p. 303.
59 The ambiguities over the kind of training that equipped exhibition designers were the subject of TNA BT57/34, Serge Chermayeff's evidence to the Board of Trade Committee on Presentation and Display of 1 February 1938.
60 David Dean, *The Architect as Stand Designer: Building Exhibitions 1895–1983* (London: Scolar, 1985), p. 71.
61 TNA BT57/34 Emberton evidence, p. 1.
62 Dean, *The Architect as Stand Designer*, p. 59.
63 The 1930 stand was commissioned for Venesta by Jack Pritchard, with encouragement from John Gloag, who worked for the advertising agency holding the Venesta account.
64 Dean, *The Architect as Stand Designer*, pp. 53, 56–7, 61–3, 70.
65 Peter Moro interview with Louise Brodie, BL National Sound Archives, part 7; Alistair Fair, *Peter Moro and Partners* (Liverpool: Liverpool University Press, 2021).
66 Black, 'Propaganda in Three Dimensions', p. 119.
67 Paul Nash, 'Finding a Living in Art To-day', *The Journal of Careers*, vol. 12, no. 130, May 1933, pp. 5–8; James Boswell, *The Artist's Dilemma* (London: The Bodley Head, 1947).
68 *Shelf Appeal*, October 1936, p. 28.
69 Jeremy Aynsley, *Graphic Design in Germany 1890–1945* (London: Thames & Hudson, 2000), p. 111.
70 *Commercial Art*, vol. 5, 1928, p. 205. Images of window displays from Reimann in Berlin had long appeared in the trade press – *Commercial Art*, June 1926, p. 113, for example.
71 Aynsley, *Graphic Design in Germany*, p. 113.
72 *Commercial Art*, vol. 2, no. 8, February 1927, pp. 41–5. The author explained that technical papers and magazines supported the profession in Germany and that the popular book on window display *Schaufensterkunst: Lehrsätze und Erläuterungen* by Elisabeth Von Stephani-Hahn had already gone into multiple editions. An article the same year in *Commercial Art* (pp. 164–6), giving a Belgian view on publicity, also reported that Germany was 'where it was at'. While another piece in *Commercial Art*, 1927, p. 118 showed French window displays to be an inspiration.
73 *Shelf Appeal*, October 1936, pp. 27–8.
74 *Modern Publicity*, 1937–8. Similar advertisements found in *Shelf Appeal*, November 1936, p. 20 and December 1936, p. 8.
75 *Shelf Appeal*, October 1936, pp. 27–32.
76 Austin Cooper, 'A Tribute to British Taste', *Display*, February 1937, p. 609.
77 *Shelf Appeal*, January 1937, p. 54. Pick quoted by Yasuko Suga in 'The Reimann School and Studios', in *Designs on Britain: An Exhibition by Jewish Museum London* (London: Jewish Museum London, 2017), p. 85.

78 The advisory council included Jack Beddington of Shell-Mex, John Grierson of GPO films, graphic artist Edward McKnight Kauffer, Fred Phillips of Baynard Press, Howard Robertson and F. R. Yerbury of the Building Centre, while Basil Marriott, late Art Director of the Empire Marketing Board, was in charge of publicity; *Shelf Appeal*, October 1936, p. 29.
79 *Modern Publicity*, 1937–8. Similar advertisements found in *Shelf Appeal*, November 1936, p. 20 and December 1936, p. 8.
80 This one 'English' (as they described him) display instructor was given as R. Ll Huws, who took charge of 'Plastics and Window Dressing'.
81 *Shelf Appeal*, January 1937, p. 54–6.
82 As Suga notes, quoted in 'The Reimann School and Studios', p. 86.
83 *Shelf Appeal*, November 1937, p. 40–1; *Shelf Appeal*, December 1938, p. 34.
84 TNA BT57/34 Board of Trade Committee on Presentation and Display, summary of evidence given by Mr. Cooper and Mr. Loew of the Reimann School, 31 December 1937, p. 1.
85 Loew wrote a text in José Schmidt, *Lehre und Arbeit am Bauhaus 1919–32* (Dusseldorf: Marzona, n.d.).
86 *Shelf Appeal*, December 1938, p. 34.
87 As noted in TNA BT57/34, Board of Trade Committee on Presentation and Display, summary of evidence given by Mr. Cooper and Mr. Loew of the Reimann School, 31 December 1937. The London Reimann was forced to close after the outbreak of the Second World War, by which time two hundred students of many nationalities were enrolled and seven hundred students had completed their studies. It never reopened. Suga, 'The Reimann School and Studios', p. 90.
88 TNA BT57/34, Board of Trade Committee on Presentation and Display, hearings with Misha Black, 1 February 1938, pp. 4–5.
89 Graphic designer Clifford Hatts, interview with the author, London, 2003.
90 UoBDA Brighton School of Art prospectus, 1935–6, p. 32. The course was largely the same in 1944–5. H. A. Rothholz went to Willesden School of Art, part of the Technical College, in the late 1930s and, in 1937, at the age of 18, won first prize for British exhibits at the International Congress on Art Education in Paris; UoBDA RHZ/3/1/2.
91 TNA BT57/34 Clive Gardiner and Kenneth Holmes evidence, 7 November 1937.
92 Kingsley Martin, *Editor: A Second Volume of Autobiography, 1931–45* (London: Hutchinson, 1968).
93 TNA BT57/34, Notes of a meeting held on Tuesday 1 February 1938, Committee on Presentation and Display, evidence of Misha Black, p. 1.
94 TNA BT57/34, February 1938, Serge Chermayeff, pp. 2–6, 7–9. Beck and Pollitzer shared this view in their evidence, p. 23.

2

Exhibitions as projection, promotion, policy and activism in three dimensions

'The task of every exhibition is to sell something, whether it is a new line of tea-pots or a plan for the regeneration of Western civilisation', wrote Misha Black, reflecting on the agility needed to practise as an exhibition designer, when demanded to turn your hand to 'selling' across extraordinarily diverse contexts.[1] Black's image of exhibitions offering the potential to sell everything from teapots to a vision of civilisation itself is apt for this chapter, which focuses on exhibitions' use in Britain from 1933 to 1939 as propaganda for promoting diverse ideas and agendas by official bodies and activist groups. Essentially, exhibitions had been identified as a potent means of 'manufacturing consent', as media theorist Noam Chomsky later described it, to control the population through non-coercive, participatory means, rather than through the imposition of penalties or conspicuously punitive methods.[2]

Exhibitions were, increasingly, operating within the complex system of communications, as an adjunct to other media forms. In his 1938 book *Propaganda* Richard S. Lambert, editor of the BBC's in-house magazine *The Listener*, argued for the value of propaganda in the dissemination of truth, observing that 'propagandist organisations' were using many types of activity as propaganda, from giving concerts and plays to arranging art exhibitions.[3] Exhibition designer Misha Black, whose work and ideas are central to this book, believed an exhibition's job 'must be based on the principle of persuasion, consent and participation'.[4] Such exhibitions could help the British population accept particular ideas, such as new government initiatives.[5]

Exhibitions were being recruited as an element in the burgeoning culture of promotion and public relations developing in Britain during the interwar years, as this chapter shows. In his account of this emerging culture, historian Scott Anthony cites Stephen Tallents and Frank Pick as central proponents. Both Tallents and Pick were firm believers in the power of exhibitions to convey ideas to the public, within a body of assorted media.[6] In the previous chapter I discussed the striking impression the German Pavilion at the 1929 Barcelona exhibition had made on Tallents. In this chapter I show how government-funded bodies such as Tallents' Empire Marketing Board (EMB), the Underground Electric Railways Company of London (UERL) and its successor the London Passenger Transport Board (LPTB) and the General Post Office (GPO) used exhibitions as part of their promotions strategy. I analyse a number of examples, including exhibitions promoting 'good design' (*Design in Modern Life*), the Postal Service (*Post Office Exhibition*), British Empire practices and products (*Timber through the Ages, Flying over the Empire* and *Peeps at the Colonial Empire)*; exhibitions amplifying government policy (*Do Not Pass by the Special Areas* and *The Highway Code)*; and exhibitions as the focus for activist campaigns (*Clear Smoke from the Air* and *New Homes for Old*).

Exhibition design in Britain, as this chapter shows, was a vital cosmopolitan practice principally shaped by refugee and migrant artists and designers from diverse contexts, who arrived in Britain in large numbers during the early twentieth century. Black had arrived in Britain from the Russian Empire as a baby, while more recent arrivals were also to become formative influences on British exhibition cultures, as this chapter discusses, including László Moholy-Nagy, Hans Schleger, E. McKnight Kauffer and F. H. K. Henrion. It was the visual sensibility of this group that shaped information design and exhibition design more generally in Britain. Many of these artists and designers reappear across the multiple exhibitionary contexts of this book.

Exhibitions as projection and promotion

Turning over public spaces to use for sharing information was becoming established practice. Indeed, this kind of public promotional material was created to be shown in spaces of transition. Filmmaker John Grierson had set up film projectors at Victoria Station in the 1920s to attract passersby to watch his documentaries, as Stephen Tallents recalled.[7] From the mid-1920s, London's Charing Cross Station became a recurrent exhibition site, with the Underground station's modest ticket hall becoming a well-used gallery as part of the liminal landscape of transport hubs and terminals. The extensive series of early Charing Cross exhibitions is recorded in the London Transport Museum archive, but the accompanying committee papers have not survived, so it is unclear who was responsible for organising and designing many of them, or what their overarching vision was. It is evident, however, that UERL started mounting exhibitions as a way to bring art to the public, to show off the progress of a design-conscious authority and to promote 'good design'. Zoological specimens were shown on the westbound platform of Mansion House Station in 1925, followed by displays of a Frigidaire company ice demonstration at Piccadilly Circus Station in 1929 and a series of showcases at Leicester Square, including Central School of Arts and Crafts students' work and glimpses of V&A collections with a label on a dress asking passersby: 'What Did Our Grandmothers Wear?'[8] Set in glass vitrines, with small informative text panels, they mirrored conventional museum installations of the day.

With Frank Pick at the helm, the LPTB (as UERL had become from 1933) was intensely focused on promotion. Architect Gerhard Kallmann commented on the way that Frank Pick allowed Charing Cross Station to be used for 'campaigns with which he was in sympathy'.[9] As well as leasing station space to other bodies, the LPTB was keen to promote its own endeavours, with exhibitions illustrating the impressive programme of rapid expansion and modernisation with the spread of the 'tube' in the 1930s northwards towards Cockfosters on the edge of Hertfordshire and west towards Uxbridge.[10] Frequent displays showed rolling stock and Underground posters, as well as flowers and vegetables grown by LPTB employees. The *London Transport Photographic Exhibition* of 1934 showcased the extensive transport stock, with photographs of the impressive trams, coaches and trolley buses, all 'at London's service'.[11] The *Design in Transport* exhibition of 1935 included photographs of transport and publicity design.

Displays at Charing Cross Station ticket hall began with conventional hangs, installed horizontally round the walls of the small space but, by the mid-1930s, these were becoming more experimental. Trade and tourism bodies like the Milk Marketing Board made displays at the station an element in their promotion programmes, installing live dairy cows and working milking machinery in the station, allowing passersby to drink fresh milk and promoting its health benefits. The British Electrical Development Association was one of many organisations that mounted displays; its *Golden Age of Electricity* in 1932 showed mock rooms using a range of technologies powered by electricity, raising the profile of their organisation and service. Commercial organisations took the space as another outlet for advertising campaigns: Shell-Mex and British Petroleum displayed products made from petrol in autumn 1932. *The Hydrogenation Exhibition*, hosted by the Anglo-American Oil Company in spring 1933, showed the 'wonders' of Essolube motor oil, from refinery to engine. Dunlop displayed the wide variety of their manufactured goods, from splayed badminton rackets to comfortable cinema chairs and latex car seats, in spring 1933 in a display reminiscent of a department store installation. The Metal Box Co. installed working machinery to showcase processes for canning English fruit and vegetables in 1935, with multiple pyramids of cans interspersed with photographs illustrating the bucolic English origins of each product, exhibitions acting as advertisements.

Promoting 'good design': *Design in Modern Life*

As well as promoting trade and commerce, throughout the 1930s Charing Cross Station was used for non-commercial exhibitions: elements in wider publicity campaigns and the corollary to radio and film broadcasts, posters and booklets. Organisations including the Housing Centre and the Council for Preservation of Rural England (CPRE) chose Charing Cross to promote their work to the passing public in exhibition form. The Design and Industries Association (DIA) chose Charing Cross for a conventionally installed exhibit of British goods 'of moderate price', *Design in British Goods*. This 1932 display illustrated the maxim 'Fitness for Purpose & Simplicity Are the Key to Good Design', with shelves of goods chosen for their 'simple and efficient design' including sections for tableware, shoes, kitchen gadgets, accessories, clocks and lamps.[12] The DIA returned to Charing Cross the following year with *Design in Modern Life*, this time moving beyond a sole focus on well-designed goods, to buildings and environments.

Showing their awareness of compounding technologies for promotional purposes by bringing objects and radio together, the DIA's exhibition accompanied the British Broadcasting Corporation (BBC)'s National Programme series of radio talks *Design in Modern Life* (1933). The series included designer Gordon Russell on 'The Living Room and Furniture', poet Francis Meynell on 'The Printed Word', transport administrator and 'good design' champion Frank Pick on 'The Street', architect Wells Coates on 'Dwellings' and housing consultant Elizabeth Denby on 'The Kitchen'. Exhibition visitors were prompted to buy a series 'syllabus' prepared by the BBC from Messrs WH Smith & Co. at a station kiosk. BBC was developing its expansive agenda to speak to the world, allying with other organisations to expand its reach and to amplify its messages, and exhibitions were a potent element in this alliance. BBC programming was also regularly focused towards supporting exhibitions on the built environment in the years up until the end of the Second World War.[13]

Promoting the postal service: *Post Office Exhibition*

Amongst the public bodies in Britain that chose exhibitions as a central means of promotion, the GPO stood out for its striking use of the medium, with exhibitions sitting alongside film in enabling the company to explain its functions in providing mail, radio and telephonic communications.[14] The company often mounted their exhibitions in newly built cinema halls, as a self-conscious embodiment of modernity, and regularly incorporated films into their displays. The GPO mounted an extensive programme of thematic exhibitions, with the Charing Cross exhibition one of twelve held in 1936. The May 1936 *Post Office Exhibition* explained the GPO's range of technologies – from telegraphs to telephones and engineering – through a series of arresting and dramatic photomontages, with a floor-to-ceiling montage of the London skyline covering the walls with huge photographic figures of postmen and customers superimposed on top and a mural on the ceiling showing clouds with messages (Figure 2.1). The front of the Charing Cross space was animated by a series of dramatic silhouettes, showing GPO transport types.

Post Office explained the mechanics of the service, with a sample postal car and maps, a telephone and other modern machinery demonstrated by a GPO worker at a desk.[15] The design of the Charing Cross installation shows the strong influence of Modernist forms at a time when designers employed by the GPO to work on exhibitions, films and other forms included László Moholy-Nagy, E. McKnight Kauffer, Tom Eckersley, Hans Schleger, F. H. K. Henrion and Austin Cooper.[16] Cable and Wireless, the telecommunications company that served the whole of the British Empire, held an exhibition of *World Communications* at Charing Cross in

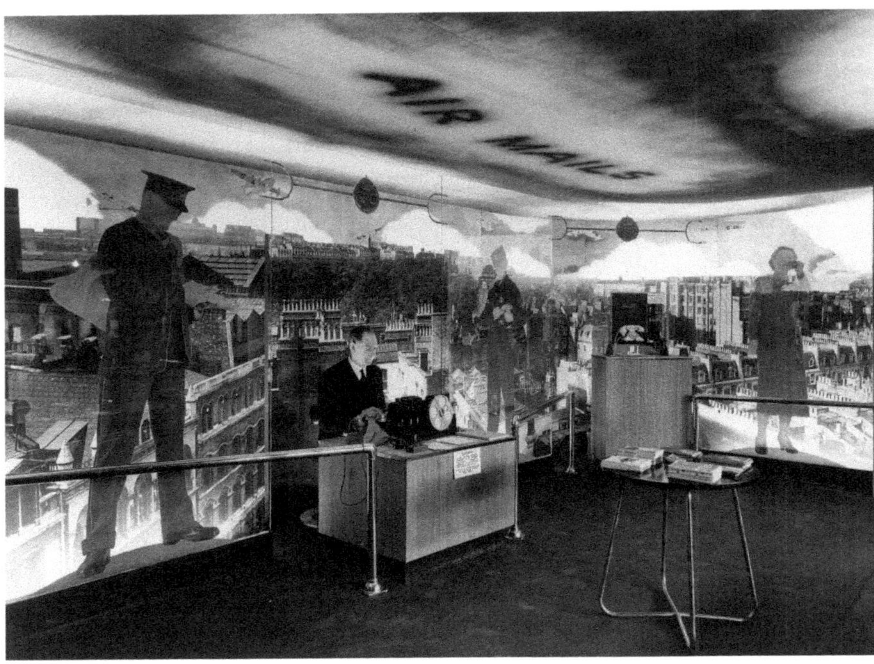

2.1 *Post Office Exhibition*, Charing Cross Underground Station, May 1936, with striking floor to ceiling murals and demonstration of new communications equipment. Topical Press. U20393 © TfL from the London Transport Museum collection. All rights reserved and permission to use the figure must be obtained from the copyright holder.

August 1936, again with a background of abstracted photomontages and demonstrating telecommunications equipment.

Promoting British Empire projects: *Timber through the Ages*

The rolling programme of trade exhibitions at Charing Cross Station included the Timber Development Association (TDA)'s *Timber through the Ages* held from February to March 1936, with artist Paul Nash as appointed designer.[17] The TDA, a trade body for timber businesses set up in 1934, was rooted in a tradition of promoting British Empire timber. Nash's appointment was likely the idea of TDA Director of Public Affairs and design writer John Gloag, who appointed Walter Gropius as judge of a timber house design competition the same year.[18] Nash was an inspired choice of artist for the TDA's Charing Cross Station display, given the romantic and lyrical associations with wood in his work; and, despite its commercial focus, *Timber through the Ages* provided Nash with an opportunity to share his artistic vision.[19]

Nash designed a large mural using eight wood veneers 'chosen for their colour, harmony and variations of grain' across the back wall of the station's space (Figure 2.2). Mural panels ranged from 'pale blue-grey, yellow and ochre, to russet brown and chestnut red'. The woods were unstained, the effect depending on their natural beauty, as *Commercial Art* enthused.[20] Splayed sidewalls allowed views to the wooden mural from outside. In addition, Nash chose a series of photographs arranged on rectilinear-shaped veneers showing wood anatomy and features, film stills illustrating the 'Life of a Tree' and ancient and modern uses of timber, constructional and decorative uses, usual and unusual uses (Figure 2.3). Images ranged from a fifteenth-century house to a spinet, a bridge to an aeroplane wing, and included photographs he had taken such as '*Gate Posts in the Making*', chosen for 'its essential, native woodenness'.[21] Despite such displays' intention to persuade and influence passersby, the lack of records of how they were received by the public makes it curiously hard to gauge what sort of impression they made.

Peeps at the Colonial Empire

The close linkages between Britain and its Empire had been shown across a century in dazzling displays. National and imperial exhibitions had been held all over Britain from 1900, transposing the imperial theme onto an epic scale, highlighting the invisible bonds holding the Empire together and selling products of Empire to the world.[22] Major exhibitions had proved efficient vehicles for reinforcing the narratives of colonising nations, in effect acting as global communications technologies for projecting and legitimising national power.

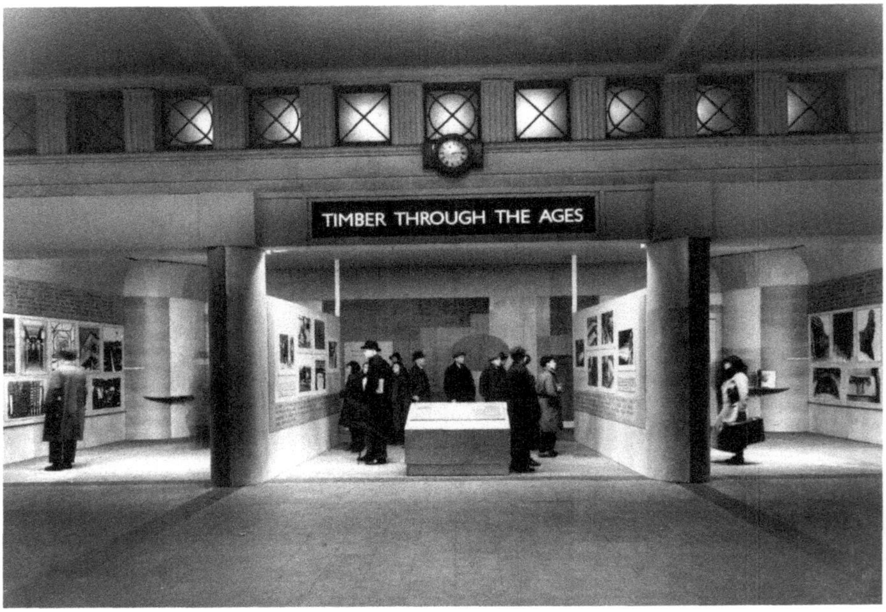

2.2 General view of *Timber through the Ages* in March 1936, at Charing Cross Underground Station, with a wood mural across the back wall, designed by Paul Nash. Topical Press. U19757 © TfL from the London Transport Museum collection. All rights reserved and permission to use the figure must be obtained from the copyright holder.

The unfolding of a tradition of spectacles of the British Empire in and from Britain, celebrations of the vast, multiracial empire, continued until the 1930s. These linkages were shown in more modest exhibitions, equally complicit in communicating messages of British difference and superiority despite their small scale and streamlined appearances. The short-lived Empire Marketing Board (1926–33), led by Stephen Tallents, consistently used exhibitions to display Empire at home and to project Britain abroad, part of a long-standing culture of commoditising non-Western cultures and geographies in Britain. Imperial Airways developed a stand at the aeronautical exhibition at London's Olympia in 1929 to promote activities and highlight imperial connections. Five years later it mounted its own exhibition, *Flying over the Empire*, at Charing Cross, comprising a large folding screen with a map mounted on it, models of Imperial aircraft, photographs and dioramas of imperial scenes. This was renamed *The Empire's Airway* and remounted at London's Science Museum in December 1935, subsequently touring to Canada, South Africa and Australia where, by the end of its run, it was estimated to have been seen by a million people.[23]

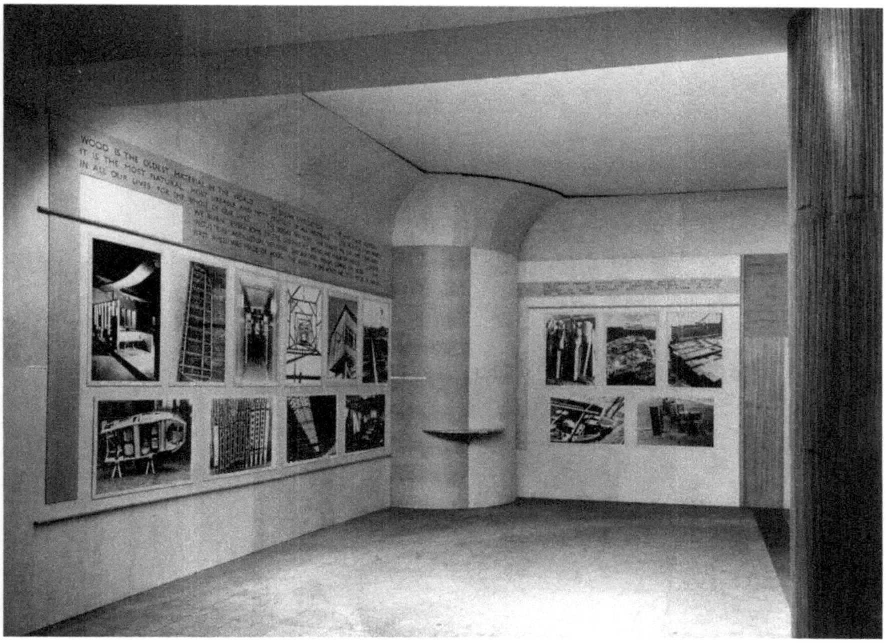

2.3 Photographs selected by Paul Nash shown to illuminate the theme of *Timber through the Ages*, Charing Cross Underground Station. Topical Press. U19760 © TfL from the London Transport Museum collection.

Imperial Airways appointed László Moholy-Nagy to design *The Empire's Airway* exhibition at Charing Cross Station in June 1936, showing the many routes flown and encouraging the transcendent benefits of flying with slogans such as 'The AEROPLANE IN THE SKY CARRIES US ABOVE MEDIOCRE THINGS', featuring cut-out photographs of planes, models of planes in section and of landscapes (Figure 2.4 and 2.5).[24] A series of peep-holes allowed glimpses through from the entrance, as well as mimicking the circular windows in planes to allow views through to illuminated displays with photographs, dioramas and statistics. The exhibition's installations were photographed in meticulous detail by news agency Topical Press.[25]

A succession of exhibitions at Charing Cross Station aimed to reinforce Britain's colonial and Dominion relationships. The government-mounted *Peeps at the Colonial Empire* of October 1936 brought together exhibits from a vastly dispersed colonial area: the British West Indies, Malaya, Ceylon, British West Africa, British East Africa, Malta and Cyprus, showing a dramatic diorama of silhouetted scenes of life from the various colonies above a series of sections showing crafts and produce such as coffee and tobacco, all accompanied by information pamphlets (Figure 2.6).[26]

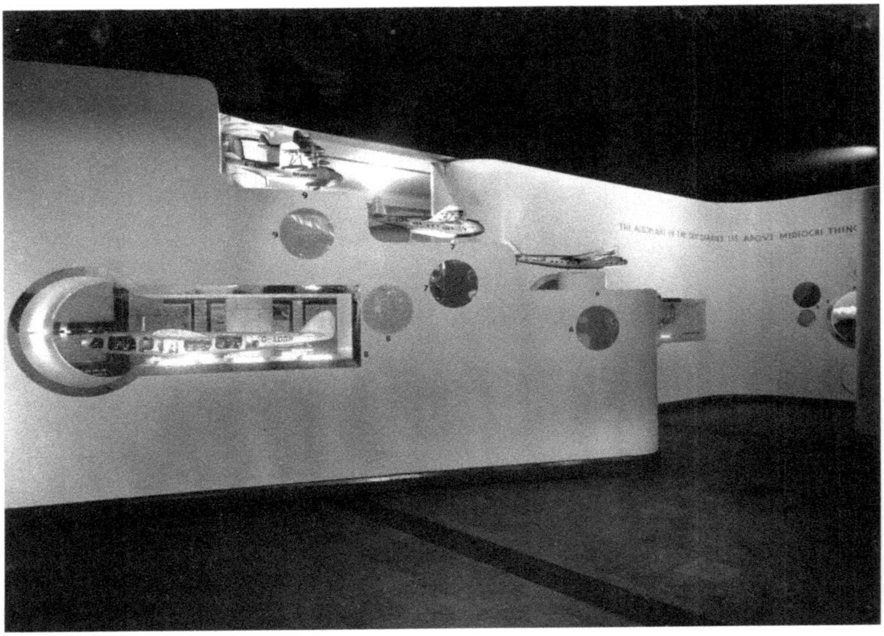

2.4 Publicity for *The Empire's Airway* at Charing Cross Underground Station, June 1936, designed by Theyre Lee-Elliott. 1983/4/4490 © TfL from the London Transport Museum collection. All rights reserved and permission to use the figure must be obtained from the copyright holder.

Empire Tea at Charing Cross of 1937 showcased Great Britain's tea industry, making a connection between the '120 thousand million cups' of tea a year drunk in the British Empire and the employment it gave to 'more than 2 million British subjects in India & Ceylon'. Photographs of tea pickers were displayed besides a diorama showing an 'Empire Tea Garden' and statistics about 'when the English family takes tea'.[27]

Policy in three dimensions: *Do Not Pass by the Special Areas* exhibition

Charing Cross Station ticket hall was used by a plethora of organisations for promotion and projection. From the mid-1930s, the ticket hall also became a site for enabling the public to visualise government policy in three dimensions. Modernist exhibition techniques – including photomontage and the new typography – were in evidence at *Do Not Pass by the Special Areas*, mounted at Charing Cross in February 1936. The exhibition was planned and designed by F. C. Pritchard, Wood & Partners, the advertising agency of the Isokon Furniture Company.[28] The small exhibition was mounted at the request of the Commissioner for the Special Areas, a

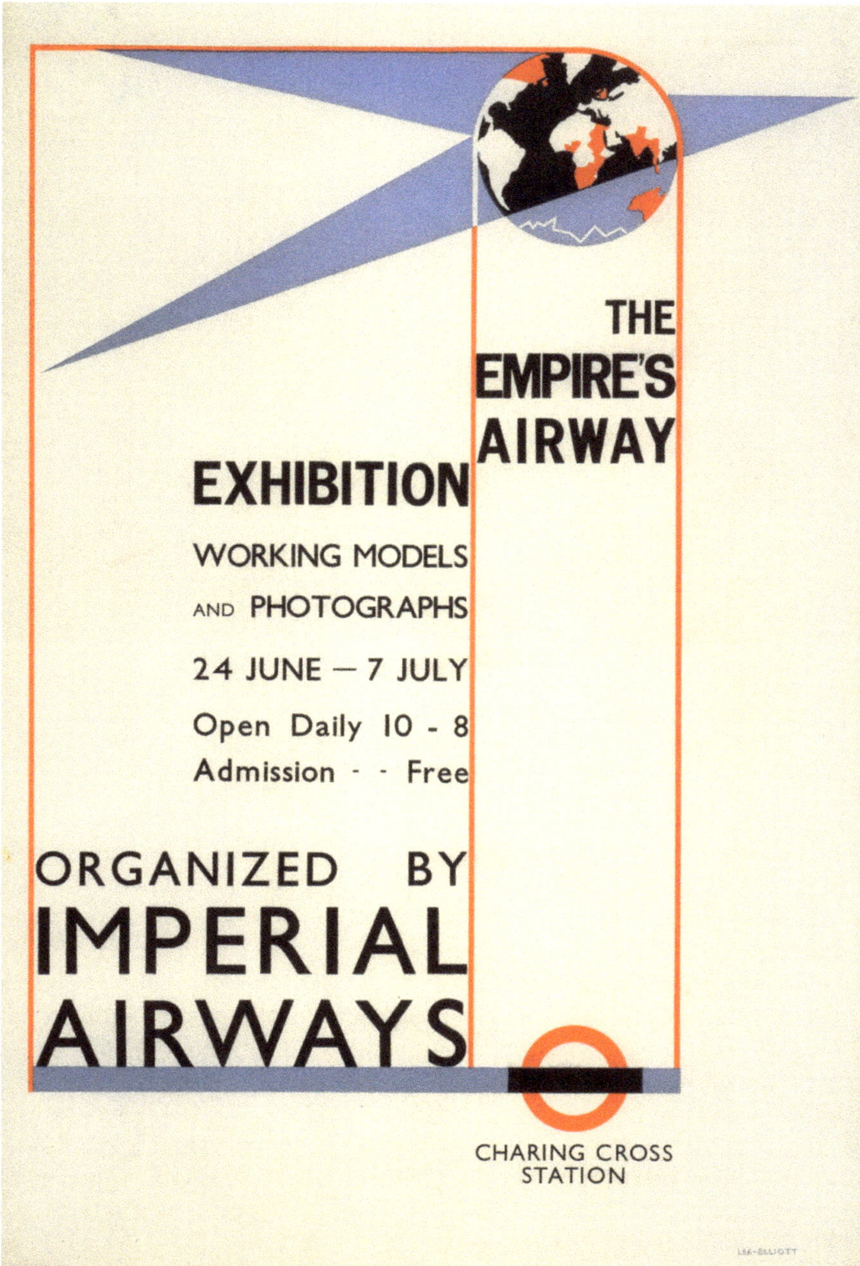

2.5 *The Empire's Airway* exhibition at Charing Cross Underground Station, designed by László Moholy-Nagy. Topical Press. 1998/42726 © TfL from the London Transport Museum collection. All rights reserved and permission to use the figure must be obtained from the copyright holder.

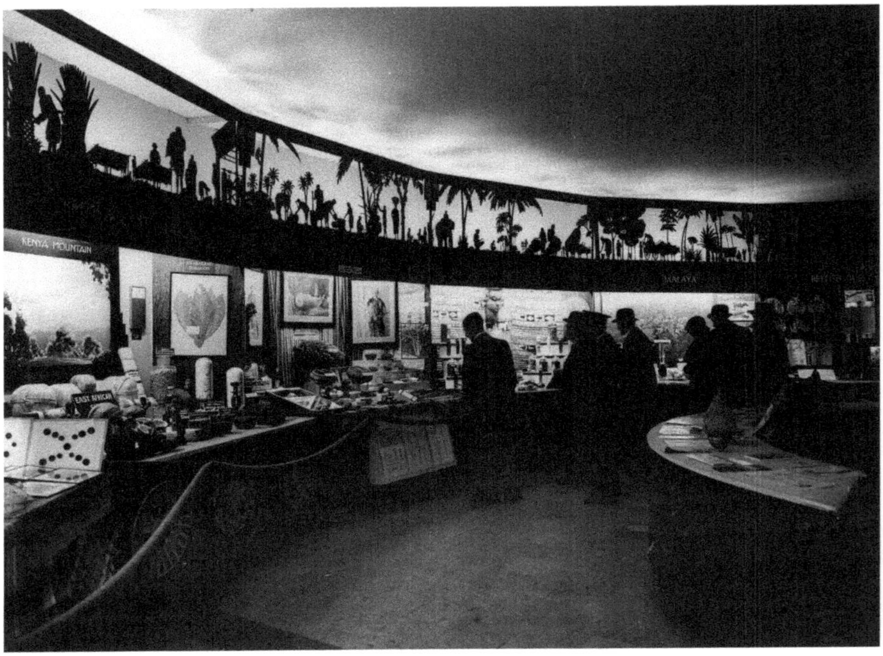

2.6 *Peeps at the Colonial Empire* exhibition, Charing Cross Underground Station, October 1936, showing products from the colonies. Topical Press. U21703

government appointee tasked with improving conditions in places stricken by the Depression, and was intended to amplify government policy. It entreated Londoners to jolt themselves out of insularity, to 'Look Beyond London – Learn How Others Live'.[29]

The exhibition's accompanying poster was designed by commercial artist E. McKnight Kauffer, its red and black script set on dynamic diagonals, centring on the cut-out photograph of an official-looking man appealing to the public's sense of social responsibility by proclaiming 'every citizen should see this' (Figure 2.7). American-born Kauffer, who had lived in England since 1914, played a key role in the emerging profession of graphic design in England. He was, by the 1930s, working for numerous clients including doing regular work for the Underground, where he was championed by Frank Pick.[30]

'Special Areas' was an official designation given by Prime Minister Ramsay MacDonald in 1934, through the Distressed Areas Act, to three districts hit particularly hard by the Depression. The Act had allocated £2 million for demolition of derelict factories and to aid self-help schemes. While the scheme was criticised as too minor an intervention, given the

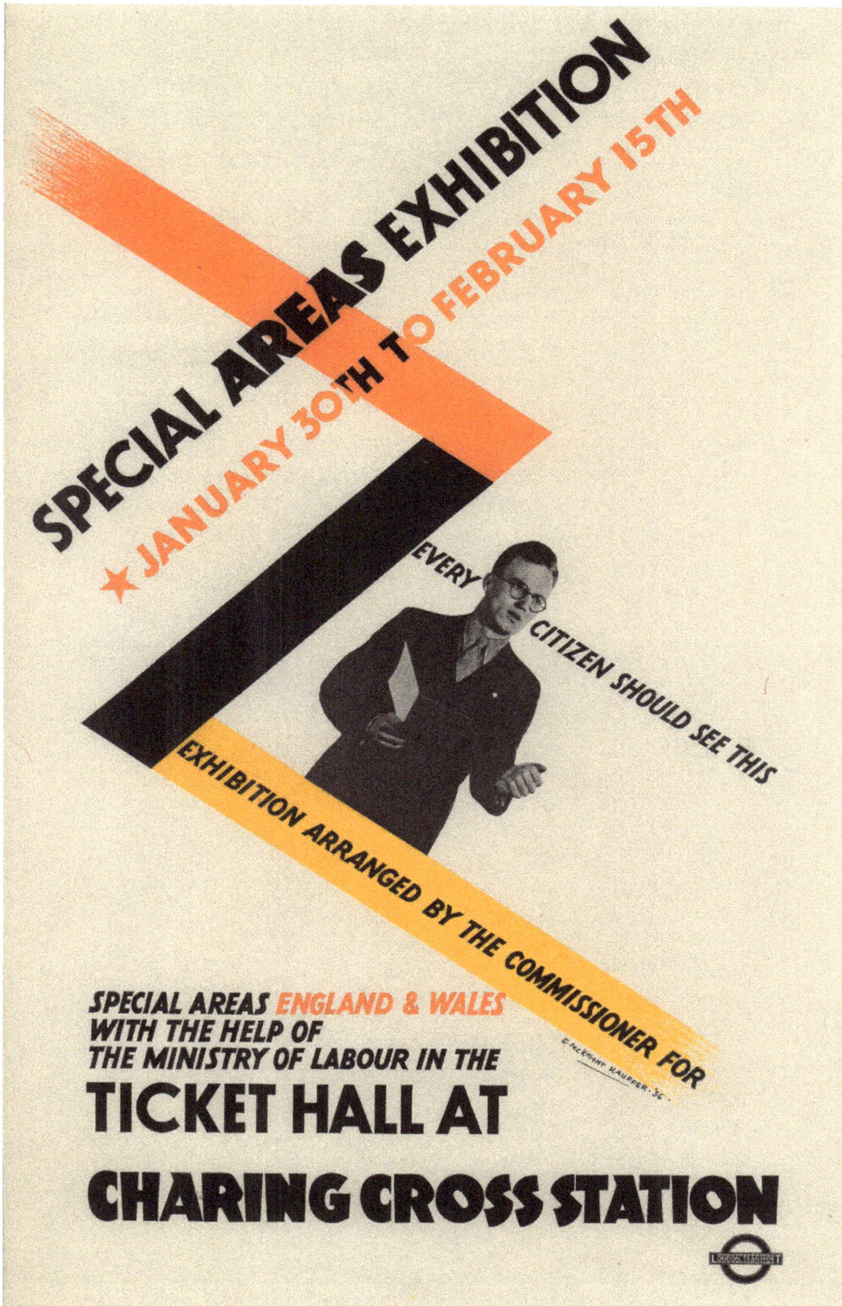

2.7 *Do Not Pass by the Special Areas*, Charing Cross Station, February 1936, poster designed by E. McKnight Kauffer.

over two million unemployed in these areas, including by MP George Lansbury who described it as 'an attempt to bale out the ocean with a spoon', the Charing Cross exhibition was part of a wider programme to promote the government scheme.[31] It showed what the government was doing to improve conditions and instructed the public on what they could do to help.

The exhibition visualised and dramatised the impoverished living conditions of the working classes beyond London, with context presented in the exhibition in documentary form, through photomontages showing the areas' rural beauty: lakes, hills and cathedrals, industrial landscapes and the people who worked and lived in them. Photographs of mills and power stations were brought together with cut-outs of sooty miners, children and babies (Figure 2.8). Maps showed the three government-designated 'Special Areas': Cumberland, Durham and South Wales (Figure 2.9). The West Cumberland and Alston section showed locations for coal, iron and steel, quarrying and weaving. Woollen, woven, metal and glass product displays entreated viewers to 'Buy the products of the special areas', while travel information showed people how to visit them and a

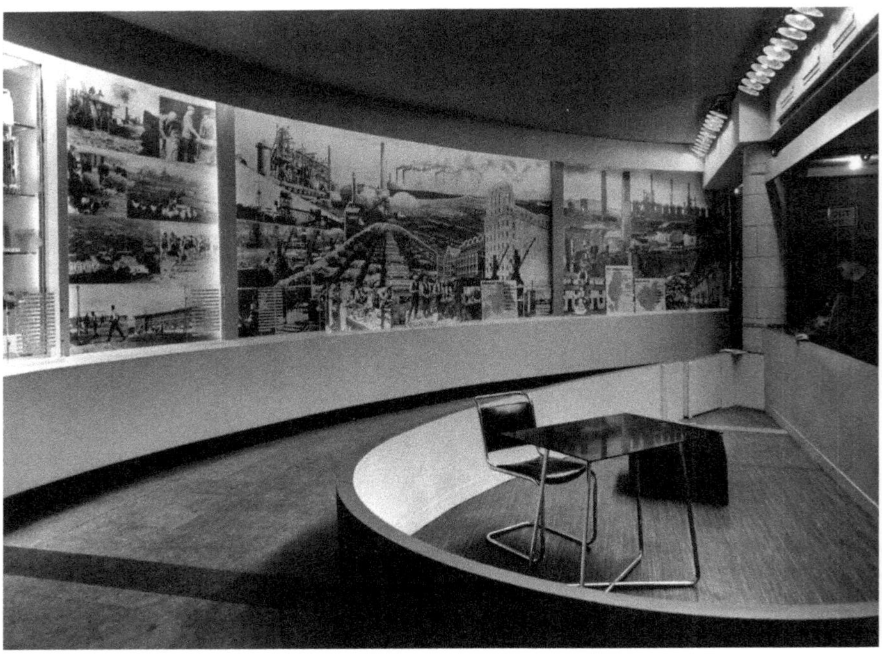

2.8 *Do Not Pass by the Special Areas*, Charing Cross Station, planned and designed by F. C. Pritchard, Wood & Partners. 1983/4/4333 © TfL from the London Transport Museum collection.

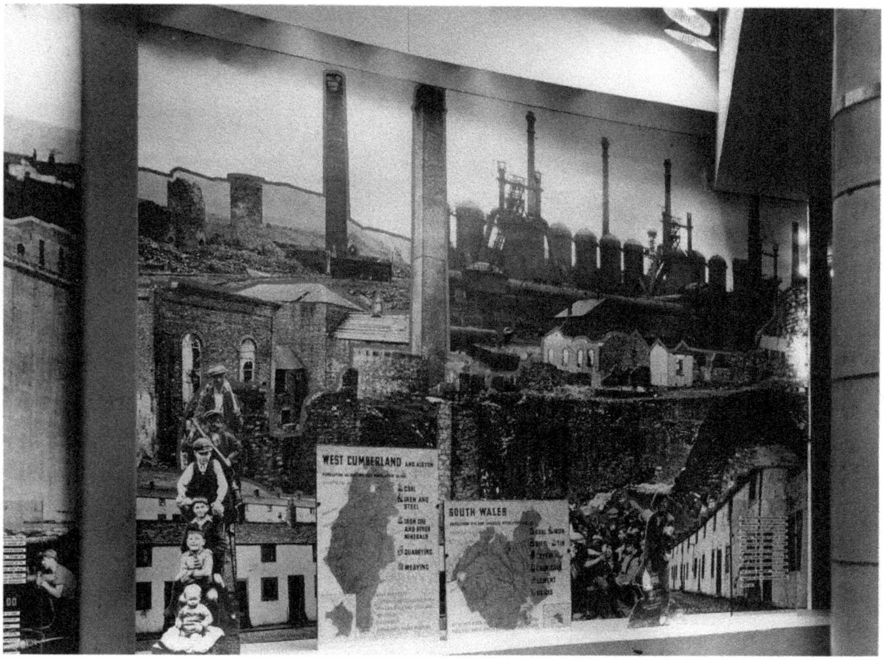

2.9 *Do Not Pass by the Special Areas*, Charing Cross Station, planned and designed by F. C. Pritchard, Wood & Partners. 1998/42360. Chancery Co. Ltd.

photocollage demonstrated government training provided for retraining men 'for whom there is little or no prospect of further regular employment in their own trades', as well as twelve recuperative courses to 'restore their physique'.[32] Such exhibitions, even in the small underground space, were ambitious elements of government communications; their installations complex and multi-faceted, combining media and script, and driven by narrative, becoming manifestations of government policy in three dimensions.

Displaying road safety: *The Highway Code: An Exhibition*

Charing Cross Station continued to be seen as a good venue for sharing official policy and information, although frustratingly little evidence of the public response remains. In November 1937, Charing Cross hosted *The Highway Code: An Exhibition*, intended to keep the public safe on roads. Developed by the Ministry of Transport, *Highway Code* was part of the campaign to increase public awareness of dangerous roads, following the appointment of modernising Transport Minister Leslie Hore-Belisha.[33] *Highway Code* was designed by Hans Schleger, who worked under the pseudonym 'Zéró'. Schleger had been born Hans Schlesinger in Poland, training at the Berlin Kunstgewerbeschule and Reimann School in Berlin, working as a designer in the US from 1924 to 1929 before joining the Berlin office of prestigious British advertising agency W. S. Crawford and moving to London with Crawford's in 1932.[34] With the support of Crawford's Director Ashley Havinden and graphic designer E. McKnight Kauffer, Schleger had secured an exhibition in the art gallery of Lund Humphries Bedford Square in 1934. He designed posters for the LPTB regularly from 1935.

Commissioned by the LPTB to create a poster to publicise *The Highway Code* exhibition, Schleger depicted ears and eyes marching (Figure 2.10). *Modern Publicity* explained that 'by simple visual analysis' the poster created 'an abstract notion of behaviour', lauding Schleger's work as 'a brilliant example of very difficult material admirably co-ordinated'. The article continued, 'the imposition of eyes and ears for heads underlines the necessity of super-awareness on the part of pedestrians in traffic'. It admired how such an 'unusual visual idea imports an interest and curiosity-appeal into a subject which would otherwise make dull poster material'.[35] Made a year after the London Surrealist exhibition, Schleger's poster showed the influence of Surrealism and of contemporaneous designs by Man Ray, Kurt Schwitters and Herbert Bayer and the work of Roland Penrose, whom Schleger had recently exhibited with in Cambridge.[36]

Signage at the Charing Cross exhibition used a Playbill font, which echoed the poster. An introductory panel declared it the work of the Ministry of Transport, 'with the Authority of Parliament', appearing like the front cover of a book complete with publisher, 'His Majesty's Stationery Office', location and date. Schleger's exhibition installation showed the influence of his German training and alliances: abstract or cut-out photographs caught in organic shapes depicted a spat between cyclists and an inconsiderate car driver; miniature model cars showed how to navigate roads safely; neat illustrative diagrams showed the dangerous times on the road and casualty numbers; drawings and cartoons, metal cut-outs overlaid photographs and illuminated signage, images glimpsed through port-holes were all accompanied by aphorisms such as 'Good Tyres, Good Brakes' and 'Drive Within the Limits of Your Lights' (Figure 2.11).[37] Beyond short reports in the trade press, there is frustratingly limited evidence as to what the public made of the installation.

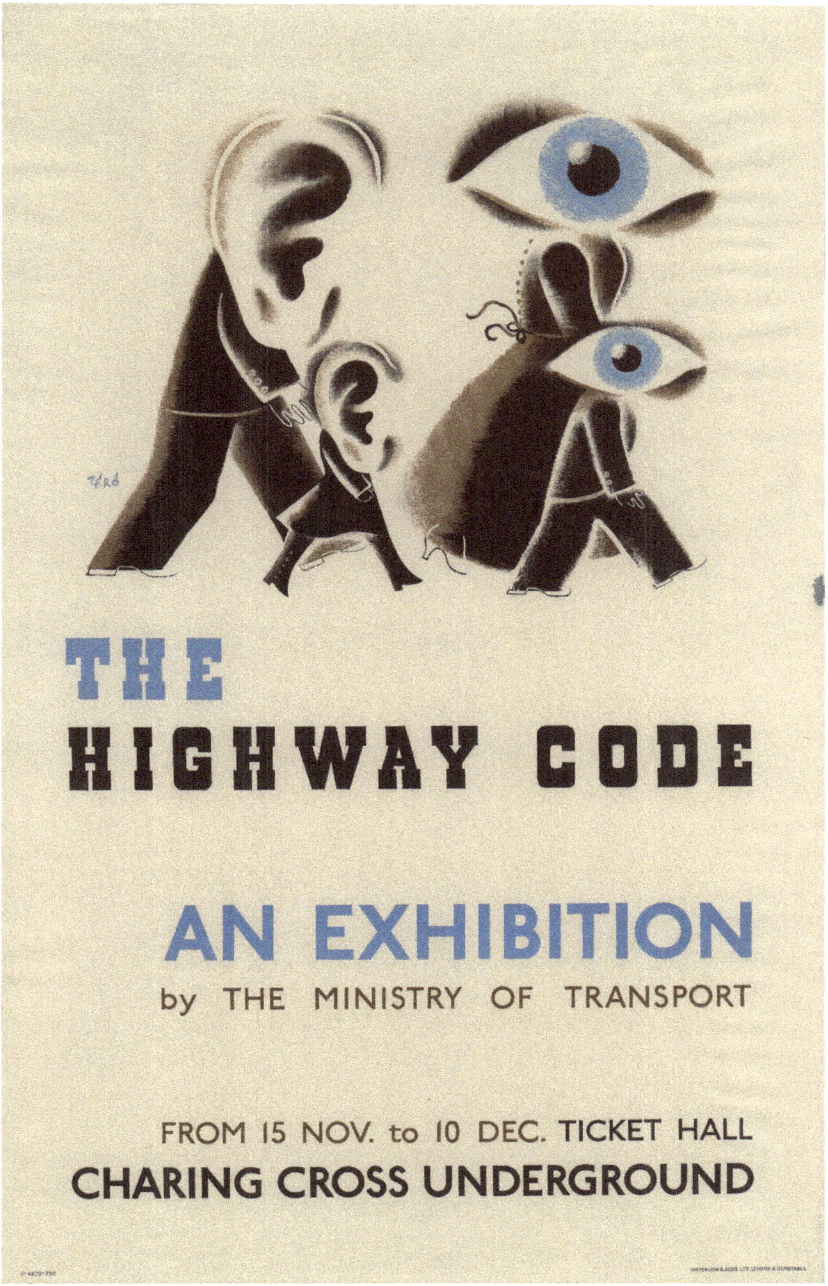

2.10 *The Highway Code* exhibition, at Charing Cross Underground Station, November 1937, poster designed by Zéró (Hans Schleger). 1983/4/4923

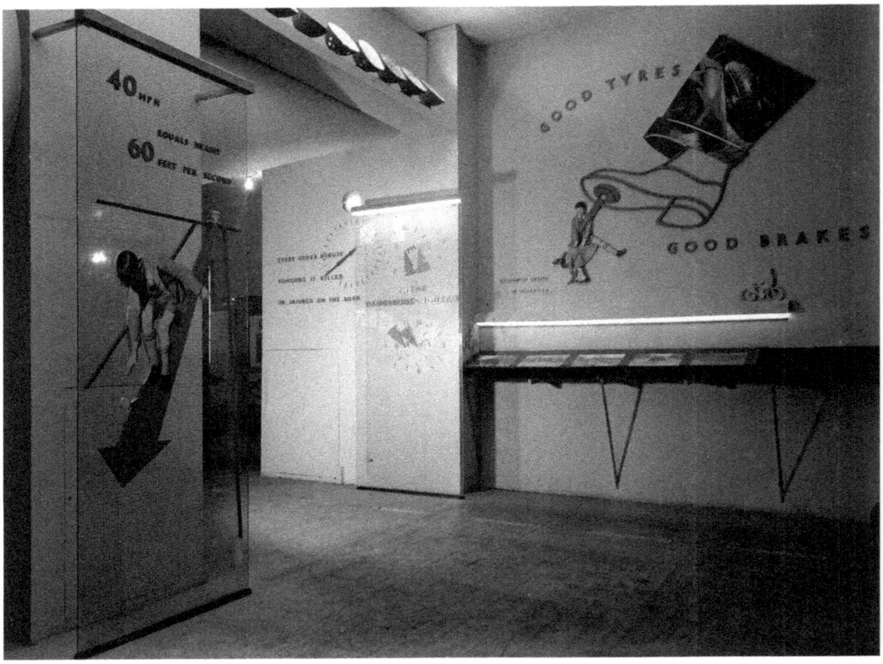

2.11 *The Highway Code* exhibition, display designed by Zéró (Hans Schleger). Dell & Wainwright. 6747-DW © TfL from the London Transport Museum collection.

Exhibitions as activism: the *Schools Exhibition* and *Food for Fitness*

At the same time as promoting organisations and goods, and giving concrete shape and form to abstracted government policy, exhibitions were being taken up by activists to promote acute social and political agendas. In the case of the examples I have chosen in this chapter, these agendas coalesced around improving health, children's education, air quality and housing standards; their audiences varied from central and local government to the general public. All were focused towards prompting social, cultural or political change. The extent to which they met their target is extremely difficult to gauge.

The Leftist *News Chronicle* newspaper had a strong social agenda to improve living and working conditions and mounted regular exhibitions to further this agenda. *News Chronicle's Schools Exhibition*, held from December 1937 to January 1938 at Charing Cross Station, gave audiences an exhibition taster before pointing them on to the more major exhibition on the same subject being held simultaneously at Dorland

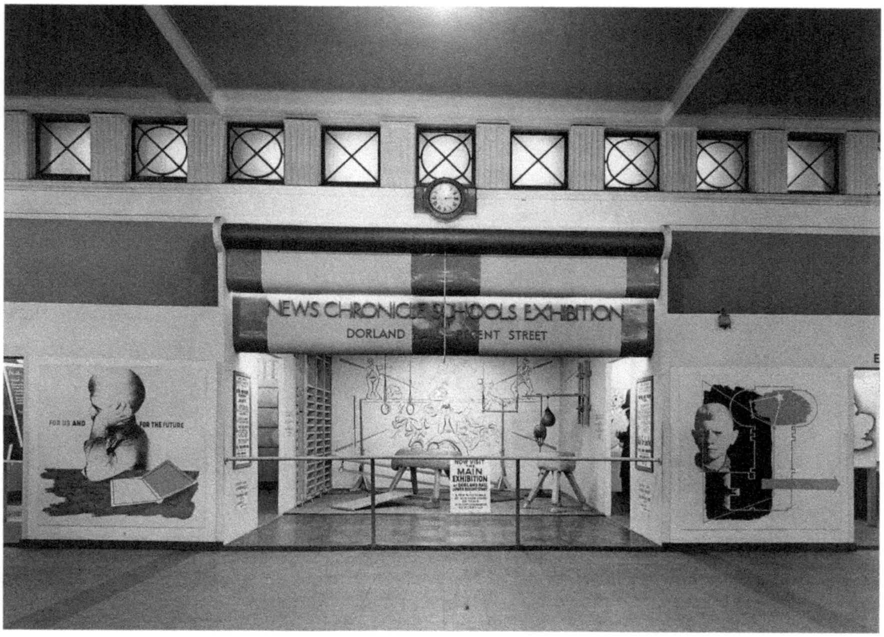

2.12 *News Chronicle Schools* exhibition at Charing Cross Station, 1938, with collages designed by E. McKnight Kauffer. U25966 © TfL from the London Transport Museum collection. All rights reserved and permission to use the figure must be obtained from the copyright holder.

Hall on Regent Street.[38] Collages by E. McKnight Kauffer flanked the Charing Cross exhibition's entrance, showing a cut-out photograph of a child leaning on a book, with the slogan 'For Us and For the Future' (Figure 2.12). A whimsical cartoon accompanied a mocked-up section of a school hall with bars, school bench, gym horse, hoops and pummel-bag. McKnight Kauffer designed the exhibition's poster, one version showing an abstracted Doric column accompanied by the cut-out photograph of a child's head, another version showing a trowel, in which the trowel's head was the drawing of a school plan (Figure 2.13).[39] Perhaps unsurprisingly, given its opposition to the Leftist politics of the exhibition's sponsor *News Chronicle*, fascist magazine *Action* wrote a scathing report of the exhibition, focusing on its use of 'photographic layouts', which aimed to typify the spirit of education in Russia and Spain but ignored Germany and Italy and hardly, they thought, showed Britain.[40]

Exhibitions addressing public welfare came in a variety of forms. *Food for Fitness* was one such exhibition. Held at Charing Cross in February 1938 and probably mounted by the government's Public Health Department, its agenda was clearly to teach the benefits of well-planned meals in terms

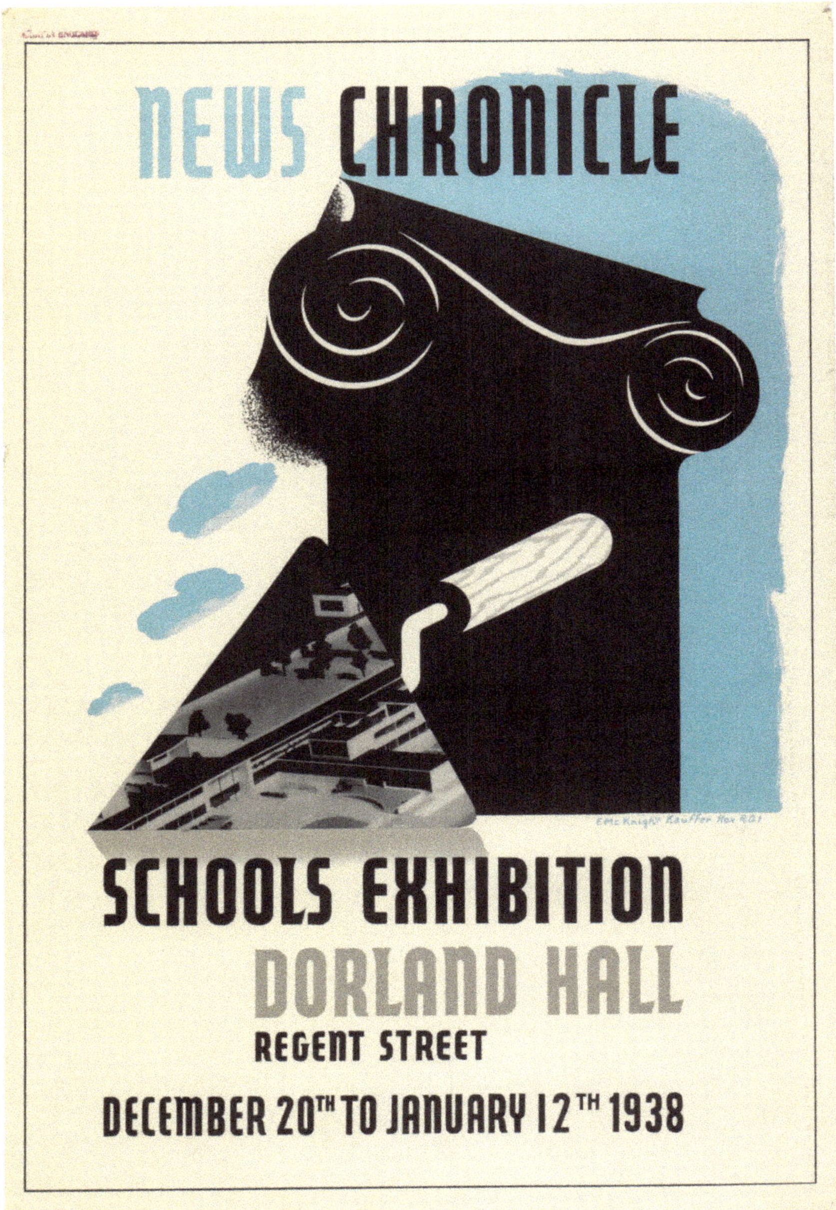

2.13 *News Chronicle Schools* exhibition at Dorland Hall, London, 1937, poster designed by E. McKnight Kauffer. Cooper Hewitt, Smithsonian Design Museum. Gift of Mrs. E. McKnight Kauffer, 1963-39-96. Photo: Matt Flynn. © Smithsonian Institution. All rights reserved and permission to use the figure must be obtained from the copyright holder.

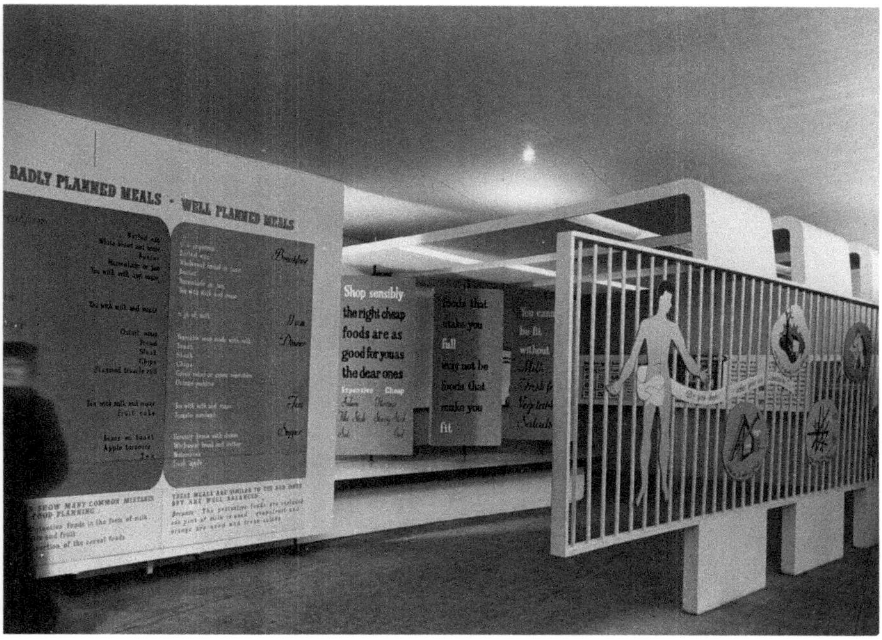

2.14 *Food for Fitness* exhibition at Charing Cross Station in January 1938, teaching the benefits of healthy eating through sample menus. Topical Press. U26173 © TfL from the London Transport Museum collection.

of budget and nutrition, as well as the negative impacts of poorly planned ones (Figure 2.14). An ideal diet for 'a working man' and a child were displayed, with sample menus showing healthy choices and blown-up photographs showed appealing looking healthy food, alongside a vast photograph of an obese man shown to caution against the 'wrong' kind of eating (Figure 2.15).

Exhibitions for air quality activism: *Clear Smoke from the Air*

A striking Charing Cross installation tackled atmospheric pollution, a subject seemingly antithetical to visual brilliance. The respiratory and pulmonary threat of the dense coal smoke that billowed from the chimneys of factories, workshops and homes had been strikingly evident for a century.[41] The *Clear Smoke from the Air* exhibition was organised by the National Smoke Abatement Society and mounted in December 1938 in co-operation with the Gas Light and Coke Company, to campaign against London's 'smoke, soot, as causes of disease, as deterrents to business and as chief obstacles to the growth of beautiful

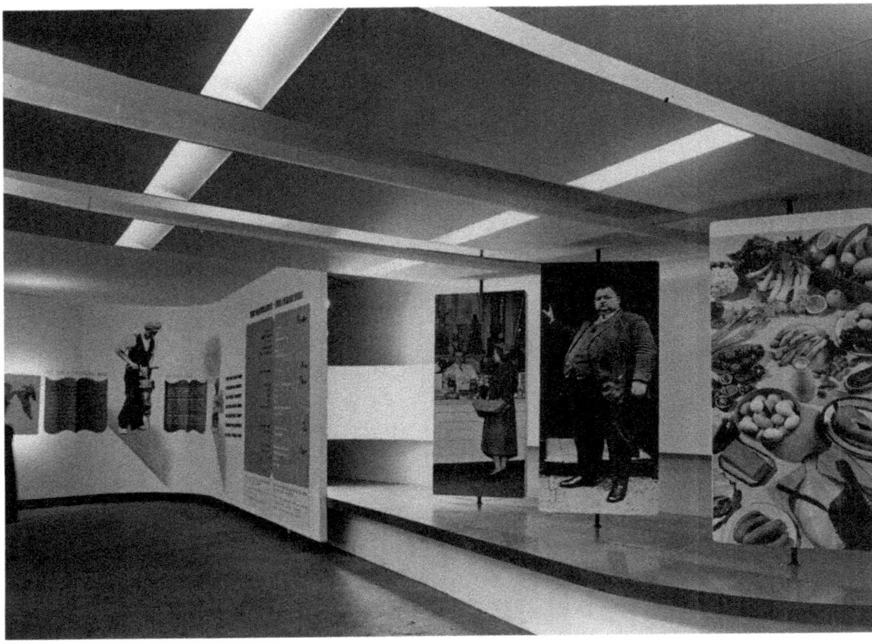

2.15 *Food for Fitness* exhibition, teaching meal planning through visual guides to healthy food, enlarged photographs and photocollages. Topical Press. U26174 © TfL from the London Transport Museum collection. All rights reserved and permission to use the figure must be obtained from the copyright holder.

cities'.[42] Herbert Morrison MP opened *Clear Smoke* and the press photographed him chatting with a sooty chimney sweep.[43] Morrison had a strong belief in exhibitions' efficacy in communicating government messages, as his regular appearances through the many phases of this book will show.

The National Smoke Abatement Society had held regular exhibitions to lobby for improved air quality through the 1920s and 1930s.[44] Aside from their success as lobbying exercises, displays had been criticised for failing to convey their visual messages effectively. A major Society exhibition at London's Science Museum in October 1936 was panned by *Shelf Appeal* as 'so magnificent a theme, so dramatic a subject' but 'so poorly handled'. Its treatment of photographs was 'crowded and under-sized' – 'half-a-dozen powerful enlargements would have smashed home the message in an unforgettable manner' – and 'charts, diagrams and maps were small, inexpertly handled, and unconvincing ... there was a lot of meat at the Exhibition, but it was served up in an indigestible manner'. 'If the Society ... had entrusted the exhibition to a display specialist – much more satisfactory results would have been obtained'. This point was echoed in the press comment, 'Never has an exhibition cried out so loudly for the directive hand of a skilled and imaginative exhibition planner'.[45]

Heeding this criticism, the Smoke Abatement Society appointed imaginative exhibition planners Misha Black and F. H. K. Henrion to design a pair of exhibitions. Russia-born Black had arrived in Britain as a child in 1912, starting his career as exhibition designer in 1927, so by the mid-1930s he had had several years' experience.[46] The Society presented Black as 'the leading exhibition designer in the country'.[47] German-born Henrion, who had arrived in Britain as a refugee in 1936, after working as a textile designer in Paris and as an exhibition designer in Tel Aviv, was praised by *Art and Industry* magazine as 'a versatile' and 'sensitive' artist who worked on press advertisements, folders, brochures, display units and photomurals.[48] Black and Henrion worked with Beck & Pollitzer Ltd as builders.[49]

Clear Smoke employed visual rhetoric and dialectic contrasts to make its point. An image of dense housing was accompanied by the label 'CENTRAL LONDON gets only half the winter sunshine', the cut-out head of a ruddy-cheeked baby and children jumping, contrasting with an image of the 187-foot Nelson's Column half buried in soot and the label 'A month's fall of soot in the County of London swept into Trafalgar Square would look like this' (Figure 2.16). A diorama was used to show the transformation of a smoky, gloomy city, plagued by 'The Soot Menace', into a clean, orderly townscape. This was accompanied by the repeated incantation: 'We CAN Work This Miracle'. The displays were at once serious and playful, visually appealing and innovative. A cartoon character, 'Sammy Soot', announced himself above the exhibition's entrance, declaring 'I make Work hard for you', lurking mock-menacingly at intervals throughout the exhibition. The inset model of a house, with a plain brass plate on the fence at the front reading 'Mr and Mrs Everyman, Makers of Smoke', underlined the message that individual homeowners had a role to play in improving the situation by changing the fuels they used to cook and heat their homes.

A cut-out photograph showing a woman struggling with a heavy sack labelled 'The Burden of Smoke' illustrated the way pollution added a hidden encumbrance to life. This was contrasted with a display about housewives' experiences of the additional washing and cleaning caused by soot entitled 'Take a Womans [*sic*] Word For It', with four photographs accompanied by textual vox pops, articulating the experiences of real women living in a smokeless London housing estate, the newly built Kensal House in London. The display playfully superimposed them on a house silhouette, with labels hanging like the outline of clothes on a washing line. 'Health from the Sun' contrasted with 'Disease and Death from Smoke', with X-rays of lungs, drawings of lines of vegetables and other devices used to illustrate the point, while a ceiling panel illustrated soot in the air and clever, angled displays showed from several sides (Figure 2.17).

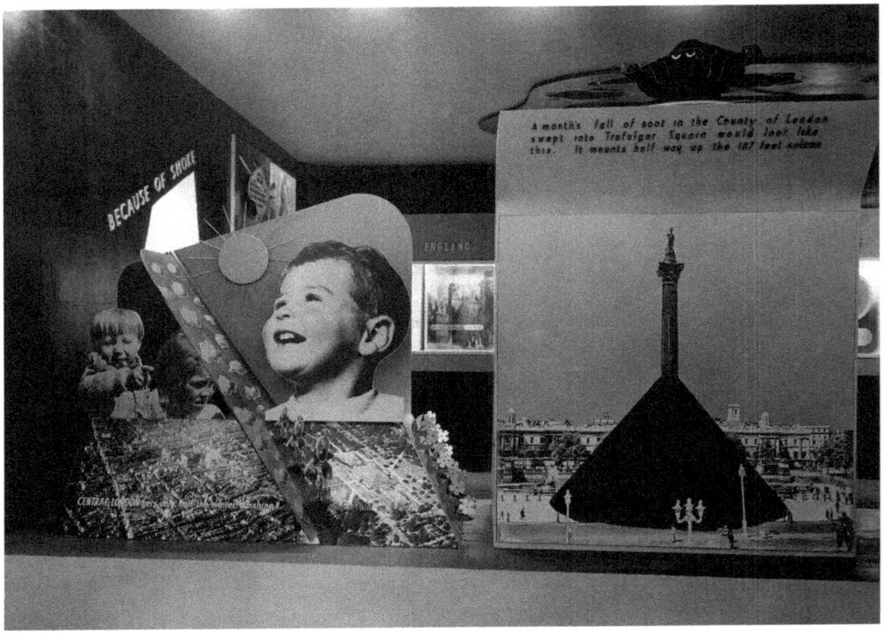

2.16 *Clear Smoke from the Air* exhibition, December 1938, organised by the National Smoke Abatement Society at Charing Cross to tackle atmospheric pollution. Topical Press. F. H. K. Henrion Archive, University of Brighton Design Archives. © Estate of F. H. K. Henrion. All rights reserved and permission to use the figure must be obtained from the copyright holder.

Display praised the Charing Cross exhibition for its 'moving effects, illuminated transparencies, striking photo work and sound design apparent in the informative displays'. It went on to note the particularly strong messaging achieved by this exhibition, saying that Charing Cross Underground Station 'promises to become the venue of London's forceful temporary exhibitions', exhibitions that did not only show objects or produce but stood to communicate a point.[50] The Society described Charing Cross as 'one of the best-known and valuable exhibition sites in London', noting with satisfaction that the exhibition had prompted an increased demand in information about smokeless fuel.[51] Exhibitions such as these were elements in strategies to raise public consciousness about the invisible threat of poor air quality across a variety of media including posters and film. In 1938 the Society ran a poster competition, judged by E. McKnight Kauffer, setting designers the task of illustrating one of three slogans such as 'Away with smoke, let in the light'.[52] And in 1939 the Society ran a film competition for films on smoke abatement, with highly respected documentary-maker Paul Rotha as judge.[53] Through their choices of judges and designers, the Society's campaigns were inextricably linked with wider Modernist

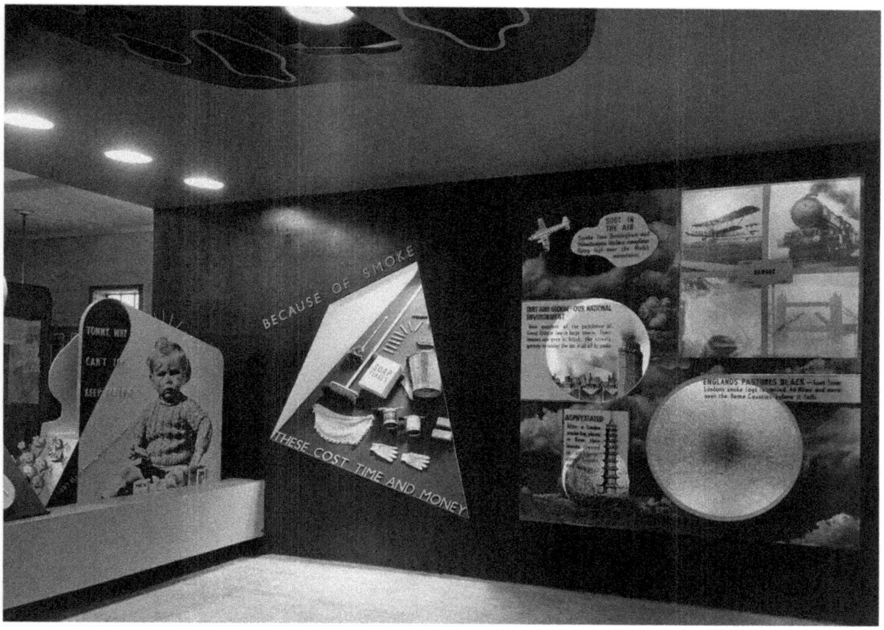

2.17 *Clear Smoke from the Air* exhibition. Topical Press. F. H. K. Henrion Archive, University of Brighton Design Archives. © Estate of F. H. K. Henrion. All rights reserved and permission to use the figure must be obtained from the copyright holder.

activism promoting social reform and new housing. It is no coincidence that another form of exhibitionary activism, that focused on housing, drew contributions from some of the same designers.

Exhibitions for housing activism: *New Homes for Old*

A series of exhibitions titled *New Homes for Old* was mounted in London between 1931 and 1938 to draw attention to overcrowding and poor housing conditions and to show that high quality cheaper housing could be built. From Ideal Home to Building Exhibitions, the exhibition had long been considered an effective medium for representing solutions to housing problems: educating the public on improved forms, selling materials to the building industry and communicating diverse planning visions.[54] By the 1930s, public interest in building was strong, something RIBA President Sir Giles Gilbert Scott attributed to the negative impacts of 'vast building schemes which have done so much to ruin England'.[55]

The 1930 Housing Act championed by Arthur Greenwood, Labour Minister of Health, had promised to bring sunshine to the slums, giving local authorities power to clear 'plague' sites. Two million children, it was estimated, lived in unfit housing; the question was how to find solutions rapidly.[56] With low inflation, this was a boom time for private housing, with 345,000 houses built annually between 1933 and 1937. But much of it was poor quality, based on piecemeal planning, and public housing programmes were faltering.[57] While commercial building exhibitions had engaged a large public by showing novel housing forms and materials, an exhibition series with a more adversarial and polemical focus was developed by the Housing Centre, a pressure group of women drawn from London's voluntary housing societies, their other activities including creating housing information, publicity and research.[58]

New Homes for Old once again demonstrated the exhibition as suited to housing activism, allowing for a mix of photographs, charts and three-dimensional elements, knitted together by a hard-hitting narrative. Housing Centre exhibitions were intended to precipitate rapid improvement by showing the ills meted out by successive governments that had invested in building accommodation without sufficient understanding of people's actual needs. The first *New Homes*, co-ordinated by housing professional Elizabeth Denby and held at Westminster Central Hall in December 1931, displayed a graphic and easily absorbed representation of the horrors of London's slums, pointing to what needed to be done.[59] Model flats and illustrations of exemplary German working-class housing were shown alongside a shock section on slums.

The pictures, intended to horrify viewers into action, included pinned-out specimens of rats, beetles and other vermin, employing a vivid use of a dualist problem/solution trope to point from the existing problem towards a preferable solution, a display device favoured by many contemporary lobbying organisations, including the DIA.[60] The exhibition was designed to tour round the country and, as architect Judith Ledeboer explained, it was successful in 'taking a large share in the rousing of public opinion to abolish the slums'.[61] The second *New Homes for Old* exhibition, held at Olympia in September 1932, again brought together a mix of media including models, photographs and statistical charts. It drew on data from the recently completed 1931 census to underpin messaging about the human cost of slums.

Sensing that such powerful material was hitting home with the public, the 1934 *New Homes* exhibition brought together visually impactful material, from documentary photographs to models and pictograms, in combination with a punchy textual narrative. Held in September 1934 and mounted at Olympia Building Trades Exhibition, the exhibition was designed across seven bays, along with a simulacrum slum alley given the title Susannah Row, Drysdale Street Clearance Area, painted by Molly MacArthur and with doorframes, windows and drainpipes brought from a demolished slum. This mock slum drew the crowds. It was contrasted with orderly sections, each designed by women, showing model housing, statistical diagrams and photographs, town planning by J. F. Abram, flat planning by Elizabeth Denby and outdoor amenities by landscape architect Lady Allen of Hurtwood. A booklet accompanying the 1934 exhibition carried a photomontage by Edith Tudor-Hart on its cover, with a slum contrasting with new housing and a round-cheeked, beaming baby rising from the roof, operating as an optimistic symbol of a healthy new future (Figure 2.18).[62]

Denby and Ledeboer spotted an opportunity to show yet more vividly the possibilities for transforming housing, by forming a collaboration with new architectural consortium the Modern Architectural Research Group (known as MARS), who they invited to collaborate on a display about the re-planning of slum areas.[63] The MARS Group, formed in 1933 as the British chapter of the Modernist architects' consortium Congrès Internationaux d'Architecture Moderne (CIAM), was an association of 'architects and allied technicians united by a common belief in the necessity for a new conception of architecture and its relation to society', formed by six members.[64] Architectural historian John Summerson, a member of MARS, described the group's name as encapsulating a sense of both their 'militancy' and a 'vision of planetary exploration'.[65] MARS, like the group around *New Homes*, had a missionary belief that they could improve people's lives through 'good' architecture. Like other artists' groups formed in the same year, including the Artists International Association and Unit One, MARS saw exhibitions as central to demonstrating their future vision and to enact this they formed an exhibitions committee early on to start planning. MARS – who identified as an embattled avant-garde – used exhibitions alongside several other methods, including press releases and articles, to campaign for modern architecture.[66]

The resulting 1934 MARS-designed section of *New Homes* centred on an analysis of slum conditions, with the organising group's own display showing what they hoped to see in terms of indoor and outdoor play and communal space, using photographs and models, while MARS contributed a didactic analysis of population density in a 'typically bad slum', Bethnal Green, through maps, 'pictorial statistics' and diagrams showing population density and types of circulation, occupations of residents, housing accommodation and rents in relation to incomes of the employed and unemployed. Understanding the importance of amplifying their messages across media, MARS reproduced the whole of their *New Homes* display in print in *Design for To-Day*, the DIA's magazine.[67] MARS ably highlighted 'the problem' of housing whilst conspicuously failing to suggest 'the solution'. Excusing this omission, the *Manchester Guardian* reported, 'They ... are determined not to form conclusions till they know a great deal more', suggesting MARS was merely in the process of research.[68]

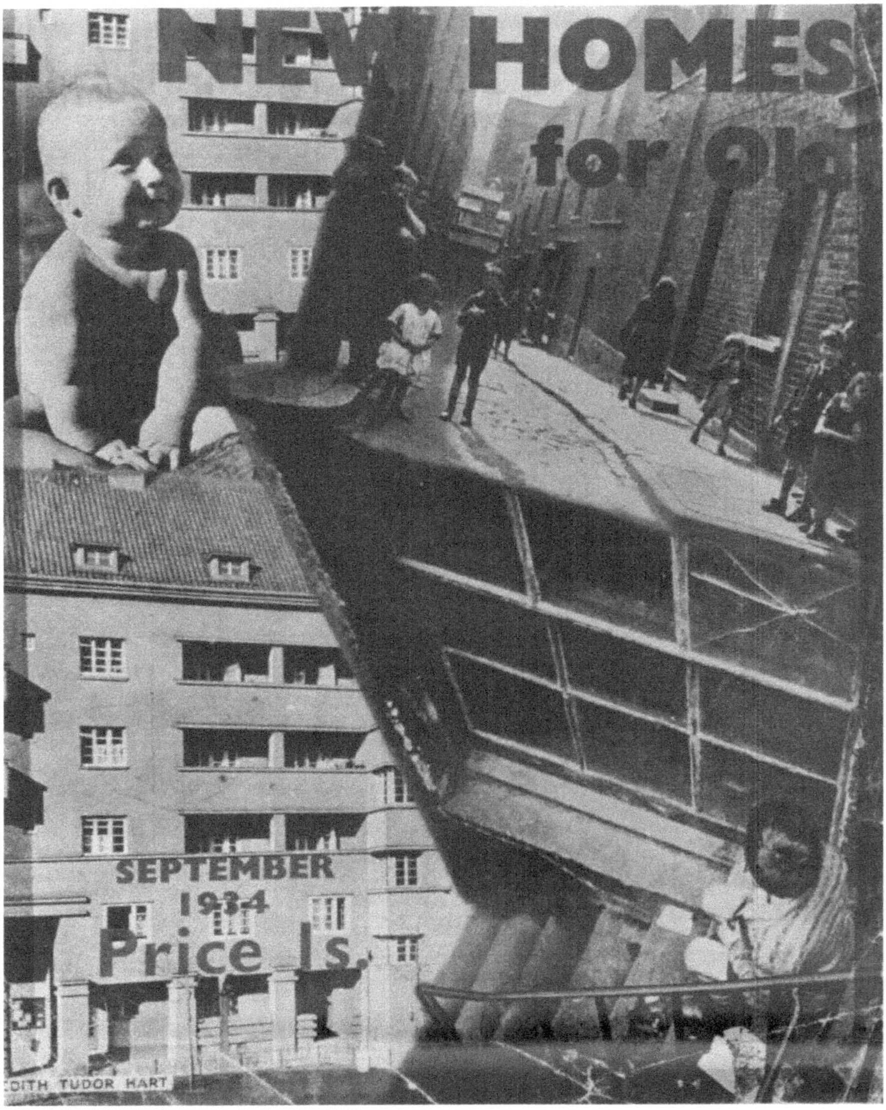

2.18 Booklet for *New Homes for Old*, 1934. The cover carried a photomontage by Edith Tudor-Hart contrasting a slum with new housing. © Edith Tudor-Hart, The Estate of Wolf Suschitzky. Image courtesy of Four Corners. All rights reserved and permission to use the figure must be obtained from the copyright holder.

The *New Homes for Old* series continued, showing the ability of its organisers to continue absorbing up-to-date visual technologies to propel its polemical housing messages. The 1935 *New Homes for Old* exhibition, held at Charing Cross Station, consisted of a photographic collage of well-planned public spaces lining the walls of the small area, accompanied by the overarching slogan 'Housing must be planned in relation to the life of the community and to centres of employment. Our cities can be made pleasant for all …', with smaller texts for each image, model flats and a nursery school.[69] The September 1936 *New Homes* saw members of the Housing Centre represented by Ledeboer and Denby collaborate again with MARS and joined by a new organisation, the Architects' & Technicians' Organisation (ATO), with the aim of urging 'better housing and more beautiful surroundings for all ages'.[70] ATO had been formed in 1935 by architects Francis Skinner, Berthold Lubetkin and other members of the Tecton practice, as a breakaway from MARS, an explicitly socialist group focused to 'support working class organisations fighting for better housing conditions'.[71] Born in Russia (now Georgia), Lubetkin had studied and worked in Paris before moving to London in 1931 and co-founding the Tecton practice. ATO's intention was to counteract the increase of slums and unplanned, badly constructed buildings, working in collaboration with the building industry. Unlike MARS, with their vague commitment to 'further an architecture which serves the needs of society', ATO was explicitly anti-fascist, with an agenda to create better housing conditions to counter, in their words, 'reactionary forces of privilege and finance'.[72] Many criticised the resulting Housing Centre display as too 'political', an accusation robustly rebutted by ATO, who believed it was necessarily political to present the issues.[73]

ATO's more robustly socialist agenda was apparent in its exhibitions. Earlier that year, in April 1936, ATO had shown their commitment to improving poor housing when they mounted the independent *Exhibition on Working-Class Housing by the Architects and Technicians Organisation* at the Housing Centre on Suffolk Street in central London, with a dramatic presentation of housing conditions and statistics and a catalogue criticising national housing policies. The ATO exhibition toured on to Blackfriars, Liverpool, Manchester, Cambridge and Poplar, each time supported by a sympathetic Leftist organisation, including the Home Counties Labour Association and ATO regional branches, with updates such as screens on the 1935 Housing (Overcrowding) Act added later.[74]

The September 1936 *New Homes for Old* was mounted in the gallery of the Building Trades Exhibition at Olympia, staffed by members of voluntary housing societies, with layout and setting by Misha Black. It took the strapline 'Britain is Being Rebuilt', with the polemical subtext 'Old mistakes must be avoided', stating that 'This exhibition is designed to point out how human needs can be more fully satisfied … A fitting background can be provided to every stage of life: to Infancy, Childhood, Manhood and Old Age'.[75] It included a focus on infants in nursery school, children at school and play, 'The Man and Woman in the Home', 'Men and Women in the Community' and 'The Old in their Leisure'. There was material on planning suitable and 'dignified' housing for older people, researched and designed by architect Godfrey Samuel, showing different models from the village to the urban local authority home in Britain and drawing on comparators from Denmark and Sweden.[76] In addition to displays of plans and photographs accompanied by written commentary, the exhibition incorporated films to draw out particular aspects of the theme. Photographs and models combined with slogans and texts, proclaiming 'bad conditions breed disease' and 'there are too many slums and out-of-date schools', and mapping the changes in housing stock since 1835, a popular device in British housing propaganda.[77]

The September 1936 *New Homes* showed a shift in ambition for the appeal of the messages within the exhibition. To reach a greater audience, the committee approached the BBC, asking if they would broadcast a programme about the exhibition and inviting architectural critic Geoffrey

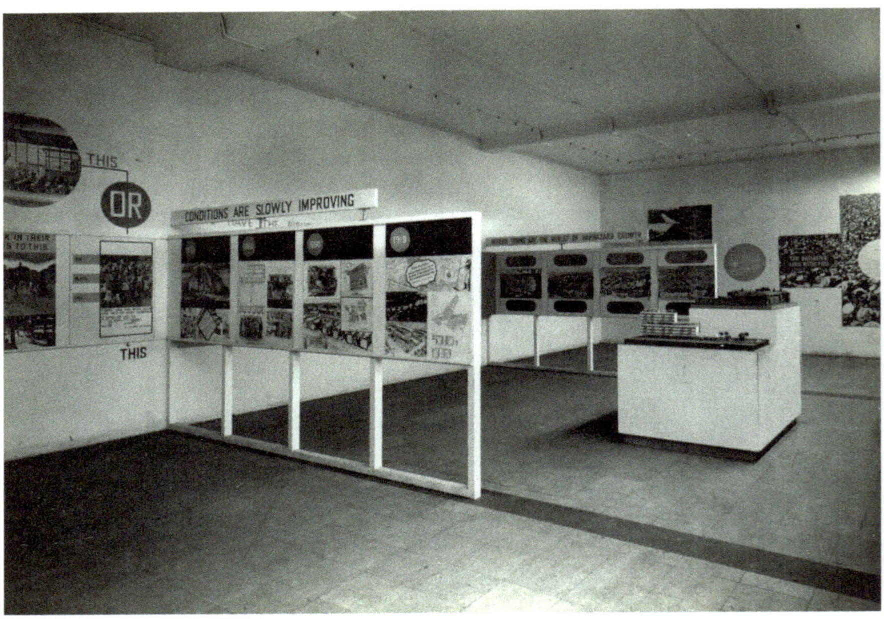

2.19 *New Homes for Old* exhibition at Charing Cross Underground Station, December 1936, with displays showing how good housing conditions improved health. Topical Press. U22228 © TfL from the London Transport Museum collection. All rights reserved and permission to use the figure must be obtained from the copyright holder.

Boumphrey to present it.[78] A smaller version of the 1936 *New Homes* was mounted at Charing Cross Station in December, again intent on activating audiences.[79] Displays focused on how good housing conditions improved health; they mapped housing improvements at intervals (from 1835 to 1860 to 1890 to 1918 to the present), showing examples of towns lacking 'order, health, convenience, comfort', all accompanied by slogans proclaiming 'BAD CONDITIONS BREED DISEASE' and 'THIS OR THIS – WE HAVE THE RIGHT TO DEMAND IN OUR HOMES' (Figure 2.19).[80]

The final *New Homes for Old* of 1938, once again organised by Ledeboer, focused on rural housing, its centrepiece a five-roomed cottage designed by Justin Blanco White displayed alongside 'an authentic rural slum house'. It had been occupied shortly before being dismantled, transported and re-erected at Olympia, including bringing its occupant.[81] The *New Homes* series showed exhibitions to be a suitable means of giving answers to housing problems through exhibitions, particularly when created through alliances between planners, architects, technicians and broadcasters and using all means of representation, from photographs, to film, to building elements.

Government and political activists had identified exhibitions as allowing scope for the presentation of problems and their proposed solutions across subjects such as areas in need of investment and housing in need of improvement. Their use, even in small spaces like Charing Cross, could then be amplified by news media and in broadcasts through the BBC network. The next chapter explores how exhibitions became three-dimensional manifestos during the 1930s in Britain.

Notes

1 Black, 'Propaganda in Three Dimensions', p. 125.
2 Chomsky, *Manufacturing Consent*.
3 Lambert, *Propaganda*, p. 115.
4 Black, 'Propaganda in Three Dimensions', p. 128.
5 Ellul, *Propaganda*, p. 70.
6 Anthony, *Public Relations*.
7 Stephen Tallents, quoting from the *Empire Marketing Board, 1927–28*, p. 42, in 'The Birth of British Documentary', *Journal of the University Film Association*, vol. 20, no. 1, 1968, pp. 15–21.
8 London Transport Museum has an excellent series of photograph albums starting from 1925. London Transport Museum Depot, Acton Album Q1 – 482Q 1925–31. See also Transport for London archives online.
9 Kallmann, 'The Wartime Exhibition', p. 97.
10 Kenneth O. Morgan, *The Oxford History of Britain* (Oxford: Oxford University Press, 2001), p. 613.
11 LTM Album, A3 1934–5, U15608 24/8/1934 *London Transport Photographic Exhibition*.
12 LTM Album, Q2 Exhibitions 1932–1933 483Q.
13 Shundana Yusaf, *Broadcasting Buildings: Architecture on the Wireless, 1927–1945* (Cambridge, MA: MIT Press, 2014).

14 Scott Anthony in Anthony and Mansell, *The Projection of Britain*, p. 10.
15 Yasuko Suga, 'State Patronage of Design? The Elitism/ Commercialism Battle in the General Post Office's Graphic Production', *Journal of Design History*, vol. 13, no. 1, 2000.
16 Yasuko Suga in Anthony and Mansell, *The Projection of Britain*, pp. 21, 23. This same distinctive Modernist style is also evident at the 1937 GPO display at the Ideal Home Exhibition.
17 'Timber Displayed by Nash', *Commercial Art and Industry*, vol. 20, April 1936, pp. 154–5. London Transport Museum, Album Q4.
18 Alan Powers, 'A Popular Modernism? Timber Architecture in Britain 1936–39', *Architectural Theory Review*, vol. 25, nos. 1–2, 2021, pp. 245–66.
19 Buckley, *Designing Modern Britain*, pp. 94–5; Lambert, *Paul Nash as Designer*; A. Causey, 'Paul Nash as Designer' in Peto and Loveday (eds), *Modern Britain, 1929–1939* (London: Design Museum, 1999); Paul Nash, *Room and Book* (London: Soncino Press, 1932).
20 *Commercial Art*, 1936, p. 154. 'Timber Display by Paul Nash', *Architectural Review*, February 1936, p. 90. Paul Nash wrote about the exhibition in 'Experiments in Wood Murals', *Wood*, April 1936, p. 195.
21 'Gate Posts' was reproduced in *Commercial Art*, 1936, p. 155.
22 Paul Greenhalgh argues that this tradition of unflinching patriotism in response to empire shows went up to the Second World War in Britain in *Ephemeral Vistas: The Expositions Universelles, Great Exhibitions and World's Fairs, 1851–1939* (Manchester: Manchester University Press, 1988), p. 58. I dispute this: by the 1920s in Britain, exhibitions begin to be a focus for opprobrium as well as celebration of Empire. See P. Kinchin and J. Kinchin, *Glasgow's Great Exhibitions* (Bicester: White Cockade, 1988). Felix Driver and David Gilbert discuss how Imperial exhibitions were an important focus for presenting the Empire in Britain, coining the phrase 'Imperial London' in 'Capital and Empire: Geographies of Imperial London', *GeoJournal*, vol. 51, nos. 1/2: *European Capital Cities* (2000), pp. 23–32. See also Jonathan Woodham, 'Images of Africa and Design at the British Empire Exhibitions between the Wars', *Journal of Design History*, vol. 2, no. 1, 1989, pp. 15–33.
23 John McAleer and John MacKenzie, *Exhibiting the Empire: Cultures of Display and the British Empire* (Manchester: Manchester University Press, 2015), p. 10.
24 Carullo, *Moholy-Nagy in Britain*, p. 36.
25 LTM archives.
26 *Imperial Institute Annual Report* (London, 1936), p. 46.
27 The text-heavy *Imperial Pilgrimage* at Charing Cross the same year attempted to tell the story of Rhodes, Raffles and Cameron, among others, although why this was the focus is not clear from the archive. Other exhibitions along these lines included *Australia* (1937), *Dominion of New Zealand* (1937) and *Empire Coffees* (1938). Photographs in LTM Album, Q5.
28 F. C. Pritchard was the brother of Jack Pritchard, Director of Isokon.
29 *Commercial Art and Industry*, vol. 20, January–June 1936, p. 153; John Mohan, 'Neglected Roots of Regionalism? The Commissioners for the Special Areas and Grants to Hospital Services in the 1930s', *Social History of Medicine*, vol. 10, issue 2, August 1997, pp. 243–62.
30 Teri J. Edelstein, 'The Underground's Alchemist of the Modern', in Caitlin Condell and Emily M. Orr (eds), *E. McKnight Kauffer, The Artist in Advertising* (New York: Rizzoli Electa, 2020).
31 *Thirties: British Art and Design before the War* (London: Arts Council of Great Britain, 1979), p. 18; Morgan, *The Oxford History of Britain*, p. 612.
32 *Commercial Art and Industry*, vol. 20, January–June 1936, p. 153.
33 Archives of the Royal Society of the Arts, London, Hans Schleger Papers.

34 *Designs on Britain: An Exhibition by Jewish Museum London* (London: Jewish Museum London, 2017), p. 20.
35 *Modern Publicity*, 1938–9, p. 70.
36 Jonathan Black, 'For the People's Good: Hans Schleger (1898–1976), Poster Design and British National Identity, 1935–60', *Visual Culture in Britain*, vol. 13, no. 2, 2012, p. 177.
37 LTM Q5.
38 LTM Q6.
39 Edward McKnight Kauffer designs for Dorland Hall, Cooper Hewitt collection, 2318796047.
40 *Action*, 13 January 1938, p. 8.
41 Peter Brimblecombe, *The Big Smoke: A History of Air Pollution since Medieval Times* (London and New York: Routledge, 1988) and Peter Thorsheim, *Inventing Pollution: Coal, Smoke and Culture in Britain since 1800* (Athens, OH: Ohio University Press, 2006).
42 According to *Display*, January 1939, p. 528. UOBDA, Henrion collection, *Smoke Abatement* at Charing Cross, 1938. Exhibitions focused towards campaigning against social ills such as noise and smoke included the *Noise Abatement Exhibition* presented by the Anti-Noise League, held at London's Science Museum in 1935, as discussed by James G. Mansell in Anthony and Mansell, *The Projection of Britain*, pp. 161–2.
43 *Smokeless Air*, vol. 1, no. 1, Spring 1939, p. 12.
44 As is clear from *The Journal of the National Smoke Abatement Society*, for example vol. 7, no. 25, February 1936, p. 3 and vol. 7, no. 28, November 1936.
45 *Shelf Appeal*, October 1936, p. 41. It referred to a previous exhibition about Imperial Airways at Science Museum being 'a wonderful example of how to dramatise a good story'.
46 Despite his attempts, Black was not granted naturalisation as a British citizen until 1950; TNA HO334/344/14923, 'Misha Black Naturalisation Certificate: Moisei Tcherny. From Russia. Resident in London. Cert BNA14923 issued 7 September 1950'. Before Black became British he had already led British state contributions as a 'stateless' person to numerous major international exhibitions, including the New York World's Fair 1939–40. See Atkinson, '"Lines of Becoming" Misha Black'.
47 *Journal of the National Smoke Abatement Society*, November 1938, p. 101.
48 Alec Davis, 'Henrion – Designer', *Art and Industry*, vol. 21, July–December 1936, pp. 238–40. Henrion was naturalised as a British citizen in 1948; TNA HO334/201/37981.
49 *Shelf Appeal*, October 1938, p. 36.
50 *Display*, January 1939, p. 528.
51 *Journal of the National Smoke Abatement Society*, November 1938, p. 101.
52 *Journal of the National Smoke Abatement Society*, May 1938, p. 45.
53 *Smokeless Air*, vol. 1., no. 1, Spring 1939, p. 54.
54 Dean, *The Architect as Stand Designer*, pp. 26–7; Janet Ward, *Weimar Surfaces: Urban Visual Vulture in 1920s Germany* (Berkeley, CA: University of California Press, 2001); Deborah Sugg Ryan, *Daily Mail Ideal Home Exhibitions* (London: Hazar Publishing, 1997).
55 'The Building Exhibition: Reconstructed Slum as a "Horrible Example"', *Manchester Guardian*, 13 September 1934, p. 18.
56 *Thirties: British Art and Design before the War*, p. 11.
57 Morgan, *The Oxford History of Britain*, p. 613.
58 Elizabeth Darling, '"Enriching and enlarging the whole sphere of human activities": The Work of the Voluntary Sector in Housing Reform in Inter-War Britain', in Christopher Lawrence and Anna-K. Mayer (eds), *Regenerating England: Science, Medicine and Culture in Inter-War Britain* (Leiden: Brill, 2000), pp. 149–78.

59 Darling, 'Enriching and enlarging', pp. 159, 161.
60 Dean, *The Architect as Stand Designer*, p. 56.
61 J. G. Ledeboer, *Design for Today*, vol. 11, no. 19, November 1934, p. 407.
62 Robert Radford, 'Edith Tudor-Hart: Photographs from the Thirties', *Camerawork*, July 1980, p. 2; Elizabeth Darling, '"To induce humanitarian sentiments in prurient Londoners": The Propaganda Activities of London's Voluntary Housing Associations in the Inter-War Period', *The London Journal*, vol. 27, no. 1, 2002, pp. 42–62.
63 J. G. Ledeboer, 'New Homes for Old', *Design for Today*, vol. 11, no. 19, November 1934, pp. 407–10 and 'MARS Exhibit at Olympia', pp. 411–14.
64 Formed by architects Wells Coates, David Pleydell-Bouverie and Maxwell Fry, writers P. Morton Shand and John Gloag and *Architectural Review* editor H. de Cronin Hastings; *The Times*, 12 September 1934, p. 10.
65 Summerson, 'The MARS Group and the Thirties', p. 305.
66 A point made by Louise Campbell in 'The MARS Group, 1933–1939', *RIBA Transactions*, issue no. 8, vol. 4, no. 2, 1985, pp. 70, 72, 76. Campbell points out that MARS members Maxwell Fry and John Gloag had come from DIA and that there was a gap between MARS's professed values and its actual engagements. See also Elizabeth Darling, 'Institutionalizing English Modernism 1924–33: From the Vers Group to MARS', *Architectural History*, vol. 55, 2012, pp. 299–320.
67 'MARS exhibit at Olympia', *Design for To-Day*, vol. 11, no. 19, November 1934, pp. 411–14.
68 'The Building Exhibition: Reconstructed Slum as a "Horrible Example"', *Manchester Guardian*, 13 September 1934, p. 18; 'New Homes for Old Housing Exhibit: The MARS Contribution', *Architects' Journal*, vol. 80, 1934, pp. 425–8.
69 LTM Album Q3 1934–1935, January 1935, U16568.
70 RIBA Godfrey Samuel papers, GS SaG/88/4/1/2, NHFO 1936 exhibition catalogue, p. 2.
71 RIBA Ove Arup papers, ArO/214/1/(v), Skinner, F. 'Memorandum for Discussion at First Meeting', 11 February 1935, p. 4.
72 Campbell, 'The MARS Group, 1933–1939', p. 72; P. Coe and M. Reading, *Lubetkin and Tecton: Architecture and Social Commitment* (London: Arts Council, 1981), p. 51.
73 Gold, *The Experience of Modernism*, p. 124.
74 RIBA OA ArO/2/14/2 (I -).
75 RIBA GS SaG/88/4/1/2. *Architects Journal*, 17 September 1936, p. 361.
76 RIBA GS SaG/88/4/1/2.
77 LTM Album Q4, January 1935, LTM Album Q4, December 1936.
78 RIBA GS SaG/88/4/2/2, minutes of NHFO exhibition committee, 17 August 1936, p. 1.
79 RIBA OA ArO/2/14/2 (ii-).
80 LTM Album, Q4, 1936.
81 Dean, *The Architect as Stand Designer*, pp. 74–5, covered in *The Architect and Building News*. Elizabeth Denby would go on to exhibit a prototype model house, the All-Europe House at the 1939 Ideal Home Exhibition, her solution to the problem of slum dwellings; see Elizabeth Darling, '"The star in the profession she invented for herself": A Brief Biography of Elizabeth Denby, Housing Consultant', *Planning Perspectives*, vol. 20, July 2005, pp. 285–91.

3

Exhibitions as manifestos

'Those were still the days of "isms", when it was possible for little groups to form for an exhibition and issue a manifesto', wrote painter Julian Trevelyan, recalling how the artists' groups he was associated with in the 1930s used exhibitions as platforms from which to enunciate views and ideals.[1] It is this confluence of exhibitions and manifestos or, indeed, exhibitions *as* manifestos that is the focus of this chapter, which explores how exhibitions operated as forms for Modernist artists and architects in Britain. These were early twentieth-century materialisations of the textual form developed by Leftist polemicists and writers of the nineteenth century as 'provocations of the modern', to use the phrase of literary historian Janet Lyon.[2] Taken up by artists and designers of the 1930s in Britain, such exhibitions were intended to draw audiences into a vivid visual and textual engagement with their work and ideas.

Urbanist Marshall Berman, in *All That Is Solid Melts into Air*, his seminal account of the experience of modernity, describes the manifesto as the 'first great modernist work of art', 'remarkable for its imaginative power, its expression and grasp of the luminous and dreadful possibilities that pervade modern life'.[3] From the 1930s in Britain, exhibitions' potential to communicate ideas and solidarities and, moreover, to act as forms of persuasion and coercion became evident, as this chapter discusses. For emerging Modernist artists' and architects' groups, including Unit One and the Modern Architectural Research Group (or MARS), both of which formed in 1933, exhibitions were central to demonstrating their future vision and in staging the transitional identities of these new groups.[4] They operated not only as frames, platforms or vehicles but, in a more complex sense, as carriers of messages through image, text and space. Exhibitions were honed as three-dimensional manifestos, hybrid visual-textual-spatial forms for promotion and self-proclamation.

I focus in this chapter on three 'manifesto exhibitions', each located in London in the 1930s but indicating the internationalist imaginations and networks of their makers. Each exhibition proclaimed the beliefs of Modernist artists', architects' and designers' groups in exhibitionary form: the *Unit One Exhibition*, held at the Mayor Gallery of April 1934; the *International Surrealist Exhibition*, held at New Burlington Galleries of June 1936; and the *Exhibition of the Elements of Modern Architecture* mounted by MARS Group in January 1938 at New Burlington Galleries. Literary critic Martin Puchner, in his book on manifestos, cites London as 'the birthplace of the genre', since Marx's manifesto had first been published there in German.[5] These London-based exhibitions of the 1930s were not designated as manifestos by their makers, but acted as such by extending and reinforcing the declarations these groups made in other forms (written, rhetorical and visual). They shared characteristics as self-referential and self-differentiating statements, and as acts of appropriation, borrowing from other contexts. Revolution – in art, design, architecture and life itself – was a subtext and pretext for these three exhibitions, which promoted their own modernity. All three were strong on remonstration (declaring what was wrong and what they would put right), focused on presenting themselves as a group challenge to what had gone before. Each was intent on generational critique, suggesting they had a perspective superior to that of their elders, which derived from their age and ability to innovate.[6]

The *Unit One Exhibition*

The Mayor Gallery, part of London's lively West End gallery culture, introduced and promoted the work of new creative groups, including the inaugural *Unit One Exhibition* of April 1934. Unit One, the short-lived but influential experiment disbanded in 1935, consisted of eleven architects and sculptors, many of whom lived within a tight geographical nucleus in north London's Hampstead area, describing the Mayor Gallery as its 'headquarters'.[7] Unit One founder artist Paul Nash explained the group's aims in *The Listener* magazine as preoccupied with 'the expression of structural purpose … in formal interaction … typified by abstract art' and 'the pursuit of the soul … psychological research',[8] while in a letter to *The Times* Nash said that 'Unit One may be said to stand for the expression of a truly contemporary spirit, for that thing which is recognised as peculiarly of to-day in painting, sculpture and architecture', rather than for a particular visual or formal style or orthodoxy.[9]

The Mayor Gallery's *Unit One Exhibition* included work by Edward Burra, Barbara Hepworth, Henry Moore, Ben Nicholson, Edward Wadsworth and the architects Wells Coates and Colin Lucas. Its installation was conventional, with works hung round the walls and sculptural works on plinths, rather than adopting the kinds of Modernist exhibitionary practices that were being pioneered by László Moholy-Nagy and others (such as the use of murals; of full-scale room models; of plywood screens with focusing peep-holes and port-holes to reveal elements of the display; or of lettering as a structural element). The *Unit One Exhibition* behaved as manifesto by identifying the need for multi-channel communication, not only in visual but in printed form. In addition to Nash's use of contemporary newspapers and magazines to explain the group, Herbert Read persuaded members of Unit One to amplify their position and focus through an accompanying book of written statements and photographs, which became *Unit 1: The Modern Movement in English Architecture, Painting and Sculpture*, edited and introduced by Read.[10] The typography and layout of the book were shaped by ideas within The New Typography movement, which had rejected the traditional arrangement of type in symmetrical columns, its striking cover carrying bold red type structured against a vast figure '1'.[11] A smaller version of the exhibition subsequently toured around the United Kingdom to Liverpool, Manchester, Hanley, Derby, Swansea and Belfast, a sign of the relaxation of hostility to Modernism in art and of the credibility of Nash and Read.[12] Nash's habit of communicating through multiple channels at once, producing art and design work alongside written proclamations, was in keeping with the habits of his Modernist contemporaries and exhibitions operated as another channel for extending ideas about the significance of his practice.[13]

New Burlington Galleries

The majority of exhibitions discussed in this book took place outside the established sites of culture, in Underground stations, workers' canteens, bombsites, parks and village halls. There are only two recurring galleries: London's Whitechapel Art Gallery and London's New Burlington Galleries. Based at 3–5 Burlington Gardens, the New Burlington Galleries were close to the capital's bustling Piccadilly and central to London's flowering inter-war art scene. Repeatedly, through the two decades of this book, the Galleries provided a base for seminal and eye-catching exhibitions. Aside from being the venue for two of the 'manifesto exhibitions' in this chapter – the *International Surrealist Exhibition*, held from June to July 1936, and the *Exhibition of the Elements of Modern Architecture*, organised by MARS in January 1938, the Galleries were the base for a number of exhibitions that feature in this book. These include the exhibition of *Twentieth Century German Art* held in July 1938 (discussed in Chapter 5); *Guernica*'s first English venue, held in October 1938 (discussed in Chapter 6); and the London venue for the Artists International Association's much publicised *The Mirror and the Square* exhibition held in December 1952 (discussed in Chapter 8). And yet due to the destruction of the institutional archive of New Burlington Galleries, frustratingly little information remains about who ran the Galleries, their interests, alignments or politics.[14]

The New Burlington Galleries' lack of adherence to the strictures and categories of contemporary art followed by neighbouring galleries was evident from its surprisingly varied programme. As well as hosting regular shows of the London Group, the prominent progressive artists' network formed in 1913 as an amalgamation of the English Cubists and the Camden Town Group, they programmed exhibitions of advertising art and work by international artists of differing political persuasions.[15] A retrospective of Colombian painter Andres de Santa Maria in May 1937 was followed by an exhibition of paintings and drawings of the King's Ships and Merchant Navy in summer 1937, a century of French caricature 1750–1850 in 1939 and an exhibition of a Franco-supporting artist, Ignacio Zuloaga, which ran in the next-door room to the display of Picasso's virulently anti-Franco statement *Guernica*.[16] Reflecting on their reputation for innovative and experimental programming, National Gallery Director Kenneth Clark described the Galleries as 'fifty yards behind the Royal Academy and fifty years ahead of it'.[17]

The *International Surrealist Exhibition*

Exhibitions provided a multiform platform – incorporating books, lectures and performances – through which artists could demonstrate their connection with artistic and political formations beyond Britain. The New Burlington *International Surrealist Exhibition*, held from June to July 1936, acted as a showcase for demonstrating connection with continental Surrealism, which had impacted on the work of British artists and writers for a decade or so, through the English translation of writings and British admiration for French Surrealist artworks. An exhibition held at London's Mayor Gallery a few years earlier had shown some works by Surrealists including Miró, Klee and Picabia, but the much more extensive 1936 show enabled British Surrealists to develop their reach and their mystique in Britain, through an esoteric collection of artworks and objects.[18]

Although the design of the exhibition's installation at New Burlington Galleries was conventional – paintings mainly hung vertically on walls through the space and sculptural works on plinths – this exhibition is worthy of discussion here for two main reasons. Firstly, because it was an act of exhibitionary showmanship, attracting a lively critical response from contemporaries and acting as a multiform artistic manifesto by combining exhibition with exhortation, published tracts, lectures, debates and performances. Secondly, it was an occasion on which the exhibition's contributors sought to provoke a serious engagement with political themes. This was most clear from the raft of related events and talks scheduled alongside the exhibition, such as a debate at Conway Hall that addressed the relative political effectiveness of realism versus surrealism.[19]

Surrealism had evolved in France as a highly literary form and the London exhibition was no different, being accompanied by a series of written works including a book edited by art critic Herbert Read, containing contributions from French and English Surrealists.[20] In its introduction, Read legitimised Surrealism in Britain as a reaffirmation of a national Romantic tradition. Read penned an introduction to the exhibition's catalogue, cautioning visitors 'Do not judge this movement kindly ... It is defiant – the desperate act of men too profoundly convinced of the rottenness of our civilization to want to save a shred of its respectability' and with the challenge 'The artists ... have only interpreted the world; the point, however, is to transform it'.[21] Read was beseeching viewers not to be distracted from underlying messages by playful forms and asserting that, despite the seemingly frivolous or even scandalous subject matter, these artists had a serious political agenda.[22]

French artist André Breton, the principal theorist of Surrealism, opened the exhibition, with luminaries in attendance, showing works by French Surrealists alongside British-based ones (Figure 3.1).[23] Contributing artist Julian Trevelyan recalled his enjoyment of the opening's carnivalesque atmosphere, with over a thousand people crammed into the galleries to see 'large canvases of Picasso, already deep in his Minotaur legends, ... great jungles of Max Ernst ... a powerful picture by Magritte ... and a huge pair of lips across a landscape by Man Ray', describing the 'feeling of richness and poetic invention' that 'pervaded the whole exhibition' and Herbert Read declaring the dawn of a new age;[24] while James Boswell's cartoon for *Left Review* mocked the genteel crowds at the exhibition's opening, its caption echoing Read's words for the catalogue 'Do not judge this movement kindly ...'[25] The *International Surrealist Exhibition* ran for only four weeks but attracted a large audience (given the gallery size) of 25,000 visitors and an extensive and varied critical response.[26]

Literary historian Martin Puchner suggests that the political programme of the French Surrealists was muddled – 'no avant garde was more devoted to the revolution than Surrealism, and none was more uncertain about what this revolution should be'.[27] However, in the British context the political agenda was more explicitly stated, with Surrealists becoming closely aligned with other Leftist groups such as the AIA, many such as Trevelyan being members of both and engaging extensively at political rallies and debates (as I discuss in Chapter 4).

MARS Group *Exhibition of the Elements of Modern Architecture*

Many excellent accounts of the activities of the Modern Architectural Research Group (or MARS) have already been written.[28] My specific interest here is in the way MARS shaped and influenced exhibition technique

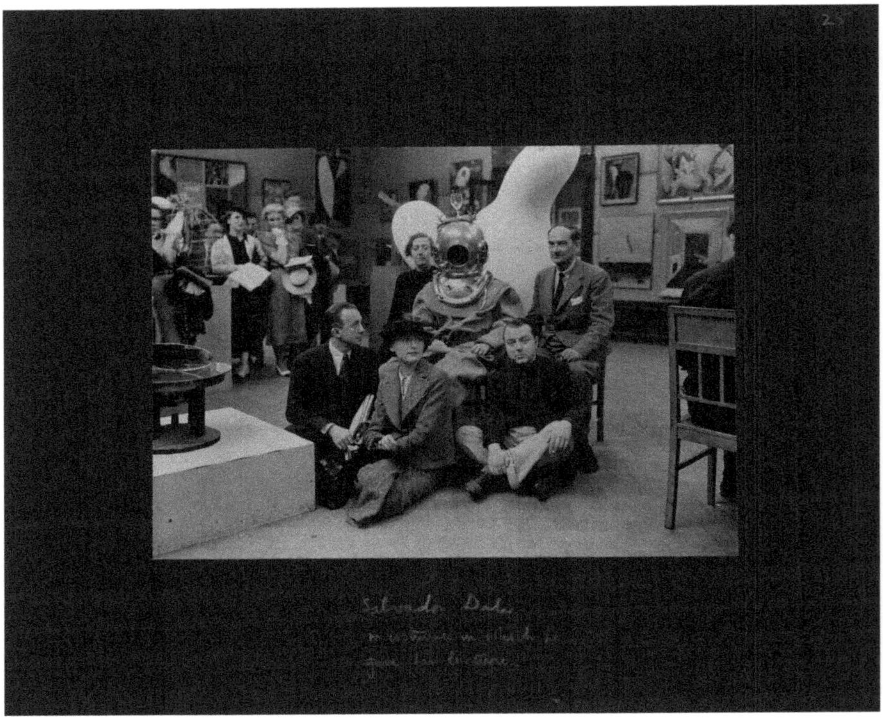

3.1 Opening of the *International Surrealist Exhibition*, 1936 with Salvador Dalí in diving suit and helmet with Paul and Nusch Éluard, E. L. T. Mesens, Diana Brinton Lee and Rupert Lee. Uncredited photographer. Images compiled and annotated by Roland Penrose. Image courtesy of the National Galleries of Scotland. All rights reserved and permission to use the figure must be obtained from the copyright holder.

in Britain and my specific claim is that MARS produced a blueprint for British exhibitions as manifestos through their 1938 exhibition at the New Burlington Galleries. This was referred to by Misha Black as 'the archetype of contemporary British exhibition design', believing the exhibition heralded an entirely 'new approach' and precipitated a vast improvement in Britain's status as exhibition designers. It proved such successful propaganda for Modernism that it provided a model for British propaganda exhibitions throughout the war, identified by many, including Black, as the first example of the 'informative and story-telling type of exhibition' in Britain.[29] Having himself acted as a co-ordinator for the exhibition, Black's claim involved some self-mythifying, but he was not alone in attributing such significance to this fairly modest exhibition, pinpointed as the model for future propaganda exhibitions in Britain, so it is worth looking more closely at the elements that made it so notable.

As signalled through their contributions to the earlier *New Homes* series (discussed in Chapter 2), MARS had a strong belief in the capacity of exhibitions to convey messages about how to improve architecture and planning and, more importantly, to prove the superiority of the Modernist architectural programme. In this spirit, they devised an ambitious programme of exhibition contributions during the period from 1933 to the outbreak of the Second World War. The 1938 MARS Group exhibition was three years in the making. The Group's exhibition committee had proposed, as early as April 1935, an exhibition 'to show good pictorial examples of new buildings, with sufficient story to arouse and maintain interest in them from a purely aesthetic point of view', that should be 'frankly propagandist' and using 'dynamic methods of presentation'. This was rhetorically in keeping with the 'good design' evangelism emerging through the exhibitions of design reform bodies such as the DIA.[30]

Over the three years of its evolution, the exhibition's key contributors changed but a notable feature was the diverse geographic origins of its makers, with contributions from, among others, Russian-born Serge Chermayeff and Misha Black; German-born Peter Moro, F. H. K. Henrion, Arthur Korn and Walter Landauer; and Hungarian-born László Moholy-Nagy, Marcel Breuer and Ernö Goldfinger. The initial lead organiser was Canadian architect Hazen Sise (temporarily living in Britain from 1929 until the eve of the Second World War), replaced first by Godfrey Samuel, then by László Moholy-Nagy and finally by Misha Black, who brought it to fruition.[31] In early 1937 its committee was led by architect Maxwell Fry, working with artist Moholy-Nagy and architects Godfrey Samuel and Serge Chermayeff.[32] Fry, Gropius and Moholy-Nagy were at the same time collaborating on designs for the interior of an electricity showroom on Regent Street, just around the corner from the venue for the 1938 exhibition. By the time the exhibition opened, Gropius and Moholy-Nagy had left Britain for the US.

Naming the MARS Group exhibition

Originally planned for 1937, it finally opened at the New Burlington Galleries in January 1938, entitled *An Exhibition of the Elements of Modern Architecture Organised by the MARS (Modern Architectural Research) Group.* Perhaps surprisingly, given its unwieldy final formulation, the exhibition's title had been the subject of hot debate within the group. An early suggestion that it be named through a competition in *Architects' Journal* was rejected as, although it was thought likely to attract good publicity, it might damage the group's prestige, showing the level to which the group was conscious of the impact of media attention. Members had instead received letters inviting a 'good, snappy, publicity title',[33] with resulting suggestions including 'Sticks and Stones', 'Towards a New Architecture' and 'Architecture for Humans'.[34] Publicising the exhibition and its messages

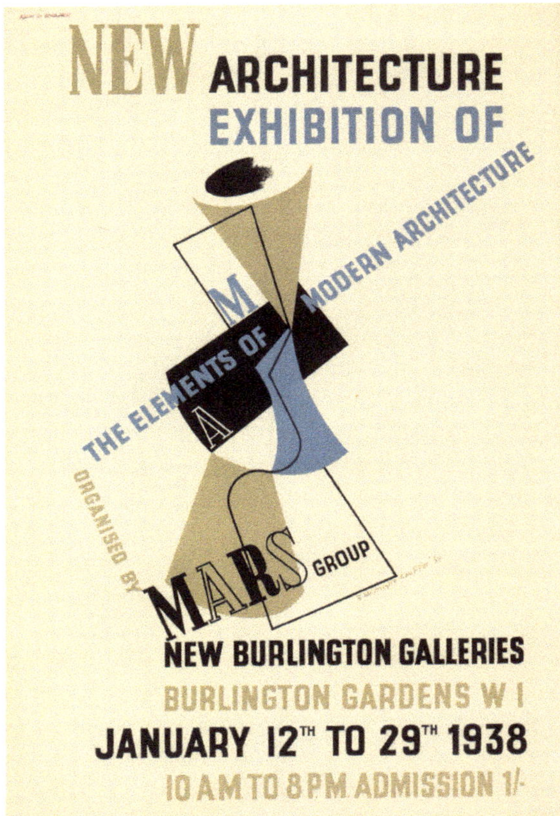

3.2 Poster for the MARS Group *Exhibition of the Elements of Modern Architecture*, designed by E. McKnight Kauffer, 1937. Cooper Hewitt, Smithsonian Design Museum. Gift of Mrs. E. McKnight Kauffer, 1963-39-97. Photo: Matt Flynn. © Smithsonian Institution. All rights reserved and permission to use the figure must be obtained from the copyright holder.

was a major focus of planning and two press officers listed in the catalogue as 'Mrs Zander' and 'Miss Bean' were part of the team. An advertising poster by E. McKnight Kauffer, centring on an abstracted cone, was circulated to Underground stations and urban billboards (Figure 3.2). The exhibition's catalogue cover, designed by Ashley Havinden of Crawford's advertising agency, referenced the show, with elements of buildings and port-holes (Figure 3.3).

The MARS Group exhibition catalogue

Exhibition of the Elements of Modern Architecture was a manifesto that enabled MARS to present themselves, to explain and represent their vision

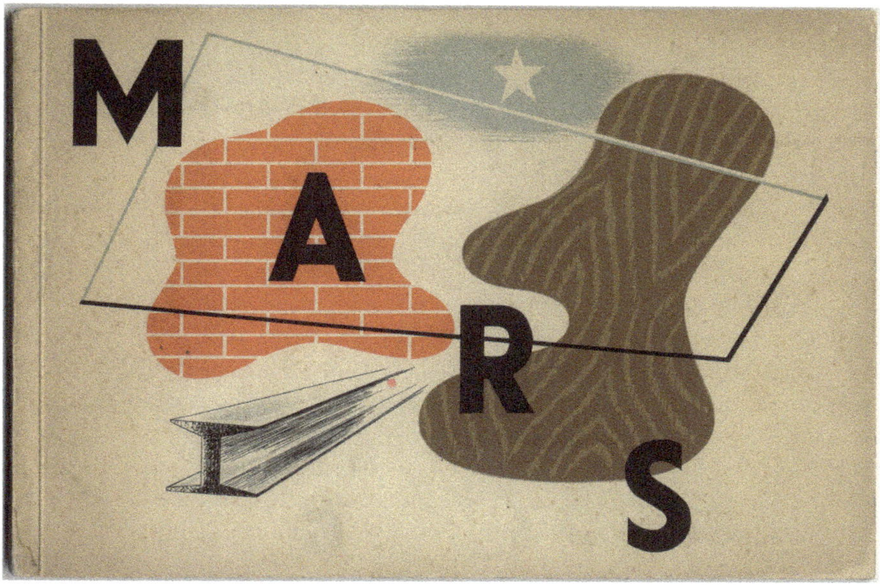

3.3 Catalogue for the MARS Group exhibition, cover designed by Ashley Havinden. Image courtesy of Ashley Havinden Estate and Webb & Webb, Design. All rights reserved and permission to use the figure must be obtained from the copyright holder.

as the future. In doing so, it fulfilled literary scholar Janet Lyon's definition of manifestos as putting a 'dramatic emphasis on *now* as a coherent moment in a well-defined political telos', as well as presenting 'a kind of political dramaturgy through which "universal" ideals are tested'.[35] CIAM and MARS, the catalogue explained, constituted 'a nucleus of practice and criticism' with 'parallels in many parts of the world', 'an international framework of opinion'. Anonymity of individual contributors to the MARS exhibition was prioritised. All were instructed that no statements by or photographs of individual members should appear in the press, that 'a story should be made out of the Group's anonymity'.[36]

This collective and unanimous approach was traced in the catalogue back to English roots. Clearly attempting to reassure its audience, it was linked to the impact of artists and engineers John Ruskin, Thomas Telford and Joseph Paxton. It then connected these with innovations before the war in Holland, Germany and Austria leading, it suggested, towards 'the ultimate realization that structural science and an exact analysis of social needs can supply a sufficient basis for the creation of an architecture of universal applicability'. Here was an attempt to cast aside English 'academic mediocrity' and insularity and to connect with the world beyond by adopting the influence of 'conspicuous landmarks' designed by architects from

Europe and North America including Charles Rennie Mackintosh, Otto Wagner, Adolf Loos, Frank Lloyd Wright, Peter Behrens, Walter Gropius and Le Corbusier. This, the catalogue emphasised, was not a 'crystallised "style"', which 'shackled' and was 'inflexible'.[37] The avowedly internationalist intention of the show was emphasised by Maxwell Fry, describing it in *Architects Journal* as drawing examples from all over the world and not over-stressing 'either the work of the Group or of Englishmen', thereby signalling a narrative of transcendent values leading to universal emancipation.[38]

For the MARS Group, written forms were closely related to other forms of Modernist creation. The MARS Group exhibition catalogue included a foreword written by socialist playwright George Bernard Shaw. Shaw's plays acted as quasi-manifestos for his political beliefs and on occasion he was even known to issue political appendixes to his plays, making him a suitable choice as contributor to this manifesto-like exhibition.[39] Initially, on being approached in spring 1937, Shaw admitted he had not 'the faintest notion of what the MARS group stands for' but was persuaded to write it.[40] His foreword stated that although the exhibition's theme was the 'New Architecture', the aim of the exhibition was not 'the promotion of novelty' but 'an approach to building problems ... still unfamiliar to the majority of the English public'. It showed an 'attitude of mind' agreed upon by architects and thinkers 'all over the world', which is 'revolutionary in a scientific and not an arbitrary or sensational meaning and which, we believe, has an invaluable contribution to offer to the life we are living today', showing his sympathy with the aims and ideas of the exhibition as being based in a specific practice (architecture) but with aspirations to share universal ideas (contributing to and improving life in general).[41] Shaw offered the MARS group advice on how to attract advertisements from the building trade, to run alongside his words. This was successful and twenty-two firms signed up to take advertising space. The exhibition was also advertised in Underground stations.[42]

Exhibition of the Elements of Modern Architecture was a three-dimensional manifesto for the Modern Movement on behalf of the sixty or so architect, engineer and writer members of MARS, its design shared between twelve architects. It was part of a wider programme of mediation and publicity for the 'new architecture' in Britain.[43] A hallmark was its reliance on the reciprocity between textual and visual elements, with MARS member architectural historian John Summerson writing the script alongside the architectural teams. The MARS Group exhibition stood, its catalogue claimed, as a 'consistent, self-explanatory statement' and it was this idea of exhibitions as acting as expanded arguments that was to be the major influence on the form of future exhibitions. The text being considered as significant as images and objects in carrying the MARS message to the public, John Summerson worked alongside all the architects leading on the section designs to script the exhibition with a coherent through-line, writing captions, hoardings and much of the catalogue. Conveying a sense of how much weight the written aspect of the MARS exhibition carried in terms of creative contribution and in transmitting the ideas of the show, Godfrey Samuel flatteringly called Summerson's draft exhibition captions 'a major contribution to English literature', asserting that the exhibition was not intended 'merely as a record of achievement' but as 'a call for public support and encouragement for modern architecture', 'put into about five words!'[44]

Although intended to convert a wide public audience to the merits of modern architecture, the exhibition's writing and referencing was highly esoteric. It was framed through a meta-interpretation of two arcane texts: the modern translation of Sir Henry Wotton's 1624 written paraphrase of Vitruvius's earlier aphorism 'In Architecture as in all other Operative Arts, the end must direct the Operation. The end is to build well'. These ideas were introduced on a curved screen standing near the entrance designed by Peter Moro and Gordon Cullen, which completed the quote: '*Well building hath three Conditions: commoditie, firmenes, and delight* …', with the exhibition structured as a visual argument around these three parts (Figure 3.4). 'Commoditie' was described as 'convenience and fluency in design', 'firmenes' as 'strength with economy and precision' and 'delight' as 'pleasure in space surface rhythm'.[45] Moro explained that the screen was to be approached from a distance so that the picture needed to be vertical but, as you approached, it twisted to become a horizontal surface.[46]

Exhibitions as mediation

Architectural historian Beatriz Colomina proposes the 'space of mass media as the true site in which architecture is produced', underlining the idea that architecture only becomes modern in its engagement with the media, and that in doing so it radically displaces the traditional sense of space and subjectivity.[47] In the presentation of the MARS exhibition, a collage combined to reference the world beyond, creating reciprocity between the design of exhibition spaces, the imaginative space of the written text and the design of streets and architectural spaces beyond the gallery.

The involvement of the BBC (an organisation with a strong agenda to advance modern architecture) in promoting the MARS Group exhibition created a multi-layered mediation of the exhibition. Exhibition caption writer John Summerson was invited on to a BBC television programme to give a commentary on modern architecture, while focusing on images of models from the exhibition.[48] In 1938 the BBC also held an exhibition about itself at Charing Cross Station, displaying publications, programmes, licences, engineering, television (as broadcast from Alexandra Palace), music and agriculture (Figure 3.5).

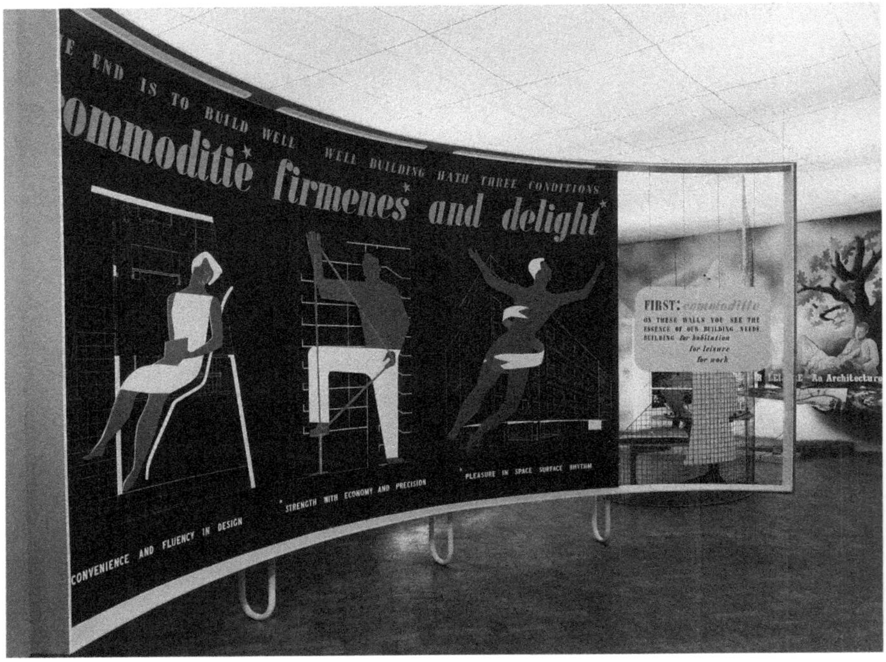

3.4 The entrance screen to the MARS Group exhibition, designed by Peter Moro and Gordon Cullen, introduced the essential conditions of architecture with the aphorism 'Well building hath three Conditions: commoditie, firmenes, and delight...'. Alfred Cracknell. RIBA24539. Image courtesy of Architectural Press Archive/ RIBA Collections. All rights reserved and permission to use the figure must be obtained from the copyright holder.

The close relationship between MARS and the BBC, signalled by its broadcast about the exhibition, may have been due to the involvement of Richard S. Lambert, who was editor of the BBC's in-house magazine *The Listener* and also a MARS member. Lambert's book *Propaganda*, published the same year as the MARS exhibition, discussed the origins of propaganda in Britain, linking it to advertising and publicity practices, mapping its intersections with entertainment and advocating for antidotes, including a free press. Lambert showed his interest in the propaganda role of exhibitions, including them in his discussion of 'indirect or "unintentional" propaganda' and referencing the Glasgow *Empire Exhibition* of 1938.[49] He saw extremist politics being played out through exhibitions, describing how the 'Marxian analysis' of art and literature in Russia had produced a 'violent reaction' from exponents of fascism, exemplified in Hitler's speech on the opening of the new art gallery in Munich in 1937.[50] He cited Stephen Tallents' book *The Projection of England*, with his idea of creating 'a school of national projection' to

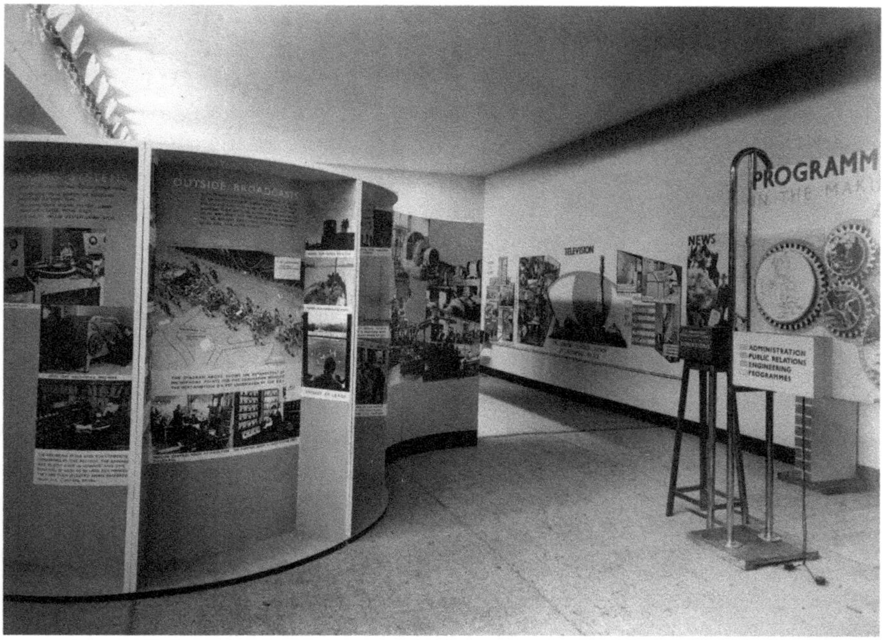

3.5 *The BBC Exhibition* at Charing Cross Station, February 1938. Topical Press. U26478 © TfL from the London Transport Museum collection. All rights reserved and permission to use the figure must be obtained from the copyright holder.

include designers of exhibitions, as one way of finding an appropriate path through tempestuous politics.[51]

The exhibition as 'story'

The *RIBA Journal* denounced the MARS exhibition as impenetrable, using 'words and phrases which are largely, if not wholly, unintelligible to the ordinary man', although without proffering evidence of the journal's consultation with 'the ordinary man' in question.[52] So how did the MARS exhibition come to be known as a 'story', perfectly suited to being a propaganda vehicle, as Misha Black was later to claim? The MARS exhibition functioned as textual and visual argument: using a dialectical approach, which invited visitors to enact an engagement with architecture by offering a succession of problems and solutions through a series of rigid, hierarchical binaries that claimed the architects as experts, with a vision that could liberate its viewers. On entering the gallery space, visitors were immediately confronted with a scene of contemporary chaos on Oxford Street, 'typifying our chaotic un-planned urbanism at its worst', according to exhibition visitor Le Corbusier. The introductory proposition was a wall-mounted cut-out photograph superimposed over a huge photograph of a tract of park-like English landscape and the striking headline 'It has come to this'. Smaller slogans running alongside read 'The mischief is done. The monstrous town enmeshes our life and wealth. We regret. We condemn. But what can we do? …'[53] This visual chaos, part of a display called 'The 20th Century Scene', set visitors up for the exhibition beyond, which would provide a restorative vision of order and orderliness.

The first large room at New Burlington Galleries posed a series of questions about what people needed from architecture – as a community, family and individuals – which were then countered in the next section on building materials, industrial method and technique. The exhibition operated as narrative, drawing visitors on a prescribed circulation route through its specially constructed spaces marked with arrows on plywood title-ribbons, 'along the path of an expanding sequence of ideas'. It used written factual statements as discursive elements interjected into the visual material and accompanied by a written brochure, which amplified its messages further (Figure 3.6).[54]

A corridor about 'Building Technique' showed ten display windows. Le Corbusier described it as 'a detailed dissertation' on building technique, which acted almost like frames in a film, as bodies moved past them. At the end of the exhibition a series of images on 'The six ages of English building' declared: 'Many times in history architecture has been

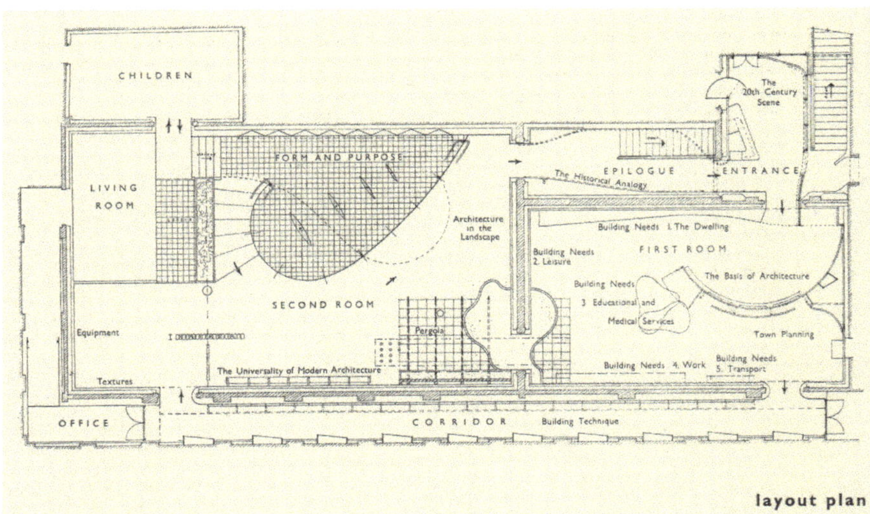

3.6 The MARS Group exhibition floor plan, New Burlington Galleries, London, 1938. Ashley Havinden. Image courtesy of Ashley Havinden Estate and Webb & Webb Design.

re-fashioned. Style gives birth to style. This is the only living continuity. To-day we recognise the beginning of a new phase in the English tradition'. Through a succession of round windows, building was shown as linked and evolutionary: from Norman Bamburgh Castle, Northumbria; to Gothic Lavenham Church, Suffolk; to Elizabethan Montacute House, Somerset; to Georgian Royal Crescent, Bath; to Victorian St George's Hall, Liverpool. All these earlier examples led to the contemporary Kensal House, London, suggesting this to be the apogee of all previous buildings and, further, the sweeping teleological path of time, from the Norman to the contemporary, echoed the line of motion of viewers stepping speedily along this galloping path of history.

Le Corbusier at the MARS Group exhibition

Architect Le Corbusier, who had long been engaged with exploring the possibilities and limitations of exhibitionary forms, wrote an extensive account of his visit to the MARS exhibition, published as a 'Pictorial Record' in *Architectural Review*.[55] The year before the MARS Group exhibition, Le Corbusier had created the *Pavillon des Temps Nouveaux* for the Exposition Internationale des Arts et des Techniques dans la Vie Moderne, held in Paris in 1937. This was an exhibitionary statement of the aims of the organisation Congrès Internationaux d'Architecture Moderne (CIAM), the International Congress of Modern Architecture, of which MARS considered itself the British section.[56] The Corbusier pavilion's interior displays covered 1,200 square metres focused on three areas – urbanism, sculpture and painting – and incorporated ramps and speakers' platforms. Alongside paintings, drawings, sculptural cut-outs, photomontages and dioramas – the stock-in-trade of exhibitions of the period – the pavilion incorporated much text as structural element.

Reviewing the Corbusier pavilion for *Architectural Review*, architect Serge Chermayeff noted that its architectural significance was not confined to formal and structural interest, but its most important function was to display 'the statements and wares of the exhibitors'.[57] An article in *Architectural Review* reproduced an image of a text-heavy stand, with many slogans including 'The world is not ending, either Europe, nor the USA, nor Asia, nor South America are at dusk – the world is reviving!' and 'for the sake of public safety, equip the country'. The pavilion was accompanied by a 'beau livre' (illustrated book) of a virtual museum: the book *Des Canons, des Munitions? Merci! Des Logis … SVP* (*Guns, Ammunition? Thank you! Homes … Please*), which told the story of the pavilion and justified the project.

Le Corbusier's article about the 1938 MARS exhibition reinforced the textual and narrative aspects of the exhibition, presented as a 'pictorial argument' with the visitor who, 'led by the hand', 'almost imperceptibly finds himself convinced by one [presentation] after another', and its hybrid elements as visual, verbal and spatial interplay.[58] Le Corbusier referred to the presence of Sir Henry Wotton's first edition volume, opened at the page which offered the quote that provided the structuring idea of the exhibition, showing his interest in the representation of ideas across text, image and architectural forms.

Forms and possibilities

Architecture was presented at the MARS Group exhibition as a plethora of forms and possibilities, rather than styles: insides and outsides of buildings, 'small equipment' and the 'large project', in photographs, models and full scale in the material itself.[59] The exhibition's visual impact came from a layering of text with photographs, statistics, montages, models, plans, built elements set on the wall or in framed dioramas. A section on 'building needs for leisure', for example, had a display focused around a central text: 'FOR LEISURE' with the cryptic words 'An Architecture – free from convention – supple gay'. This was overlaid on a background collage incorporating photographic cut-outs of people at leisure, superimposed on a Surrealistic painted scene with vignettes of sail boats, a tree, an actor, a sports person, with a series of photographs of buildings and landscapes mounted on a frame and a small, paved area with grassy verge in front.

A garden pergola formed of four right-angle fins supporting a plywood canopy, designed by Christopher Nicholson, incorporated a leafless birch tree, paving, planters with jaunty flowers, climbing plant and plywood furniture. The possibilities offered by new materials such as plywood were billed as a main attraction of the show.[60] The exhibition's interpretation of 'delight' – one of its main themes – had hallmarks of the picturesque that *Architectural Review* had been developing, in its use of light and dark and its modest attempts to vary perspective and to provide visual variety.[61]

The exhibition's installation was playful: a revolving wheel illustrating transport spun to uncover glimpsed photographs, each accompanied by written aphorisms. The end of the show was appropriately, for this text-heavy exhibition, named the 'epilogue'. The *RIBA Journal* reported that in the epilogue a 'gramophonic voice' echoed the exhibition's key principles (described above) by intoning '"Commoditie, firmenes, and delight", followed by a very short (but not too clear) sermon'.[62] This summed up the eccentric and esoteric impact of the exhibition. Some of these elements had featured in earlier British exhibitions: commercial exhibition stands had used lettering in their structure, expanded photographs, peepholes and cut-outs and room models. Modernist exhibitionary influence was conspicuously clear. *RIBA Journal* described it as using 'the exhibition technique of modern times', much of it echoing elements of exhibitions earlier in the decade by Bauhauslers: murals; full-scale room models;

plywood screens with focusing peep-holes and port-holes to reveal elements of the display; and the use of lettering as a structural element.[63] All of this amounted to giving the audience a sense of solving a puzzle through optical toys, cryptic and aphoristic captions, allowing them to experience a satisfying sense of engagement and revelation.

Exhibition as 'interior landscape'

The exhibition design's similarity to shop display distracted some critics who wanted more instruction and less suggestion or, perhaps, more education and less entertainment. *Architectural Association Journal* complained it was 'impossible to find a straight photograph of a building clearly displayed with some explanation of its function and planning; the visitor had to peer through distracting holes in an atmosphere of flippancy more appropriate to a display of DAKS than of modern architecture', adding that 'the content of the exhibition was submerged in the form, and the form seemed to explain nothing'.[64] This observation that content and form were vying in this kind of exhibition was to be prophetic: an exhibition's messages were in danger of being lost in its multiform presentation if not spelt out.

The exhibition mobilised the idea of exhibition as 'interior landscape', echoing the 1931 *Ausstellungsstand der Baugewerkschaften* (*Exhibition of the Building Workers Unions*) in Berlin by Bauhaus faculty, with shifts of perception effected through expanded photographs and photomontages. The 1938 exhibition's interactive route through New Burlington Galleries included changes of floor finish, from wooden blocks to grass and paving, and changes of height and perspective, which allowed visitors to see models both at eye level and from below and above, as if from the air, employing 'New Vision' optics.

The MARS Group exhibition owed a debt to Gropius's ideas, by pioneering exhibition as a reproducible, technical medium for the machine age. It naturalised 'the new architecture' by framing the elements on show as the most logical, scientifically evidenced and evolved responses to contemporary issues, part of the rational universalist gospel characteristic of the manifesto form. The catalogue explained that the exhibition 'is, if you like, propaganda. But we do not seek converts. All we hope to do is to win the loyalty of those who have not already made up their minds'.[65] *RIBA Journal* extolled the exhibition's virtues as 'undoubtedly the most brilliantly presented statement of an architectural idea that has yet been offered to the public or the profession here in London, or probably anywhere else', recognising 'the qualities of the show as a show' and as a demonstration of good exhibition technique. The *Journal* ran an article specifically scrutinising the story-telling qualities of the exhibition, which 'aims at putting architecture across to the public' through telling a story to the public with a 'propagandist aim'.[66]

The exhibition was expensive to mount and John Summerson remembered the level of visitors was so large as to be 'startling'; it attracted seven thousand visitors in eighteen days, and elements were re-used at the Building Centre exhibition later in 1938.[67] The exhibition was – as I have already shown – well covered in the architectural press and continued to resonate afterwards. In his major analysis of British wartime propaganda exhibitions in *Architectural Review*, AA-trained German refugee architect Gerhard Kallmann declared MARS 1938 the last word in 'up-to-date' exhibitions.[68] A few years later, eminent industrial designer Milner Gray repeated this sentiment saying, 'The circulation was informal, the display structure was colourful and fluid, using photomontage, captions and actual objects to put over the story. There was a coherent tale to be told, and for the first time a script was used to tell it'.[69] Identifying that exhibitions could tell stories and communicate messages had precipitated a huge growth in this form. Misha Black included the MARS exhibition as a key moment in the evolution: 'the use of exhibitions as a method of propaganda for ideas has suddenly blossomed from a frail plant ... into a vigorous growth which now spreads its tendrils from Oxford Street and Piccadilly to provincial towns, remote villages and isolated army camps'.[70]

Exhibition of the Elements of Modern Architecture at New Burlington Galleries 1938 fulfilled the aspiration of it makers to create an exhibition as argument: it acted as a manifesto for the MARS Group, it was a manifestation of Bauhaus ideas about exhibitions as reproducible forms for the machine age and it allowed for extensive experimentation with the 'New Vision' ideas of Moholy-Nagy, including directly by him while still based in Britain, through its incorporation of photomontages and collages alongside built elements and infographics, all shaped by a polemical narrative. Several of the designers who were associated with the exhibition – including Misha Black and Peter Moro – continued to be central to making propaganda exhibitions during war and its aftermath. Misha Black and a number of other MARS Group members went on to be employed by the Ministry of Information, allowing him regular and extensive experiments with this format at the heart of government, as I will discuss in Chapter 7.

Notes

1 Julian Trevelyan, *Indigo Days: The Art and Memoirs of Julian Trevelyan* (London: MacGibbon & Kee, 1957), p. 54.
2 Janet Lyon, *Manifestoes: Provocations of the Modern* (Ithaca, NY and London: Cornell University Press, 1999), p. 2.
3 Marshall Berman, *All That Is Solid Melts into Air* (London: Verso, 1983), p. 89.
4 Lyon, *Manifestoes*, p. 37.
5 Martin Puchner, *Poetry of the Revolution: Marx, Manifestos, and the Avant-Garde* (Princeton, NJ and Oxford: Princeton University Press, 2006), p. 109.

6 Alex Danchev (ed.), *100 Artists' Manifestos* (London: Penguin, 2011).
7 Catalogue of 50-year anniversary show at the Mayor Gallery of Unit 1, *Unit One: Spirit of the 30s*, May–June 1984.
8 *The Listener*, 5 July 1933, p. 16.
9 Paul Nash, 'Going Modern and Being British', *Weekend Review*, 12 March 1932, pp. 322–3; Paul Nash, letter to *The Times*, 12 June 1933.
10 *Unit 1: The Modern Movement in English Architecture, Painting and Sculpture* (London: Cassell, 1934).
11 Charles Harrison, *English Art and Modernism* (New Haven, CT: Yale University Press, 1994), p. 249.
12 Juliet Gardiner, *The Thirties: An Intimate History* (London: HarperPress, 2010), p. 331.
13 Myfanwy Evans (ed.), *The Painter's Object* (London: Gerald Howe, 1937).
14 There is no institutional archive for the New Burlington Galleries, which were demolished for development in the 1970s (thank you to Sue Breakell for confirming this). London Metropolitan Archives, LCC Architects' Department collection ref GLC/ AR/ BR/ 13/ 152199 has photographs of the New Burlington Galleries, 3–5 Burlington Gardens.
15 TGA London Group GB 70 7713/6/1/26 and *Journal of the Royal Society of Arts*, vol. 82, no. 4257, 22 June 1934, p. 838.
16 TGA 806/1/630, letter from E. L. T. Mesens to James Bolivar Manson, 15 May 1937.
17 Kenneth Clark, *Another Part of the Wood: A Self Portrait* (London: Hamish Hamilton, 1985), quoting Tom Driberg MP, p. 253.
18 Michael Remy, *Towards a Dictionary of Surrealism in England* (Nancy: Groupe-Édition Marges, 1978), p. 15. Held in 1933.
19 Remy, *Towards a Dictionary*, p. 18 includes a list of all the talks and associated events. Herbert Read's account of the Conway Hall debate, *International Surrealist Bulletin*, no. 4, 1936, is mentioned in Robert Radford, *Art for a Purpose: The Artists' International Association 1933–1953* (Winchester: Winchester School of Art Press, 1987), pp. 87–9. For debate over surrealism's political potential see supplement in *Left Review*, vol. 2, July 1936, and Anthony Blunt, 'The Realism Quarrel', *Left Review*, vol. 3, April 1937, p. 170.
20 Herbert Read (ed.), *Surrealism* (London: Faber & Faber, 1936) included contributions from André Breton, Hugh Sykes Davies, Paul Éluard and Georges Hugnet.
21 Read's first article on surrealism was 'Beyond Realism', *The Listener*, 16 April 1930, p. 29. Read, *Surrealism*, p. 28.
22 *Modern Publicity 1936–7*, p. 108, catalogue for *Surrealism* exhibition cover and poster designed by Max Ernst.
23 Including Paul Nash, John Armstrong, Tristram Hillier, Roland Penrose, John Banting, Eileen Agar and documentary filmmakers Humphrey Jennings and Len Lye.
24 Trevelyan, *Indigo Days*, pp. 68–9.
25 'Surrealism Supplement', *Left Review*, vol. 2, July 1936. Boswell's cartoon is reproduced in Radford, *Art for a Purpose*, p. 88.
26 As noted by *Shelf Appeal* in its impact on Shell and Guinness advertisements, June 1936, pp. 35, 39. Critical response is discussed in Gardiner, *The Thirties*, p. 336. It was dismissed by the *Daily Worker*, the Communist Party newspaper, and *Action*, newspaper of the BUF: 'Exhibition – or Exhibitionism?', *Action*, 2 July 1936, p. 11.
27 Puchner, *Poetry of the Revolution*, p. 179.
28 Among others: Campbell, 'The MARS Group, 1933–1939'; Darling, 'Institutionalizing English Modernism 1924–33'; and Gold, *The Experience of Modernism*.

29 Black, *Exhibition Design*, pp. 2, 77. Black also described the MARS Group exhibition as one of the two greatest influences on building styles in Britain in 'Notes on 1951', Misha Black collection, V&A AAD/ 1980/3/4/file 2/ 3. The other was the Festival of Britain 1951.
30 RIBA OA Ar0/2/6/9 and ArO/2/6/5 (i-). According to Campbell, 'The MARS Group, 1933–1939', p. 76, the MARS exhibition committee's summary of the 'characteristics and principles' of modern architecture annoyed many MARS members.
31 Gold, *The Experience of Modernism*, p. 125.
32 Iain Jackson and Jessica Holland, *The Architecture of Edwin Maxwell Fry and Jane Drew* (Farnham: Ashgate, 2014), p. 86.
33 RIBA GS SaG/88/4 2/2, Box 88, minutes of MARS designers' meeting, 25 October 1937 and letter from M. B. Bradshaw.
34 RIBA GS SaG/88/4 2/2, Box 88, recommendations to Exhibition Committee, 16 November 1938.
35 Lyon, *Manifestoes*, pp. 30, 37.
36 RIBA GS Papers, recommendations from Executive Committee to Publicity Committee, 16 November 1937.
37 Foreword to exhibition catalogue, reproduced in *Architects Journal*, 20 January 1938, p. 121.
38 *Architects Journal*, 20 January 1938, p. 124.
39 Puchner, *Poetry of the Revolution*, p. 109.
40 RIBA GS SaG/88/4 2/2, letters to Godfrey Samuel from GBS, 7 April 1937 and 30 April 1937.
41 Foreword to MARS Group exhibition catalogue, p. 121.
42 RIBA GS SaG/88/4/2/2, as per letter of 15 February 1938.
43 Elizabeth Darling, *Re-forming Britain: Narratives of Modernity before Reconstruction* (London: Routledge, 2007) discusses the role of mediation in British Modernism.
44 RIBA GS SaG/88/4/2/2, letter 28 December 1937, from Godfrey Samuel to John Summerson.
45 John Gold, 'New Architecture and the Search for Modernity', in Robert Freestone and Marco Amati (eds), *Exhibitions and the Development of Modern Planning Culture* (Farnham: Ashgate, 2014), p. 89.
46 Peter Moro interview with Louise Brodie (part 7 of 15), British Library National Life Story Collection: Architects' Lives.
47 Beatriz Colomina, *Privacy and Publicity: Modern Architecture as Mass Media* (Cambridge, MA and London: MIT Press, 1996).
48 'New Architecture', BBC Television, 11 January 1938 https://genome.ch.bbc.co.uk/schedules/service_bbc_television_service/1938-01-11 (accessed 16 May 23).
49 Lambert, *Propaganda*, p. 75.
50 Lambert, *Propaganda*, p. 118.
51 Lambert, *Propaganda*, pp. 150–1.
52 *RIBA Journal*, 24 January 1938, p. 290.
53 Le Corbusier, 'The MARS Group Exhibition of the Elements of Modern Architecture: A Pictorial Record', *Architectural Review*, no. 496, 1938, p. 109.
54 *Journal of the RIBA*, 24 January 1938, p. 292.
55 Le Corbusier, 'The MARS Group Exhibition', pp. 109–16.
56 Danilo Udovicki-Selb, 'Le Corbusier and the Paris Exhibition of 1937: The Temps Nouveaux Pavilion', *Journal of the Society of Architectural Historians*, vol. 56, no. 1, 1997, pp. 42–63; Catherine de Smet, *Le Corbusier Penseur du Musée* (Paris: Flammarion, 2019), p. 59.
57 Serge Chermayeff, 'Circulation: Design: Display: The Architect at the Exhibition', *Architectural Review*, September 1937, p. 93.
58 de Smet, *Le Corbusier*, pp. 109–10.
59 *Architects Journal*, 20 January 1938, p. 122.

60 *RIBA Journal*, 6 December 1937, p. 130.
61 Harriet Atkinson, '"The First Modern Townscape"? The Festival of Britain, Townscape and New Picturesque', in John Pendlebury et al. (eds), *Alternative Visions of Post-War Reconstruction: Creating the Modern Townscape* (Abingdon: Routledge, 2014).
62 *RIBA Journal*, 24 January 1938, p. 292.
63 *RIBA Journal*, 10 January 1938, p. 242.
64 Unit 15, 'The MARS exhibition', *Architectural Association Journal*, 53, February 1938, pp. 386–8.
65 Foreword to catalogue, reproduced in *Architects Journal*, 20 January 1938, p. 121.
66 'The MARS Group Exhibition: Some Considerations of Exhibition Technique', *Journal of the Royal Institute of British Architecture*, vol. 150, 24 January 1938, pp. 270, 290–2.
67 RIBA OA Ar 0/2/13, account book of MARS exhibition, the exhibition cost £2,000 (approximately £100,000 in today's value), over and above the materials gifted by manufacturers; Summerson, 'The MARS Group and the Thirties', p. 309.
68 Kallmann, 'The Wartime Exhibition', p. 95.
69 Milner Gray, 'A Century of Commercial Design 1851–1951', in *Designers in Britain*, vol. 3 (London: Allan Wingate, 1951), p. 17.
70 Peter Larkham, 'Exhibiting Planning in Wartime Britain', in Freestone and Amati, *Exhibitions and the Development of Modern Planning Culture*, p. 132 quotes Black from 1947, p. 97.

4
Exhibitions as demonstrations

On Mayday 1938, at an anti-fascist gathering at London's Hyde Park, amongst the assorted banners was one that read:

> *A WARLIKE STATE CANNOT CREATE*
> – William Blake
> SURREALIST GROUP (Figure 4.1)

This pithy quotation, an adaptation of words from one of England's most celebrated Romantic poets, critiqued contemporary government policy and referenced British Prime Minister Neville Chamberlain's policy of appeasement, which, they believed, was allowing Hitler to expand German territory unchecked.[1] The London Surrealist Group's Blake banner typified the playful, politically engaged and below-the-radar nature of contemporary artistic interventions. This chapter explores such interventions, showing how the definition of 'exhibitions' was stretched during the 1930s, so that they morphed to operate as strategic forms of public 'demonstration', intersecting with wider protest cultures in Britain, at a time when permissible public activities were severely curtailed.

With the rise of political extremism, laws and byelaws increasingly controlled activities allowed in public spaces and were intended to curtail protests and other behaviours viewed as disruptive. In this context, exhibitions became expedient vehicles for politically engaged artist- and designer-activists. They allowed political ideas to be shared publicly, in a form less conspicuously challenging to authority, and so were more permissible spaces for raising political issues to the public, acting as another form of public communication that went under the wire. In order to consider the varied activities discussed in this chapter as forms of 'exhibition', we need temporarily to remove our main focus from their specific, material forms and, instead, to consider them as varied 'acts of exposure', to draw

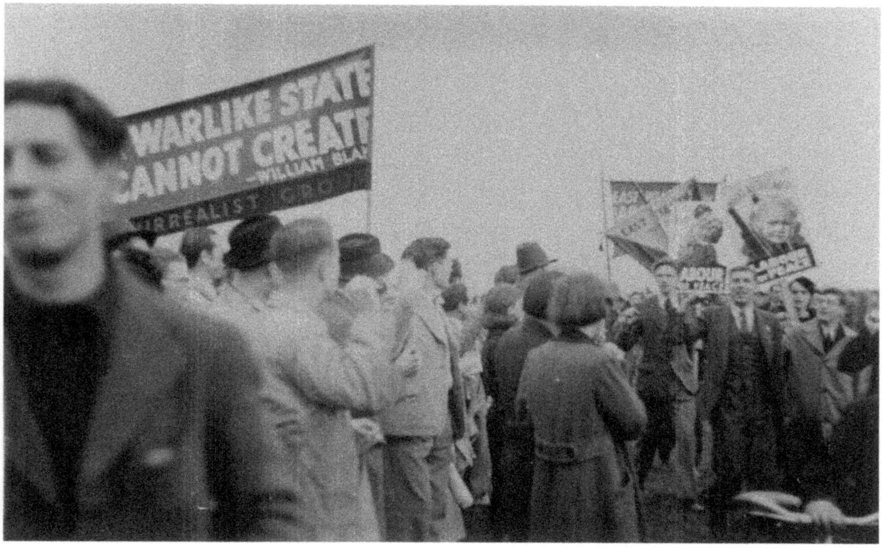

4.1 Surrealist group banner at Hyde Park rally, May Day, 1938. Unknown photographer. © reserved. Image courtesy of TATE. TGA 7043/14. All rights reserved and permission to use the figure must be obtained from the copyright holder.

from cultural theorist Mieke Bal's proposition for understanding exhibition's rhetorical power as public exposition and public presentation. If we can focus temporarily on the discourse within which exhibitions evolved in this period and the ideas and rhetoric deployed in and through them, rather than on their particular forms, we are able to see the playful and provocative ways in which they were being rethought to become available for political means.[2]

Although exhibitions were being used to striking effect in totalitarian states across Europe, both on the Left and Right during the 1930s, in Britain publicly mounted exhibitions were the preserve of artists and designers on the Left. My primary focus in this chapter is on the exhibitionary 'demonstrations' held by two groups in particular. The first group most actively pioneering exhibitions for sharing political ideas in early twentieth-century Britain was the Artists International Association (AIA), their innovation as exhibition makers being in conceptualising urban space as a platform for exhibitions performed 'live', as well as temporary assemblages of billboards and banners. The second group was the pacifist Cambridge Anti-War Council who created impactful exhibitions for demonstrating peace.

The groups had overlapping protagonists, with Misha Black a prime mover across both contexts. Drawing from ideas in philosopher Henri

Lefebvre's *The Production of Space*, I will suggest that exhibitions operated for the AIA in particular during this period as a means of creating multi-faceted and interpenetrating 'social space': space created by the relationships they supported, space created from ideas and solidarities; where ambiguous continuities existed regardless of whether there appeared to be visible boundaries and forms of spatial separation.[3]

The formation of the AIA and exhibitions as 'demonstrations'

The founding meeting of Artists International (as they were initially named, echoing the Communist International) was held in 1933 at the London Covent Garden flat of 23-year-old designer Misha Black, a group of young artists – including James Fitton, Clifford Rowe, James Boswell, James Lucas and Pearl Binder – sitting on fruit boxes, by candlelight. The group, which a few months later became the Artists International Association (or AIA), styled themselves as anti-fascist and anti-imperialist activists: artists and designers who wanted to use their practice as a political weapon. They declared themselves 'The International Unity of Artists Against Imperialist War on the Soviet Union, Fascism and Colonial Oppression' who would achieve their ends through creating 'working units' of artists to make 'propaganda' in the form of 'posters, illustrations, cartoons, book jackets, banners, tableaux, stage decorations' spread through the press, lectures and meetings and, crucially, through mounting exhibitions.[4]

The formation of the Leftist AIA came in 1933, the year Hitler took power in Germany, in the wake of the Depression and with half the British population living below the poverty line. With the rise of extremist groups on the Left and Right in Britain, including the formation of the British Union of Fascists in 1932, marches held to protest poor living and working conditions led to regular clashes, causing public authorities to pass a flurry of laws and byelaws curtailing public freedoms to gather or demonstrate in public space. The Incitement to Disaffection Act of 1934, popularly known as the 'Sedition Act', included a prohibition against public gathering, while the Public Order Act, passed after the 1936 clashes popularly known as the Battle of Cable Street, regulated use of public political symbols such as flags, banners and emblems, banned political uniforms and gave police powers to regulate public processions or to ban them altogether.

The AIA's work made a direct response to such prohibitions. They affiliated themselves to organisations like the National Council for Civil Liberties (NCCL), formed to oppose such legislation and to defend free speech and assembly, setting up a system for reporting 'irregularities' including banning or interference with meetings, processions and propaganda, use of leaflets, chalking or loud speakers, and police action at open-air meetings.[5] The NCCL made clear to its network the legal limits of police powers in curtailing moving processions, while public meetings could be broken up more easily.

Taking up this language of politics, the AIA had a strong sense of their exhibitions as politically engaged protests, regularly describing them as 'demonstrations' and evoking exhibitions as performative, active, provocative and participative vehicles for propaganda and persuasion. For the AIA, exhibitions were more than a vehicle through which to display their work as artists and designers: they were an urgent form of political engagement; a significant affective form; a means of expression for making manifest and visible issues that were abstract or invisible; a way of communicating common beliefs; and vehicles for provoking public awareness for issues of concern.

The AIA's spatial negotiation through exhibitions can be described, in Lefebvrian terms, as the interplay of 'dominated space', controlled and regulated by the British authorities, and 'appropriated space', in which the AIA sought to create ever more playful means through which to connect with others.[6] With the articulation in British legislation of definitions of prohibited public behaviours, the AIA's sense of the possibilities of exhibitionary space became ever more expansive and fluid: at different times exhibitions became representations, performances, platforms, meeting points and events. In making this argument, I am complicating Tony Bennett's use of Michel Foucault's panoptic impulse to describe the control that museum-based exhibitions have over audiences, by suggesting that in the context of 1930s Britain, exhibitionary spaces outside of institutions of the kind I am discussing were less surveilled and circumscribed than other spaces.[7] AIA member artists had seemingly contradictory aspirations: at once to bring activities and representations of the street into their work, while pursuing the liberation of exhibitions from established institutions and spaces, by taking them out to the streets.

Early work by AIA members satirised government anxieties about public gatherings and behaviours considered seditious. James Holland's cartoon for the *Left Review* titled *Incitement to Disaffection* (c.1934) saw lumbering police racing to their next site of public concern, while Edith Tudor-Hart's photograph *Sedition?* (1935) showed a crowd as an unspecified object of suspicion and Pearl Binder's lithograph *Chalking Squad* (c.1932) had a sedate woman in the process of enacting a supposedly seditious activity: writing a political message in chalk on a wall. Lithography as a form created a kind of spatial porosity for AIA artists. Several founder members had met at a lithography evening class at the Central School. Many AIA lithographs took their subjects from the street, and lithographs, being affordable and reproducible, allowed the AIA to remove pictures from gallery spaces and to send them on tour. The AIA's Everyman Prints series, launched in 1939, allowed work usually seen only in galleries to become dispersed across all kinds of sites around the country.

AIA members had a recurring interest in bringing street art into exhibitions. The assimilation of such work was highly uneasy, laying bare conspicuous class inequalities. North London pavement artist David Burton, given an exhibition at the AIA's Charlotte Street gallery space in 1945, was patronisingly dubbed the 'Hampstead Primitive' by *Picture Post.*[8] AIA members acknowledged discomfort at seeing fellow members taking inspiration from the subjects of the street purely as visual experience, a merely anthropological interest, rather than born of a more genuine working experience of their subjects.

If the AIA sought to bring the imagery and activities of the street into their exhibitions, central to their work was the seemingly contrary aspiration towards taking exhibitions out, to the street, creating social space as 'encounter, assembly, simultaneity', to quote Lefebvre.[9] The AIA's innovation as exhibition makers was in their perception of public urban spaces as sites for exhibitionary 'demonstrations' that allowed people to come together in public when this fundamental right was being threatened. Central to AIA exhibitions was a professed belief in democratisation; their literature declared, 'it has always been our aim … to make our exhibitions accessible to the widest possible public'.[10] In order to do this art must be taken out of the studio and the museum, to be shown on sites that would attract a public not drawn to galleries, such as 'Underground Stations, factory canteens and working men's settlements', the AIA's *Bulletin* explained. Art should 'come to the people and not be simply a form of luxury goods'; it should 'perform a social function'.[11]

Early AIA exhibitions: *The Social Scene* and *Artists Against Fascism and War*

The AIA's aspiration to take exhibitions ever further out, ever more removed from galleries as contained and closed, was enacted most convincingly as symbolic and rhetorical. Despite their aspiration to take art to the people, the AIA's exhibitions in the first five years were largely restricted to central London, on sites in Charlotte Street, Soho Square and Mayfair's Grosvenor Square. What the group lacked in the diversity of its sites and spaces, however, it achieved in stylistic variety, the geographical diversity of its vision and the internationalism of its contributing artists. Its vision of the world, and its real and imagined networks, encompassed a vast geography. A lack of adherence to particular stylistic orthodoxies was a key tenet of AIA exhibitions, including accepting works from trained and untrained painters alike (though not without some distinction being drawn when the latter were regularly singled out as 'amateur' or 'unprofessional').

The AIA exhibited much painting and sculpture containing overt political and social critique. Their inaugural exhibition, *The Social Scene*, was held at a former motorcycle showroom on Charlotte Street, behind

London's busy Tottenham Court Road, in October 1934. It caused criticism and disagreement from critics, which the AIA countered by saying that these people 'disliked the presumption of a number of artists co-operating to criticise the society in which they lived'.[12]

The AIA's second exhibition, *Artists Against Fascism and War*, was held in November 1935, a few weeks after the Italian fascist invasion of Abyssinia of October 1935, at 36 Soho Square in a 'splendid Georgian house' that the AIA rented for the purpose, as lead member Betty Rea described it.[13] No images of the exhibition's installation are held in the archive, but it was likely a conventional hang, with paintings shown at eye level round the walls. An influential committee of artists – including Eric Gill, Augustus John, Laura Knight, Henry Moore and Paul Nash – selected works by six hundred artists. These included abstract painting and sculpture by Henry Moore, Barbara Hepworth, Ben Nicholson, John Piper, John Tunnard and László Moholy-Nagy and figurative work by Ethel Walker, Harold Knight (*Cottage Bedroom*) and Charles Cundall (of the *Miners' Gala, Durham*).[14] One section of *Artists Against Fascism and War* was devoted to photographs of working-class life, including work by Edith Tudor-Hart (whose photographs feature in several contexts through this book) and other members of the Leftist English Workers' Film and Photo League of which she was part.

The AIA received submissions for *Artists Against Fascism and War* from France, Holland, Poland and Russia, from organisations sympathetic to AIA's aims, indicating the reach of members' reputations and affiliations. Many works took up Leftist subjects, such as Clifford Rowe's painting *Canvassing the Daily Worker* and Peter László Peri's sculpture *Against War and Fascism*. The use of social realism was considered an effective way of raising problems up the political agenda, and socially engaged subject matter was seen particularly in prints and drawings, with titles like *Prostitution*, *South Wales Tubercular Miner* and a satirical work by Peggy Angus called *Poison Gas*, attacking the Jubilee celebrations of that year.

As spaces that allowed gatherings, AIA exhibitions provided an occasion for exploration of wider social and political themes and ideas, with articles published alongside talks from prestigious speakers. Although AIA members espoused clear ideas about the artistic styles most suited to political engagement, early AIA exhibitions did not dictate formal or stylistic orthodoxy, and *Artists Against Fascism* was no exception. It was accompanied by lectures on 'Marxism and Aesthetics' by art critic Alick West, and on 'The Crisis in Culture' by politician and writer John Strachey. It was opened by writer and philosopher Aldous Huxley whose 1932 dystopian novel *Brave New World* had warned against ominous trends in politics and technology.[15] In his foreword to the exhibition catalogue, Huxley proposed the case for the artist 'as a special case of the good citizen … while painting, he is controlled, scrupulous, conscientious'.[16]

Visited by six thousand people, the exhibition was praised by critic Montagu Slater in *Left Review* for its impressive scale and ambition, for attracting people who did not normally visit art shows and for bringing war and anti-fascism together, recognising it as 'something between a demonstration and a national gallery', with 'no "market" to speak of', operating as he saw it outside the commercial considerations of West End art galleries. Despite praising its ambition, Slater regretted that *Artists Against Fascism* mainly failed to achieve its intention, that only in the abstract room had he found the 'tendency and direction'. Essentially he felt the show gave no answer to the question 'where do we go from here?' and, he thought, 'lacked a positive'.[17] *Artists Against Fascism* allowed AIA members to demonstrate relationships and solidarities between the many groups with common interest in these issues and to provide a platform from which to express them. But it failed to build a consensus around how to respond. Moreover, the class differences between the exhibitors were painfully manifest, specifically distinctions between those attempting to sustain themselves through artistic practice, perhaps with private income to supplement, and its working contributors, such as the miners of the Ashington Group, whose other 'working' identities were exposed while apparently championed.

'Exhibiting' and 'demonstrating' at Trafalgar Square

Aside from AIA exhibitions mounted in central London houses temporarily given over for the purpose, the AIA were creating exhibitions 'live', in the open air, as a way of reclaiming and appropriating public spaces dominated by official rules. This was an astute method for circumnavigating prohibitions imposed by the Office of Works, which had banned the exhibition of art works in the street, only permitting banners at demonstrations.[18] On one occasion, AIA painters Rodrigo Moynihan and Victor Pasmore, attending an 'Arms for Spain' rally in Trafalgar Square in February 1939, applied sketches to blank banners while the rally was in progress, with speeches being given by biologist J. B. S. Haldane and trade unionist Tom Mann. Some banners bore motifs drawn from Goya's condemnation of the universal evils of warfare *Disasters of War* (1810–20) – ironic given that the rally was in fact focused towards arming Spain.[19] These were then constructed behind the speakers to create an impromptu temporary exhibition. AIA member Nan Youngman's sketches of an earlier rally at Trafalgar Square convey this sense of banners surrounding the base of Nelson's Column, creating an impromptu 'structure' as backdrop (Figure 4.2).

Such 'live exhibitions', created in front of the Trafalgar Square rally, exemplified the plural performativity and carnivalesque quality of AIA activities.[20] This same attitude was apparent in May 1938 when two hundred artists, including many AIA members, took to the streets for a May

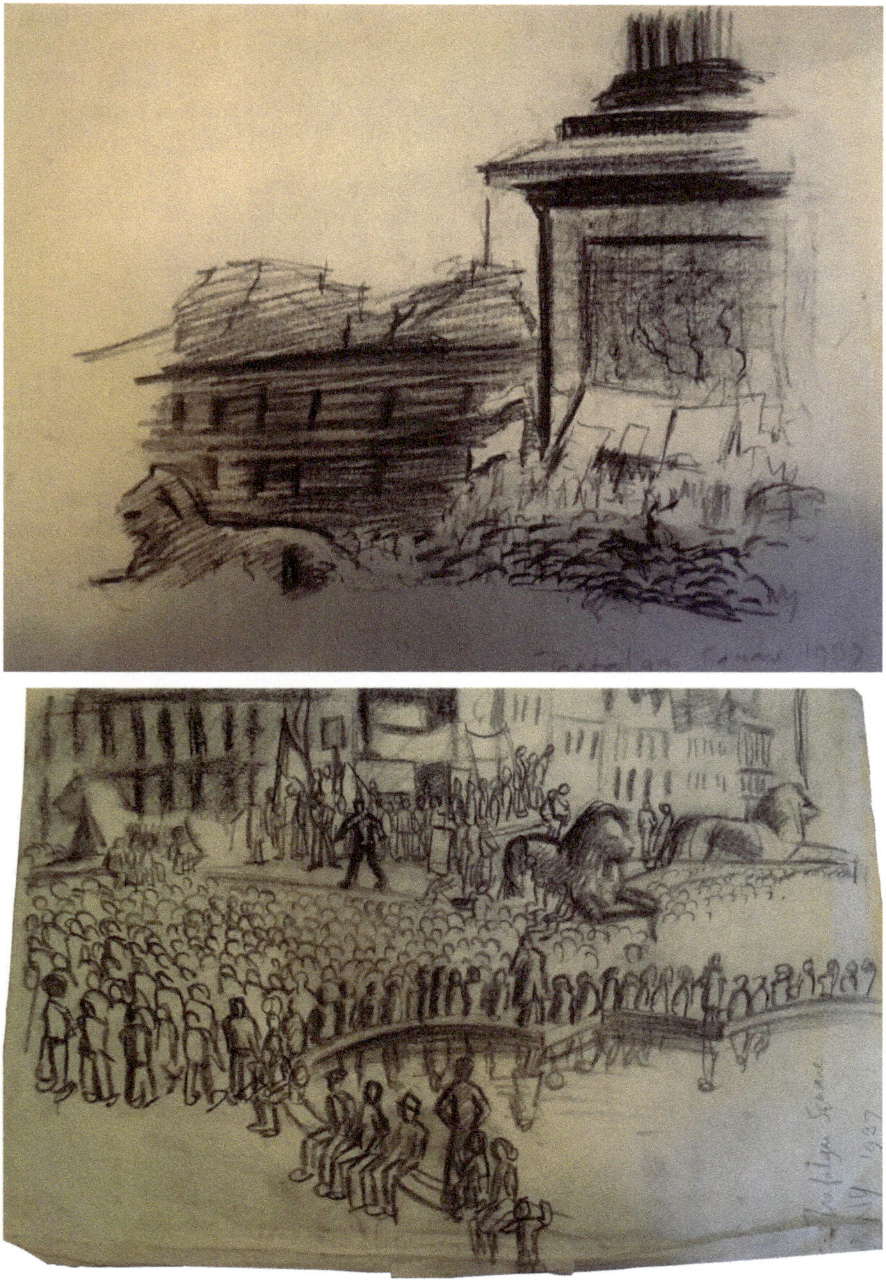

4.2 AIA member Nan Youngman's sketches of a political rally at Trafalgar Square in 1937. Nan Youngman.

Day procession protesting the government's policy of appeasement in response to the rise of Nazism. Surrealist artists Roland Penrose, James Cant, Julian Trevelyan and F. E. McWilliam marched wearing masks of Neville Chamberlain and carrying placards with the slogan 'Chamberlain Must Go!', with occasional ironic Nazi salutes. The Surrealist group's van appropriated public space by carrying a loudspeaker issuing the tunes of 'The Internationale' and the 'United Front', both recognisable anthems of the Left, with a great gilded birdcage perched atop with a whitened skeleton inside, captioned 'Present from the Dictators', a commentary on the Prime Minister's dangerously uncritical relationship with European fascist leaders, as they perceived it (Figure 4.3).[21]

Advertising hoardings as 'exhibition' sites

The AIA's extension of the boundaries of exhibitions as public spaces and 'live' experiences, acts in defiance of the police and other authorities, was to the fore when in February 1939 they took over twenty public advertising hoardings around London as sites through which to raise Spanish relief, painting illustrated slogans calling for support for the Popular Front

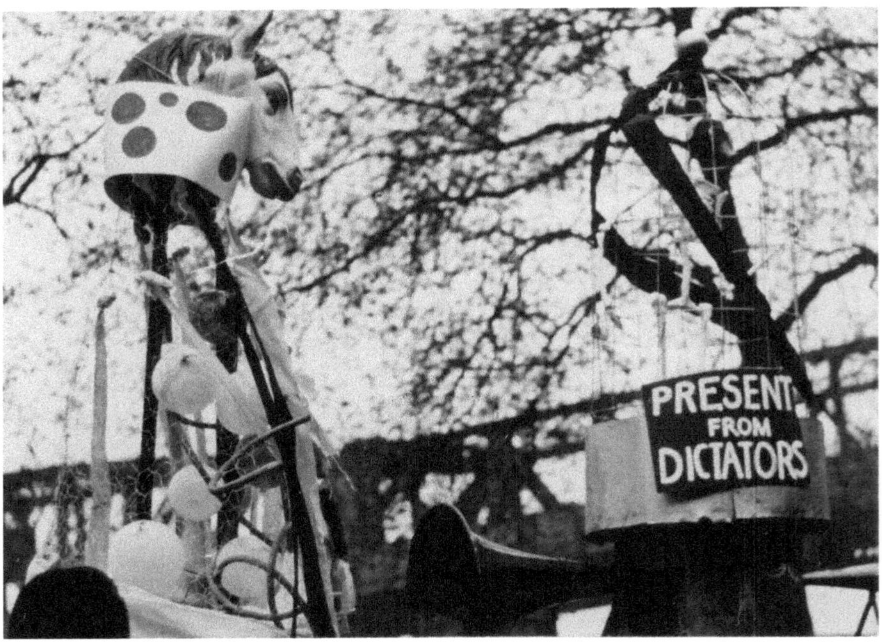

4.3 Artists International Association photograph of a rally in London, 1938. Unknown photographer. © reserved. Image courtesy of TATE. TGA 7043/14.

in Spain live in front of a crowd. This was a bold act of appropriation of commercial space, which we might describe, in Lefebvrian terms, as 're-appropriation, diversion, detournement'.[22] The *Manchester Guardian* showed a photograph of AIA member artist Julian Trevelyan up a ladder by a billboard at the intersection of Bouverie Street and Fleet Street, in the process of painting Isotype-style warships and submarines and the slogan 'Send Food to Spain Now!', with a large crowd of onlookers, which, the newspaper reported, had 'quite flustered' the police.

The same day artists painted eight other hoardings for Spanish relief. One of the best, according to the *Manchester Guardian*, was that by James Holland just beside Tower Bridge, which 'neatly included the bridge itself – a symbol of London's solid security – in a design showing war-stricken Spain'. Intrigued by the novelty of this new kind of public display, the newspaper remarked, 'The AIA, whose exhibitions have already contributed to the cause of Spanish relief, to-day tried a new sort of exhibition – if exhibition it can be called – for the same cause' and likening this kind of work to 'novel bill-stickers' or 'pavement art'. The newspaper continued approvingly, 'If this practice spread it would certainly be one way of improving London's hoardings'.[23] Such painting was in the spirit of the Soviet windows displayed by the TASS News Agency in Moscow, being painted by artists on the spot and appearing at the same time as the news items they illustrated, as topical propaganda.[24] In these activities exhibitionary cultures converged with street theatre and agitprop.

The AIA showed through their hoarding painting their prowess at courting media attention to their campaigns. AIA members demonstrated clear understanding of the potency and potential spectacle of gathering in public space to paint and create 'live' exhibitions. AIA member Priscilla Thornycroft notified newsreel cameramen that she and Nan Youngman would be painting a hoarding on 23 February 1939. They were then surprised when, instead of giving them time to start their painting, the crew arrived almost immediately to film them painting the slogan '*Spain Fights On – Send Food Now*'. Thornycroft recalled:

> [We] … painted frantically, 'Food for Spain', terribly badly, because we thought the message was more important than the art. Yes, it was one of those embarrassing things … I never saw the film but other people rang me up and said they'd taken a proper film and it really was in the Gaumont News or Pathé perhaps.[25]

A few days later, the AIA's billboards, with their 'Aid to Spain' slogans, were photographed in the *Evening Standard* daubed with fascist slogans after a messy attack from Mosleyite supporters. Under 'Send Food to Spain Now!' someone had painted 'Mosley Will Win' and a fascist symbol. *Action*, the magazine of the BUF, picking up on the story, reported the hand-painted work of 'Spain savers' as having been 'altered for the better', 'evidently by some sympathiser of British Union'.[26] This simple alteration showed such sites' vulnerability to having their meanings overturned and redirected, becoming public information battlegrounds.

The AIA's anti-fascist *Art for the People* at Whitechapel Art Gallery

A landmark exhibition-as-demonstration of the AIA's first decade was the popular display of members' work, the anti-fascist *Art for the People* at Whitechapel Art Gallery from February to March 1939. This claimed to be 'A cross-section of every form of contemporary art in Great Britain exhibited as a demonstration of the Unity of Artists for Peace, Democracy and Cultural Progress'. Works ranged from a cartoon by Augustus John to sculptures by Jacob Epstein, Frank Dobson and Peter László Peri.[27] Peri's large concrete piece was called *Save Spain.* Noting the contribution of Surrealists to the exhibition, journalist Derek Stanford declared 'Surrealism has come to the East End'.[28] A provocative lecture series ran alongside, the first titled 'THEY LIKE WHAT THEY KNOW: Criticisms of the Present Exhibition', given by *News Chronicle* art critic Frederick Laws, and another by *Sunday Times* and *Manchester Guardian* art critic Eric Newton.[29]

It is unclear how the exhibition was hung. The exhibition's most notable aspect was its attempt at outreach: of giving people who did not usually see contemporary art an experience of it, a missionary instinct central both to AIA and Whitechapel Art Gallery work of the time. Intending to respond to the snobbery of the art world, by moving its exhibitions away from Mayfair and the West End to the Whitechapel, AIA and Whitechapel interactions with the public often made class inequalities and condescension all the more conspicuous. On Sundays, admission to the AIA's exhibition at Whitechapel was to be free, the exhibition invitation noting in red at the top: 'The exhibition will be opened by THE MAN IN THE STREET' (original emphasis), 'intended as a symbol of the relation of anti-fascism to the art of the people'.[30] This became *Art for the People*'s most widely reported aspect. The *Star* enthused:

> When five women artists went outside the Whitechapel Art Gallery to choose A Man in the Street to open the exhibition of the Artists' International Society, they chose Mr James O'Brien because, of the passing crowd, he seemed interested.[31]

This gimmick paid off: East End butcher O'Brien was considered perfectly emblematic of the AIA's intended working-class audience and the press showed much interest in the images O'Brien preferred. The *Manchester Guardian* opined, 'He seemed a typical decent East End workman in the thirties ... He had no particular interest in art, but the Surrealist sculpture and paintings in the gallery did not deter him'. It went on, 'The Association is seeking to break down the barriers between artists and the people, and no better place for their efforts could be than the Whitechapel Art Gallery, which has always sought in its forty years to procure the best pictures for the poorer art-lovers'.[32] Meanwhile the *Star* tabloid carried the story of artist miner L. F. Smith's contribution to the show. Smith had not been able to afford to join the AIA but the other members had done 'a whip-round', it reported, presenting him with the cost of a subscription.[33]

The exhibition's anti-fascist message added to the clamour of voices calling for peace. In its appraisal, *The Times* praised the Whitechapel exhibition for demonstrating the importance of peace. 'No doubt it is tempting to the outraged man to further peace by representing the horrors of war' – partly, the paper suggested, because 'scenes of cruelty' gave some pleasure – but 'to saddle art with propaganda is to show lack of confidence in its intrinsic power'. It concluded, *Art for the People* 'may be hailed as a solid contribution to the causes of peace, democracy, and cultural progress'.[34] Forty thousand people visited the show.[35] The exhibition's one-month run coincided with the British government recognising Franco's Nationalist government in Spain.

Art institutions' sensitivity to becoming platforms for political messaging and potentially falling foul of public authorities is indicated by an exchange with painter and Mass Observation founder Julian Trevelyan. Following the success of the 1939 Whitechapel AIA exhibition, Trevelyan wrote to Whitechapel Art Gallery proposing a new exhibition of works he described as 'by working class artists, chiefly pictures from all over the world by unprofessional painters that Mr Tom Harrisson and myself collected together'. An internal Whitechapel memo showed nervousness, stating that 'So many of these bodies are political, that we really must exercise considerable care in ensuring that we do not allow our Gallery to be used for propaganda by any political or semi political artistic organisation'.[36] The exhibition was rejected. The rebuff did not deter the AIA from writing back a few weeks later to propose yet another exhibition: of art produced under the US Works Project Administration, never before seen in England and already enjoying a high reputation among artists for its social commitment and as the first example of major government funding for artists. The outbreak of the Second World War put paid to this exhibition idea.

Demonstrating for peace: Cambridge Anti-War Council exhibitions

As the threat of fascism became ever more present across Europe, newly formed British pacifist organisations, looking for ways to share anxieties about the possibility of war, identified exhibitions as a way to represent and explain their values. One such was the Cambridge Anti-War Council, which mounted a series of exhibitions from 1933, with a focus on 'effective *action*, collective and organised, to prevent or to stop the conduct of war', starting at home and beginning 'now', a way of attracting recruits to the cause.[37] Although attracting illustrious contributions to the cause, the remaining evidence of these exhibitions is fragmentary: a small advertisement in a limited circulation bulletin or flyer, usually text-heavy and frustratingly image-light, forcing me to piece scattered information together. Despite this, these pacifist exhibitions were significant for two particular reasons: firstly, because they attracted the involvement of many of the same cast of designers, artists and writers who were developing exhibitions in other contexts across this book; secondly, because they show how political organisations in Britain were using exhibitions as arguments, platforms and recruiting grounds.

The Cambridge Anti-War Council, formed at a conference at Cambridge's Co-operative Hall in summer 1933, was focused against British rearmament and to organise action 'for peace and international co-operation of all people and all workers'. Their range of activities included regular demonstrations, such as one against fascist activist Oswald Mosley during his visit to Cambridge in March 1935. Despite the evident importance of imagery to the Cambridge exhibitions, as is clear from those involved whose *modus operandi* was visual presentation, extremely limited visual material remains to show what this series looked like. Much of the discussion of anti-war exhibitions in this section therefore comes from textual material including catalogues, letters and news reports. The first *Anti-War Exhibition* organised by the Cambridge Anti-War Council was held in November 1933 at St Andrew's Hall, Cambridge, touring extensively, to forty venues around Britain, with a second showing in Cambridge a year later.[38]

This free exhibition, mounted by designer Misha Black (in the year he founded the AIA) and actor and political activist Barbara Nixon, innovated in using an exhibition as a space through which to share a political perspective on the perils of war. It provided an immersive three-dimensional environment through which audiences would absorb visual and textual material. Its structure was created through vertical display screens created by W. Doel, chair of the Council. It was accompanied by a striking catalogue designed by AIA member artist Clifford Rowe, its cover image a gas mask in green, red, black and white (Figure 4.4), inside incorporating lithographs by Paul Nash. The exhibition's seven sections explained the lead into the First World War, showed the unfolding horror of war itself and outlined how the anti-war movement was addressing current dangers with photographs (some official, some forbidden by censorship), news-cuttings, cartoons, maps, charts and war recruitment posters.

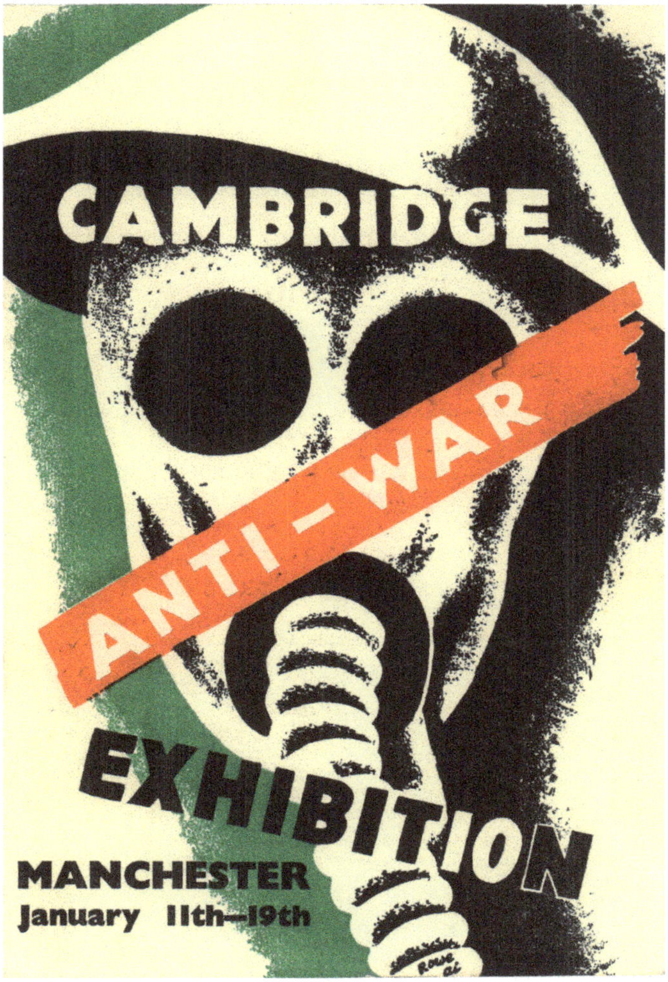

4.4 Cover for the *Cambridge Anti-War Exhibition* pamphlet designed by Clifford Rowe, held originally at St Andrew's Hall Cambridge in 1933, touring to Manchester in 1934. Clifford Rowe. © Anna Sandra Thornberry, daughter. Image courtesy of Working Class Movement Library.

Information about war and anti-war was amplified through work by Flemish painter and graphic artist Frans Masereel, German painter George Grosz and Hungarian painter and illustrator I. Szegedi-Szuts. Photographs of paintings by John Nash, stills of war films and a map by radical cartoonist J. F. Horrabin were also used in the exhibition. Posters by graphic designer Ashley Havinden, artist Pearl Binder and others carried slogans declaring 'War? No! Strike', 'War Means Workers Fighting Workers – Smash the War

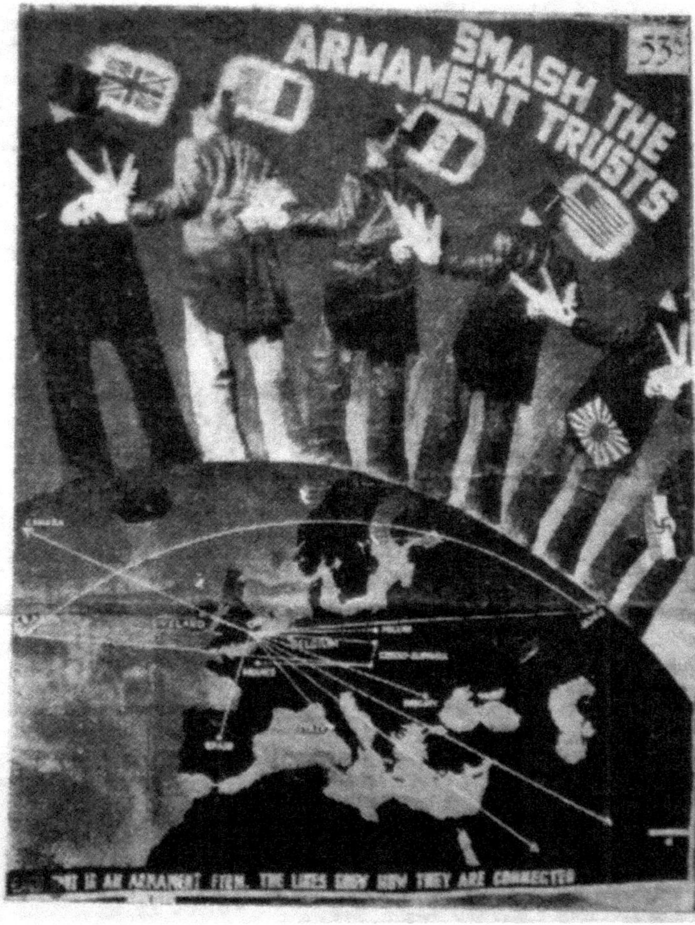

4.5 Misha Black poster, 'Smash the Armaments Trust', reproduced in *Bulletin of the British Anti-War Movement*, December 1933. Author photo. All rights reserved and permission to use the figure must be obtained from the copyright holder.

Plots!' and 'Against War – Bread! Not Battleships'.[39] Misha Black's poster, centring on a photomontage bearing the slogan 'Smash the Armament Trusts', explained in small print: 'Each spot is an armament firm. The lines show how they are connected' (Figure 4.5). Black's poster was reproduced on the front cover of the *Bulletin of the British Anti-War Movement*.[40] Talks were programmed alongside the *Anti-War Exhibition* on subjects including 'Women and War', 'Pacifism in Germany and France, 1914–18', 'War: a Personal Experience' and popular science writer J. D. Bernal speaking on 'Science and War'. Paul Nash donated a set of war lithographs to the exhibition, to be sold in aid of the Council.

Isotype at the Cambridge *Anti-War Exhibition*

Viennese polymath Otto Neurath created Isotype diagrams especially for the 1933 Cambridge *Anti-War Exhibition*. At the time, the Neuraths were still based in Vienna. After the outbreak of the Second World War, Otto and Marie Neurath fled to Britain, establishing the Isotype Institute in Oxford in 1941, which supplied pictograms to Ministry of Information exhibitions, with Isotype taken up in these wartime contexts as a succinct and powerful universal language, as I discuss in Chapter 7. Isotypes elucidated key statistics in visual form, aiming to simplify and universalise knowledge. These were mentioned in the Cambridge exhibition's programme but, to my knowledge, no visual records remain of the specific diagrams.

Isotype, a language of 'informative pictures', as its originator Neurath described it, was a system of symbols for figures and objects that had been developed for public information exhibitions, initially to explain housing and gardening in an exhibition that became the basis for Vienna's Museum for Housing and City Planning in 1923.[41] The pictogram language of Isotype, regularly incorporated as charts and posters, became an integral element of the visual lexicon of documentary exhibitions across Britain and Europe. In his 1931 book *Gesellschaft und Wirtschaft* (*Society and Economy*), museum director Neurath developed a new visual language for quantitative information using interpretable icons, to elucidate details of issues as varied as global motor manufacture and the world's energy consumption (Figure 4.6).[42] Neurath developed the lexicon to stop museum visitors from being overwhelmed by viewing vast quantities of material that lacked a system of arrangement. 'Many go away [from an exhibition]', Neurath explained, 'blaming themselves for not having gained a better grasp of the information it is intended to convey but those who analyse exhibitions seriously as a means of communication consider that the way in which they are set out is often a visual offence'. Isotype was offered as a universal language, purely factual and seemingly neutral. 'The Isotype maker is therefore bound to be as "neutral" as a map-maker and to provide material for free discussion from any point of view', Neurath wrote.[43] Having been developed to extend the communicative reach of exhibitions, Isotype was perfectly suited for use in the *Anti-War Exhibition*.

Neurath distinguished between exhibitions and other contexts where information was being shared, such as lectures or films, explaining:

> Visitors can stand around an exhibit and discuss it freely ... they can walk backwards and forwards ... to collect their knowledge. An exhibition gives more freedom and is a stimulus to community life; if some people need more time than others there is nothing to prevent them examining the exhibits again.[44]

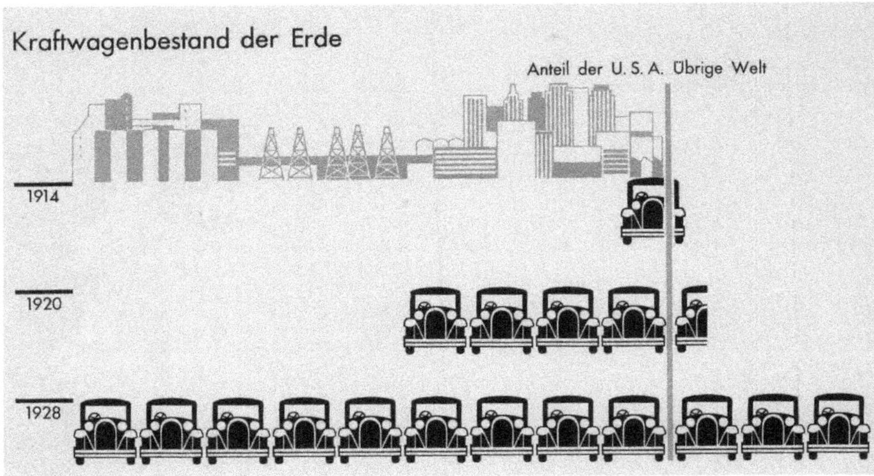

4.6 Isotype diagrams used to show global motor manufacture in Otto Neurath, *Gesellschaft und Wirtschaft: Bilstatistches Elementarwerk* (Leipzig: Bibliographisches Institut, 1931), p. 56.

The specific experience of visitors walking around exhibitions at their own speed and being able to revisit areas in order to clarify was key to providing them with 'permanent information', as Neurath described it: information that stayed with them. Neurath explained that 'Isotype leads to the presentation of events stripped of superfluous details by means of an international language-like technique. Training people to deal with the mass of material in a documentary photograph leads them to the international visual environment of modern man'.[45] Neurath's interest in interrogating how people receive information, in three-dimensional environments, shaped the serious regard with which such exhibitions were viewed as spaces for sharing knowledge.

Neurath's driving interest was in interrogating ways that information shared in exhibitions could be made to remain in the mind of the public. Its actual impact was hard to evaluate. Despite this, the idea continued to be a central concern in contexts where exhibitions formed one possible route for sharing information amongst many, including those mounting official information exhibitions during the wartime in Britain.

The second Cambridge *Anti-War Exhibition*

The Cambridge 1933 exhibition, with its display boards, illustrations and graphics, pictograms and text, enabled visitors to be immersed in a multiform political presentation. Pleased by the positive response, the Cambridge

Anti-War Council started preparing a second *Anti-War Exhibition* early in 1935; the high-profile committee included experimental physicist P. M. S. Blackett, in the process of doing pioneering cosmic ray research work and lending weight and credibility to the subject.[46] The committee intended this new version to draw out more clearly the linkages between fascism, nationalism, militarism and war, to highlight forms of anti-fascism inside Germany, Italy and Austria; and to be mobile, allowing it to tour easily. The exhibition had originated as the anti-fascist *L'Exposition Internationale sur le Fascisme* shown at Galerie La Boetie in Paris from March to April 1935, organised by a group including, once again, artist Frans Masereel and writer and art historian André Malraux. Malraux would later introduce the phrase 'musée imaginaire' ('museum without walls'), an acknowledgement of the museum's expansion beyond physical spaces, through photographic reproductions and in the space of memory and imagination.[47] This was an idea that likely took shape through contributions to projects like the Galerie La Boetie show.

The Paris *L'Exposition Internationale sur le Fascisme* addressed 'the promises of fascism', fascism and young people, fascists and women, anti-Semitism and myths about race. Anticipating its transfer to London, William Gillies, first International Secretary of the Labour Party, asked Victor Schiff, Paris Correspondent of British Labour-supporting *Daily Herald* newspaper, to send a report on the exhibition. Schiff responded by bemoaning the exhibition as 'purely Communist propaganda ... anti-Socialist and anti-democratic', which set out to prove that '"democratic illusions" of Social democrats, in Germany and in Austria as well, are responsible for the triumph of Fascism'. In summary, he declared the exhibition 'a scandal'.[48] Despite this, it transferred to London in 1935.

In its London iteration, efforts were focused towards fundraising, intending any profit to support anti-fascists inside Germany, and attracting a donation from novelist E. M. Forster. Writer Virginia Woolf joined the exhibition's committee, agreeing to canvass financial support for the exhibition; other supportive luminaries included sculptor Frank Dobson, painter Augustus John and architect Wells Coates.[49] Several of Woolf's friends, including Clive Bell, criticised her involvement with what they rightly perceived to be a communist organisation. Poet R. C. Trevelyan wrote to Woolf saying he would not sponsor it on this basis: 'I do not believe this is the right way to set about counteracting Fascism ... It can only irritate Fascists abroad'. Trevelyan went on, 'It seems to me to expose the evils of Fascism and Nazism, and to say nothing whatever of the similar evils of the Russian regime, is completely wrong-headed'.[50]

The resulting *Cambridge Exhibition on Fascism & War* was shown, like the earlier one, at St Andrew's Hall, Cambridge in November 1935. It used 'documentary evidence' – photographs, newspaper cuttings, sketches and documents – to explain the origins and history of fascism in Italy and Germany, 'Semi Fascism' in Austria and Spain, 'Embryo-Fascism' in Britain, the relationship between fascism and militarism and, finally, the anti-fascist movement and was accompanied by a set of cartoons by AIA artist and London *Evening Standard* cartoonist David Low. Designed by artist Paxton Chadwick, the exhibition had an upright display screen constructed by Anti-War Council chair W. G. Doel, accompanied by a pamphlet entitled 'Explanation', the cover designed by graphic designer E. McKnight Kauffer (Figure 4.7). In addition to carrying out work for the Cambridge Anti-War Council, during the 1930s Kauffer aligned himself

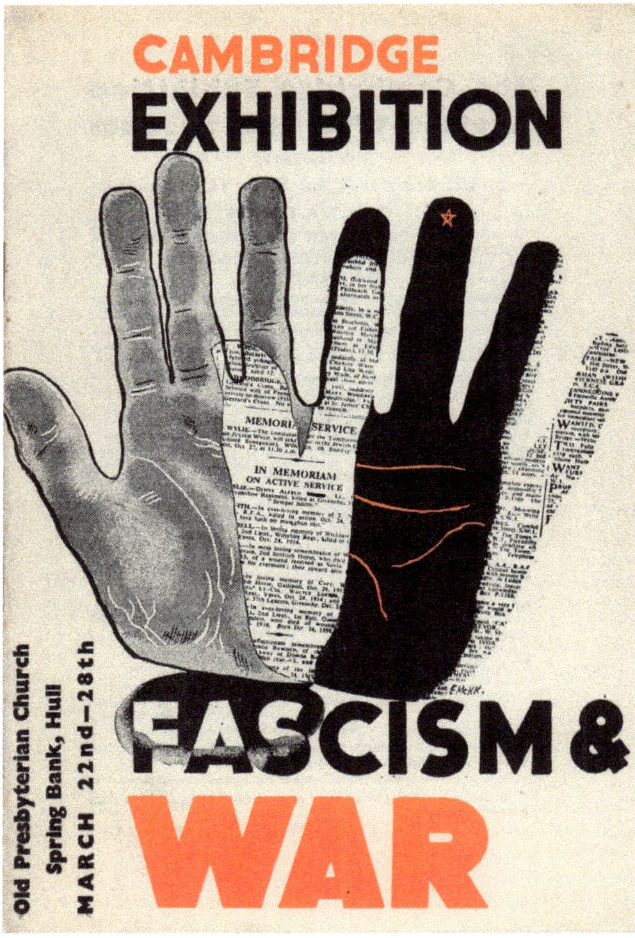

4.7 *Cambridge Exhibition on Fascism & War* brochure for its showing at Hull, with cover image by E. McKnight Kauffer. © Simon Rendall. Image courtesy of Hugh Gilbert and Webb & Webb, Design.

with a range of Leftist political organisations. He had designed sets and costumes for the Arts League of Service (ALS), formed in 1919 to democratise art and theatre; for the pacifist Peace Poster Service and Peace Pledge Union; for the AIA; he designed the brochure *In Defence of Freedom, Writers Declare against Fascism* and, by 1937, in solidarity with the Republican side in the Spanish Civil War, designed an aid to Spain poster, combining dynamic lettering in primary colours set against a head by Spanish Renaissance painter El Greco, entitled 'Help to Send Medical Aid to Spain' (Figure 4.8). Kauffer's voracious appetite for taking

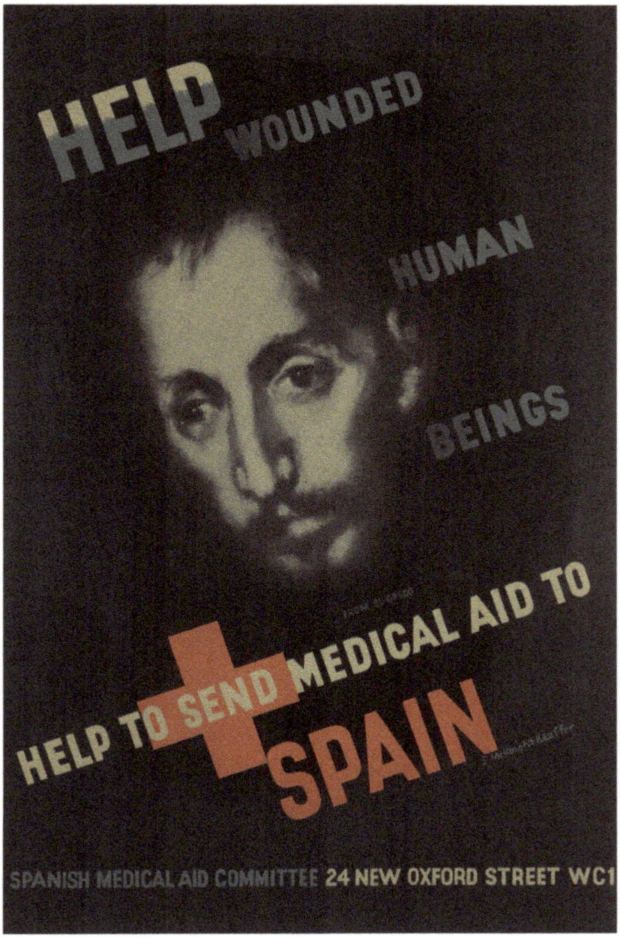

4.8 E. McKnight Kauffer poster for the Spanish Medical Aid Committee, 1937. Cooper Hewitt, Smithsonian Design Museum. Gift of Mrs. E. McKnight Kauffer, 1963-39-107. Photo: Matt Flynn. © Smithsonian Institution. All rights reserved and permission to use the figure must be obtained from the copyright holder.

commissions from a plethora of organisations led him also to work in direct contradiction to his stated alignments. His cover illustrations for Oswald Mosley's treatise *The Greater Britain* of 1932, which marked the formation of Mosley's British Union of Fascists and led to mass recruitment for fascism in Britain, is just one example.[51]

Following a visit to the 1935 *Anti-War Exhibition*, Woolf confessed in her diary to being 'plagued by a sudden wish to write an Anti-fascist Pamphlet' and, following a failed attempt to write a more politically engaged novel, which became *The Years* of 1937, she wrote *Three Guineas*, published in 1938.[52] The book, suggestive of being a text accompanying an exhibition, played with modes of exhibition, voice and presentation. Woolf's narrator evoked the atrocities of war, inviting the reader to examine Spanish Civil War photographs of 'dead bodies and ruined houses', 'piling up on the table', at the fictional exhibit. Woolf referred in the book to embryonic fascism in Britain as 'the egg of the very same worm that we know under other names in other countries'.[53]

Aside from Woolf's involvement with the Cambridge Anti-War Council, her most active political involvement was her membership of the anti-fascist group For Intellectual Liberty (for the 'defence of peace, liberty, and culture'), with Margaret Gardiner as secretary, the British affiliate to the French group Comité de Vigilance des Intellectuels Antifascistes.[54] Woolf and other Bloomsbury Group members were regularly involved with AIA activities during this period. At the AIA's request, Woolf wrote 'Why Art Today Follows Politics' for Communist newspaper the *Daily Worker*.[55] Woolf wrote in her diary that she had been shocked and inspired in this by a packet of photographs sent from Spain 'of dead children, killed by bombs'.[56] Her piece opened, 'I have been asked by the Artists' International Association to explain as shortly as I can why it is that the artist at present is interested, actively and genuinely, in politics'.

The article closed by explaining why, in the present circumstances, it was necessary for artists to organise themselves into groups like the AIA. Bloomsbury Group painter Duncan Grant was at that point an AIA member, calling for submissions to the AIA exhibition. By the outbreak of the Spanish Civil War eight months later, Woolf's sister Vanessa Bell was an AIA member.[57] An editorial published with Woolf's piece sought to distance itself by clarifying that Woolf's view was not that of the newspaper, adding 'we doubt whether artists in the past have been so peacefully immune from the conditions and issues of the society in which they live as she suggests'. After its showing in Cambridge, the *Anti-War Exhibition* travelled to London, accompanied by a lecture series, shown for just four days in November 1935 at 27 Soho Square, a site across from the Soho Square premises being used simultaneously by the AIA for *Artists Against Fascism and War*.[58]

Anti-war exhibitions in Manchester

The Cambridge Anti-War Council was not the only pacifist group demonstrating their beliefs by creating anti-war exhibitions. A number of other exhibitions were mounted around the country on the same theme. Remaining evidence exists in the form of brief advertisements and small newspaper articles. The Manchester and District Anti-War Council arranged a programme of exhibitions in 1935, one of which opened in January at the Friends Meeting House, with eight sections looking at the causes, conduct and aftermath of the First World War and how preparations for another war were being made.[59] *Woman To-Day*, the magazine of the British section of the Women's World Committee Against War and Fascism, reported in 1937 on a *Women's Anti-War Exhibition* arranged 'by Manchester Women'. 'Simple, attractive and not without humour', the exhibition showed through drawings, photographs, graphs and press cuttings how warfare impacted on women (although I have not found visual records, so the mode of installation is unclear).[60]

The same year a *Peace Exhibition* was held at Manchester's Central Hall and the year after, in January 1938, the Manchester & District Anti-War Council again hosted the *Cambridge Anti-War Exhibition* at two venues. I have not found visual records but know from flyers that Maurice Dobb, a Marxist economist from Cambridge, opened the exhibition. Art and lighting direction was by E. G. Barlow, who lent six of his own drawings, with design and mounting by Misha Black and Barbara Nixon.[61] Exhibitions had been recruited to the cause of peace, acting to create immersive, multimedia arguments in three dimensions. But this powerful, pacifist vision shared by many on the Left in Britain was severely challenged by Italian fascist leader Benito Mussolini's invasion of Abyssinia in October 1935, further disrupted by the increasingly hostile Nazi regime in Germany after Hitler's rise to power in 1933, by the Spanish Civil War's call to arms, by increasing knowledge of Stalin's abuses in the Soviet Union, by Germany's annexation of Austria in 1938 and by British Prime Minister Neville Chamberlain's official policy of appeasement and accommodation of these developments.[62] The next chapter explores the way in which counter-exhibitions were developed as political arguments in Britain during the late 1930s.

Notes

1 The actual quote from William Blake's *On Virgil* was 'A warlike State never can produce Art'.
2 Bal, *Double Exposures*.
3 Henri Lefebvre, *The Production of Space* (Oxford: Blackwell, 1984), p. 86.
4 The first AIA statement of aims was published in *International Literature* in 1934, according to Lynda Morris and Robert Radford, *The Story of the Artists International*

Association 1933–1953 (Oxford: MOMA Oxford, 1983), p. 11. Ian Grosvenor and Sian Roberts, 'Art, Anti-fascism, and the Evolution of a "Propaganda of the Imagination": The Artists International Association 1933–1945', in Frederik Herman, Sjaak Braster, and Maria del Mar del Pozo Andres (eds), *Exhibiting the Past: Public Histories of Education* (Berlin: De Gruyter, 2022), pp. 217–38.

5 British Library WP7937, Miscellaneous pamphlets, The National Council for Civil Liberties.

6 Lefebvre, *The Production of Space*, pp. 164–5.

7 Tony Bennett, 'The Exhibitionary Complex'.

8 'The Hampstead Primitive', *Picture Post* (3 February 1945).

9 Lefebvre, *The Production of Space*, p. 101.

10 AIA Archive TGA 7043, leaflet entitled 'Full Employment for the Artist? A Programme', 1945, on six contact sheets, photographs of some of the material used in the AIA exhibition 1979–80, and four sheets of negatives.

11 TGA 7043/20 *AIA Bulletin* No. 81, January 1944 and *AIA Bulletin*, 1945 p. x.

12 TGA 7043/20 'AIA: The First Five Years, 1933–1938'.

13 Betty Rea quoted in article by Charles Morris, 'On the Side of Humanity' marking the AIA's first twenty-five years, held in private archive.

14 Montagu Slater's review of the exhibition in *Left Review*, January 1936, p. 161.

15 Aldous Huxley, *Brave New World* (London: Chatto & Windus, 1932).

16 Foreword to *Artists Against Fascism* catalogue TGA 7043.

17 Montagu Slater reviewed the exhibition in *Left Review*, January 1936, pp. 161–4.

18 Harm Kaal and Casper Kirkels, 'Public Order Acts and Their Effects on Street Politics in 1930s Europe: A Case Study of Britain and Netherlands', *Société française d'histoire urbaine*, vol. 2, no. 55, 2019, pp. 125–40.

19 Gill Clarke, *Conflicting Views: Pacifist Artists* (Bristol: Sansom & Co., 2018), p. 86.

20 Judith Butler discusses the 'plural performativity' in the act of gathering in public space, drawing on the linguistic work of J. L. Austin's *How to Do Things with Words* (1962) and his concept of 'performative utterances'; *Notes Toward a Performative Theory of Assembly* (Cambridge, MA: Harvard University Press, 2015), p. 8.

21 Morris and Radford, *The Story of the Artists International Association*, p. 48.

22 Lefebvre, *The Production of Space*, p. 167.

23 *Manchester Guardian*, 18 February 1939.

24 Peter Kort Zegers and Douglas Druick, *Windows on the War, Soviet TASS Posters at Home and Abroad 1941–1945* (New Haven, CT: Yale University Press, 2011).

25 Angela Jackson, *British Women and the Spanish Civil War* (Abingdon: Routledge, 2002). Footnote 182 says the newsreel was British Paramount News, Issue no. 833, 23 February 1939. 'Artists aid refugees by painting hoardings: London. Well known artists turn London streets into studios for vivid publicity scheme for Spanish refugee fund'. ITN Archive, London, tape V619 (1): https://reuters.screenocean.com/record/203946 (accessed 20 December 2023).

26 'Wave of Unrest in Britain', *Action*, no. 157, 25 February 1939, p. 7.

27 Donald Drew Egbert, *Social Radicalism in the Arts: Western Europe; A Cultural History from the French Revolution to 1968* (London: Duckworth, 1970), p. 504.

28 Remy, *Dictionary of Surrealism*, p. 20.

29 Newton was a popular critic who later appeared in Jill Craigie's film *Out of Chaos* (1944) explaining the meaning of art to a group of men.

30 Egbert, *Social Radicalism*, p. 504.

31 *The Star*, 10 February 1939.

32 'The Man in the Street's Exhibition', *Manchester Guardian*, 10 February 1939. Egbert named him as a butcher; *Social Radicalism*, p. 503.

33 *The Star*, 10 February 1939.

34 *The Times*, 15 February 1939.

35 BL *AIA Bulletin*, no. 81, January 1944.

36 Trevelyan wrote in July 1939 according to Martin Rewcastle and Nicholas Serota, 'Art and the Social Purpose of the Whitechapel Art Gallery', in *Art for Society: Contemporary British Art with a Social or Political Purpose* (London: Whitechapel Art Gallery, 1978), p. 7.
37 BL WP7937, [Miscellaneous pamphlets] 'What is The Cambridge Anti-War Council?', Cambridge Anti-War Council (Cambridge, England) Cambridge, [1935–].
38 BL 'What is the Cambridge Anti–War Exhibition', 13–18 November 1933, p. 3. In another leaflet at the BL, 'A Cambridge Exhibition on Fascism and War', the venues are listed as including Oxford, Manchester and Birmingham.
39 Morris and Radford, *The Story of the Artists International Association*, p. 11. Photograph of 'Posters in the Anti-War Exhibit arranged at Cambridge University by the Artists International "revolutionary group of England"' from *International Literature*, vol. 1, no. 7, 1934, p. 152.
40 V&A AAD, Misha Black collection, *Bulletin of the British Anti-War Movement*, no. 12, December 1933.
41 Otto Neurath, *From Hieroglyphics to Isotype* (London: Hyphen Press, 2010), p. 100; C. Burke, E. Kindel and S. Walker, *Isotype Design and Contexts, 1925–1971* (London: Hyphen Press, 2014).
42 Michelle Henning, 'The Pig in the Bath: New Materialisms and Cultural Studies', *Radical Philosophy*, vol. 145, September/ October 2007.
43 Neurath, *From Hieroglyphics*, pp. 5, 125.
44 Neurath, *From Hieroglyphics*, pp. 118–19. See also Marie Neurath, 'An Isotype Exhibition on Housing', *Journal of RIBA*, vol. 54, 1947, pp. 600–3; Otto Neurath, 'Health Education through Isotype', *The Lancet*, 25 August 1945, pp. 236–43; Sue Walker, 'Effective Antimicrobial Resistance Communication: The Role of Information Design', *Palgrave Communications*, 5, 2019, p. 24; and Burke, Kindel and Walker, *Isotype: Design and Contexts*.
45 Neurath, *From Hieroglyphics*, p. 119.
46 P. M. S. Blackett went on to win the Nobel Prize for Physics in 1948: www.nobelprize.org/prizes/physics/1948/blackett/biographical/ (accessed 15 March 2023).
47 André Malraux, *Le Musée imaginaire de la sculpture mondiale* (*Museum Without Walls*) introduced in his first book, the three-volume *Psychology of Art* (published in French, 1947–9), later revised and published as one volume in *The Voices of Silence* (published in French, 1951).
48 People's History Museum archive ID/CI/22/8i-iii.
49 The 'Explanation' pamphlet listed contributors; Amy M. Lilly, 'Three Guineas, Two Exhibits: Woolf's Politics of Display', *Woolf Studies Annual*, no. 9, part 1, 2003, p. 43. People's History Museum archive ID/CI/22/18, Schiff_015. Naomi Black, *Virginia Woolf as Feminist* (Ithaca, NY: Cornell University Press, 2004), p. 25. Eve Colpus, *Female Philanthropy in the Interwar World* (London: Bloomsbury, 2018) discusses the long history of philanthropic cultural activities led by women in Britain.
50 As Hermione Lee notes in her biography of Woolf, *Virginia Woolf* (London: Vintage, 1996), p. 685.
51 Caitlin Condell and Emily M. Orr (eds), *E. McKnight Kauffer: The Artist in Advertising* (New York: Rizzoli Electa, 2020), p. 26.
52 Virginia Woolf, *Three Guineas* (New York: Harcourt, Brace & Co., 1938); Lilly, 'Three Guineas, Two Exhibits'; David Bradshaw, 'British Writers and Anti-Fascism in the 1930s', Part I, *Woolf Studies Annual*, vol. 3, 1997, pp. 3–27 and Part II, *Woolf Studies Annual*, vol. 4, 1998, pp. 41–66.
53 Woolf, *Three Guineas*, p. 96.
54 Hermione Lee and Joanna P. Gardner-Huggett, 'Margaret Gardiner: Collecting as Activism', *The British Art Journal*, vol. 6, no. 2, Autumn 2005, pp. 76–82; Lee, *Virginia Woolf*, p. 686.

55 Ben Harker, '"On Different Levels Ourselves went Forward": Pageantry, Class Politics and Narrative Form in Virginia Woolf's Late Writing', *ELH*, vol. 78, no. 2, Summer 2011, pp. 433–56, footnote 7, p. 452; 'Why Art To-Day Follows Politics', *Daily Worker*, 14 December 1936, p. 6.
56 Lee, *Virginia Woolf*, p. 675, quoting from Woolf's diaries.
57 Lilly, 'Three Guineas, Two Exhibits', p. 40.
58 Lilly, 'Three Guineas, Two Exhibits', pp. 32–3, although whether the exhibition was held in the end at 27 Soho Square is disputed. Advertisement for Left Front cultural activities in 1935 reproduced in Morris and Radford, *The Story of the Artists International Association*, p. 30. In March 1936 the exhibition showed at Hull; see programme in Condell and Orr, *E. McKnight Kauffer*, p. 226. In May 1936 the exhibition showed at Hackney Wick Workers' Club: Cambridge exhibition against fascism and war programme, held in Working Class Movement Library Fascism Box 1.
59 WMRC, TUC archives MSS.15X/2/373/1 Oxford District Peace, Council Workers' Educational Trade Union Committee leaflet, promoting 'Trade unionism, democracy, dictatorship' by Dr Franz Neumann.
60 *Woman Today*, October 1937, p. 6.
61 https://radicalmanchester.wordpress.com/2011/10/19/peace-and-antiwar-activiti-in-1930s-manchester/ (accessed 15 March 2023).
62 R. A. C. Parker, *Chamberlain and Appeasement: British Policy and the Coming of the Second World War* (Basingstoke: Red Globe Press, 1993) and Martin Gilbert, *The Roots of Appeasement* (London: Weidenfeld and Nicolson, 1966).

5

Counter-exhibitions

Angered by reports of a new exhibition of German art currently being held in London, German Chancellor Adolf Hitler seethed: 'There is no room for any Neanderthal culture in the twentieth century, no room at least in National Socialist Germany!'[1] Hitler's words about the London exhibition and its 'Neanderthal culture' – as he described paintings by many of the most well-regarded German Modernist painters of their day – was exactly the angry response to the exhibitionary provocation that those mounting the London exhibition had hoped for. This chapter focuses on the culture of counter-exhibitions that developed in Britain as political arguments during the 1930s. It might be possible to claim several exhibitions in this book as counter-exhibitions but my specific definition here is exhibitions used as arguments or justifications, which mounted a direct riposte.

The two counter-exhibitions that this chapter centres on were held in 1938. The first was the *Workers' Empire Exhibition* in Glasgow, mounted by the Communist-aligned Independent Labour Party, a direct critique of the Glasgow *Empire Exhibition* at Bellahouston's propaganda in support of the British Empire. The second was *Twentieth Century German Art* held in London, mounted by a group of prominent art historians and curators with the Freier Künstlerbund (Free Association of Artists) in direct response to the German degenerate art exhibitions. In both cases, exhibitions acted as a vital, cosmopolitan practice, a mode that offered the potential to make a forceful denouncement of cultures of fascism and imperialism in a shifting ecology of multimedia communication.

Counter-exhibitions take shape

Inspiration for such exhibitionary rebuttals likely came from France, where Surrealists were increasingly using counter-exhibitionary tactics to show solidarity.[2] From May to December 1931, the French government had held *L'Exposition Coloniale Internationale* (*International Colonial Exhibition*) in Paris's Bois de Vincennes, showcasing its colonial relationships.[3] In response, the Ligue Anti-Imperialiste (Anti-Imperialist League) and, at the request of the Comintern, the Surrealists organised the counter-exhibition *L'Exposition Anti-Impérialiste: La Vérité sur les colonies* (*The Anti-Imperialist Exhibition: The Truth about the Colonies*) from September 1931 to February 1932 at the former Soviet pavilion of the Paris 1925 Expo.[4] The Paris counter-exhibition's intended audience included French and colonial workers who had not yet developed a revolutionary consciousness.[5] Its poster showed three exaggerated and stereotyped illustrations representing colonial subjects buckling under backbreaking work, set against an exoticised temple structure, with no sign of Surrealist influence.

In the Paris *L'Exposition Anti-Impérialiste*, a section on the ground floor curated by historian André Thirion tried to expose the brutality of the colonisers. It featured maps, documents, photographs of the indigenous way of life and charts recording the abuses committed by imperialism, beside text panels celebrating the 'good life' in the USSR, with its 'wonderful ethnic diversity'.[6] French civil rights organisation Ligue de Défense de la Race Nègre (League for the Defence of the Negro Race) worked with the Surrealists to contribute a room devoted to visual art, working with Louis Aragon, Paul Éluard, Georges Sadoul and Yves Tanguy. Two Surrealist tracts amplified ideas in the exhibition: the first, of May 1931, 'Ne visitez pas l'Exposition Coloniale', ('Do not visit the Colonial Exhibition') signed by twelve Surrealists, attacked the French government for its exploitation of colonised peoples and portrayed the Colonial Exhibition as a denigrating ideological force; while the second, in July 1931, decried colonialism, focusing on the hypocrisy of missionary practice.[7]

The trend for counter-exhibitions as political responses to major exhibitions was catching on elsewhere. In the Netherlands in summer 1932 the Anti-Koloniale Tentoonstellings Actie (Anti-Colonial Exhibition Action) organised themselves to oppose the *Indonesian Exhibition* staged in The Hague.[8] While in 1936 the Dutch counter-exhibition titled *D.O.O.D.*, doubling both as the Dutch word for 'death' and acronym for *De Olympiade Onder Dictatuur* (*The Olympiad under Dictatorship*), was held in Amsterdam as a direct riposte to Nazi propaganda chief Goebbels' Nazi Art Olympiad. British painter Jessica Dismorr, who had showed with the AIA in 1937 and 1939, was one of only seven British women with works at *D.O.O.D*, which also included two works by AIA sculptor Betty Rea.[9]

Fascism in London: Mussolini's 'infamous' *The Italian Exhibition*

The British establishment's belief in exhibitions held on home soil as consummate displays of universal, civilising culture somehow unblemished by contemporary political concerns was called into question in 1930 when Italy's fascist leader Benito Mussolini shaped an exhibition of Italian treasures at London's Burlington House. Mussolini's proposal, enthusiastically taken up by Sir Austen Chamberlain MP and developed over five meetings, became *The Italian Exhibition.*[10] To ensure the exhibition was as eye-catching as possible, Mussolini had issued a diktat that no treasures from Italian museums would be off-limits to Burlington House and went out of his way to promote the exhibition.[11] The fascist leader's huge personal investment in the development of *The Italian Exhibition* showed his awareness of exhibitions' value for his regime's collective self-fashioning, as well as their potential to attract public approval. As desired, the exhibition courted much attention: UK's Pathé News captured the Italian treasures' arrival in Britain, with Sir Austen and Lady Chamberlain filmed greeting the ship at the docks and the treasures, worth a noted £14 million, being off-loaded into a van theatrically labelled 'Transport for Italian Art Exhibition – Anglo-Italian Express'.[12]

While the Italian Ambassador to London declared *The Italian Exhibition* 'the greatest, most effective propaganda one could imagine or wish for on behalf of Italy', onlookers in Britain recognised that the institution had been manipulated from afar by the Italian dictator; used for self-promotion, to court approval and to signal his support within Britain. Soon after its closure, many reacted with horror to the Chamberlains' complicity in enabling Mussolini to promote himself in London. Art historian and museum director Kenneth Clark echoed them when he recalled the 'infamous' exhibition in his autobiography.[13] Exhibitions were being used as instruments of fascist states in their own countries, but their manipulation from afar in Britain, as a tool of remote propaganda for a foreign dictator, was a novel, shocking and altogether unpalatable development.

Subverting a propaganda tradition: counter-empire exhibitions in Britain

This book's main focus is on exhibitions from 1933 onwards. But in the context of exhibitions of empire in Britain, the threads were longer: from the 1920s conspicuous displays of imperial might in Britain were starting to attract opprobrium, as well as praise.[14] Exhibitions mounted as the focus for displays of imperialist propaganda in Britain were increasingly subverted and eclipsed. They were used as platforms for showing the opposite of what their commissioners had intended, diverted to a focus on foregrounding anti-imperialist discourses through 'counter-exhibitions', a potent strategy for opposing ideas and political positions. From 1919 onwards, oppositional voices in Britain began speaking up, and shadow gatherings, organised by increasingly extensive anti-colonial networks, drew strength and strategies from one another.

The oppositional voices that started to be heard after 1919 were shaped by the increasing hardening of racial barriers in Britain and the colonies, meaning everyday manifestations of racism increased, such as denying people of African descent service at hotels and restaurants.[15] This opposition was partly inspired by the declared commitment of the Communist International to liberating people of African descent and colonial peoples. Many British-based artists and designers were affiliated to the Communist International, founded in 1919, which sponsored conferences; others were associated with it indirectly through groups such as the League Against Imperialism, some of which ran exhibitions alongside them, bringing anti-colonial activists to Moscow to study and exchange ideas.[16] Other internationalist organisations in Britain attracted mass participation to anti-slavery, feminist and pacifist causes such as the British Commonwealth League, League of Nations Union, Anti-slavery and Aborigines Protection Society. They became the focus for anti-colonial dissent. Exhibitions, which were focused towards strengthening and consolidating the imperial body, became an increasingly potent vehicle for opposition.[17]

A major focus for support, as well as increasing opposition to empire, was the *British Empire Exhibition* at London's Wembley of April 1924. This had attracted more than 27 million visitors to see the empire 'reproduced in miniature'.[18] In his opening oration King George V described the exhibition as a 'living picture of the history and structure of the British Empire'.[19] Wembley enabled racist narratives and beliefs to be reinforced, conspicuously through its 'native villages' and 'native workshops', housing roughly sixty Hausa, Yoruba, Mendi, Asante and Fanti speakers who lived and worked on site for the duration of the exhibition.[20] In the months leading up to and during the exhibition, Felix Oladipo (Ladipo) Solanke, a Nigerian law student from Abeokuta and a member of Union of Students of African Descent (USAD), had gained notoriety for publishing a series of letters in *West Africa* criticising the exhibition's racist and salacious depictions of Africans.[21] According to Solanke, the outrage over his letters and other coverage of West Africans at Wembley led directly to the creation of the Nigerian Progress Union (NPU) and the West African Students' Union the following year, thrusting him to the forefront of Black activism in London.[22] Wembley sparked renewed attempts by Black intellectuals in London to organise London-based West Africans across colonial divisions.[23]

The *British Empire Exhibition* activated public figures who had not previously taken a public stance on British colonialism. Novelist Virginia

Woolf wrote the essay 'Thunder at Wembley' after her visit, describing the ominous sky above the exhibition and imagining a force more powerful than empire that would cause it all to tumble down: an imperial apocalypse, showing Woolf's ambivalent mix of nostalgia and disgust.[24] Wembley's representations of Africa in London activated Africans living in London to mount their own shadow or counter-exhibition to counter it. Galvanised by the letters of Solanke and Joseph Boakye Danquah from the Gold Coast and other members of USAD, they initiated an extensive letter-writing campaign and passed a resolution denouncing representations of Africans that 'hold up to public ridicule citizens of countries whose money has been voted in large sums for the purpose of the exhibition'.[25]

The *Workers' Empire Exhibition*

Exhibitions offered an excellent vehicle for conspicuous displays of political opposition to colonialism. In the interwar years, criticism of the British Empire was a minority discourse outside Black radical circles, only occasionally percolating through to public and parliamentary debates.[26] Some exponents of anti-fascism openly defended the empire, distinguishing British and French colonialism from the Nazi racial politics of the 1930s, while at the same time Black intellectuals and activists in London and elsewhere attempted to force an anti-imperialist cultural front.[27] Despite the fact that the Artists International Association's founding idea was to oppose 'Imperialist War' and 'Colonial Oppression', as articulated in 1934, in practice the AIA was more focused towards opposing the fascism occurring in Europe, while its particular formulation of an oppositional politics often remained woolly and undefined, especially when it came to the British Empire. An exception was the AIA's exhibition *Artists Against Fascism and War* held in November 1935, which, in standing as a protest against the rise of fascism, included Mussolini's recent invasion of Abyssinia.

Internationalist groupings proliferated in Britain during the 1920s, supporting colonial dissidents, opposing empire and white supremacy. Mussolini's invasion of Abyssinia in October 1935 had brought into focus the largely unacknowledged correlation between European fascism and European colonial endeavours, and galvanised these groupings towards a global Black coalition of resistance, with the founding of the International African Friends of Ethiopia (IAFE).[28] London became an important locus of Black resistance to racism and empire, with the emergence of new organisations and publications to engage with British and imperial publics and to create a common platform. Exhibitions were, once again, identified as tools of dissidence. Anti-fascist exhibitions developed in multiple contexts during this period, with campaigns in support of Ethiopia including exhibitions. The Friends of Abyssinia, who formed the Friends of Abyssinia [Ethiopia] League of Service in 1935 in response to Mussolini's invasion of

Ethiopia, held the painting exhibition *Ethiopia House* in 1936 at 20 Ludgate Hill EC4, for example, accompanied by a 'bureau of information' to raise awareness of the plight of Ethiopians.[29]

Anti-colonial sentiment was once again piqued by the mounting of Britain's final major *Empire Exhibition* of 1938 at Glasgow, when crowds thronged Glasgow's Bellahouston Park to wander down Dominion Avenue, watch fountains play by day and night and look out over the extensive expo from the Tower of Empire designed by Scottish architect Thomas Tait of London-based practice Burnet Tait and Lorne. At the same time, across the city, an exhibition of a rather different kind was taking shape. This 'other' exhibition – the *Workers' Empire Exhibition* – was being developed as a forceful counter-exhibition to the larger and showier *Empire Exhibition*. This smaller exhibition, created to critique its major counterpart, highlighted the appalling conditions of working people in Glasgow and across the British Empire, allowing political activists to develop and enunciate their adversarial anti-imperialist agenda. Compared to the much-photographed Bellahouston Park show, the other was little documented.

The *Empire Exhibition* was an extremely ambitious venture for a population recovering from the dire economic slump, with 1.9 million still unemployed.[30] It cost £11 million and was led by Scottish industrialists with British government support and the King as Patron. It was intended as overt pro-Empire propaganda: to showcase Scottish industry and to provide visually persuasive evidence of 'the progress of the British Empire at home and overseas', as well as showing off the Empire to future generations. The UK Government Pavilion, the largest national building, was designed by architect Herbert Rowse and devoted to showcasing British industries. Misha Black, despite his oppositional politics, worked for the British government at Glasgow designing displays for the Steel, Coal and Public Welfare halls with a section on 'Fitter Britain'. Flow-charts, photomontage, text, models, dramatic cascading lighting and illustrative vignettes were set in streamlined cabinets to illuminate scientific and industrial research. Echoing recent display stands that had been lauded in the trade press by designers such as Richard Levin (introduced in Chapter 1), Black incorporated type as a structural element through titles and foregrounded explanatory text to create a narrative arc. In Shipbuilding, backlit photographs showed the shipmaker's craft and abstracted elements of the ship (Figure 5.1). These were accompanied by titles with filmic qualities that propelled the visitor through the exhibit with multimedia displays including newsreels.[31]

The major fractures already appearing across the Empire were barely hidden within the Glasgow *Empire Exhibition*'s colonial narratives. For the first time, India – nearing independence – refused to participate.[32] Displays included the West African Colonies, Southern Rhodesia, Victoria Falls and East Africa, Malaya and the West Indies. In the Colonial Pavilion

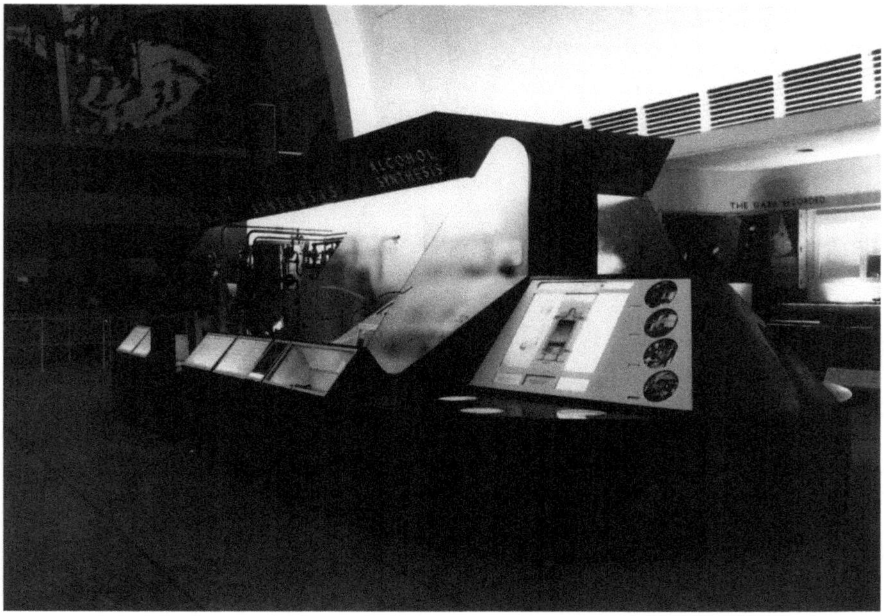

5.1 Misha Black displays for interior of the Shipbuilding pavilion, Glasgow *Empire Exhibition*, 1938. Image courtesy of the University of Brighton Design Archives.

a strangely disparate grouping brought together Malta, Somaliland, Hong Kong and the Falkland Islands. South Africa, already a 'sovereign independent state' by 1938, was inappropriately represented by a pavilion in the form of a Cape Dutch house, a style of rounded gables resonant of the Amsterdam town houses associated with Dutch colonial settlers. There were pavilions for the 'Dominions' of New Zealand, Canada, Australia and Ireland, which had seceded from the United Kingdom several years earlier.[33]

There was fierce opposition to the exhibition's representation of Empire. Socialist newspaper *Forward* said in an editorial, 'Anyone who has wandered about the Empire Exhibition at Bellahouston would think that the British Empire was a federation of happy and prosperous and contented nations and that everything in the imperial garden was lovely'. It continued, 'hundreds of millions of natives' live 'under the heel of the exploiter'.[34] Another socialist newspaper explained that, at the *Empire Exhibition*, 'Ideal conditions are being created for enticing your men to become murderous robots on behalf of the ruling caste';[35] while journalist George Padmore, writing in a Pan-Africanist journal, mocked the *Empire Exhibition* for 'informing their Imperial Majesties what a glorious contribution to the peace and prosperity of the people of the Empire this Exhibition represents', even 'as the working masses of the West Indian island of Jamaica [are] being shot and bayoneted for demanding betterment of their miserable working conditions'.[36] The *Empire Exhibition*'s condensed presentation served to crystallise colonial opposition, just as it had hoped to achieve the reverse.

Beyond the *Empire Exhibition*'s colonial misrepresentations, conditions for workers at the Glasgow Bellahouston site were also controversial. Sources described expo workers being made to pay the same inflated prices for meals as the visiting public and the casual workers in the Amusement Park working 16-hour days and attendants working 68-hour weeks.[37] Ironically, the Scottish Trades Union Congress had a small pavilion at the *Empire Exhibition*, focused on the need to shorten the working week.[38] The *Empire Exhibition*'s vision of commonwealth cosmopolitanism was predominately white and the overt racism of contemporary Britain was conspicuous to the expo's visitors: a Scottish missionary complained she and an African colleague had been refused service in the exhibition's most luxurious restaurant, the Atlantic, where the King and Queen had recently been entertained.[39] Many Glasgow hotels refused non-white guests, something that might have gone unreported had it not been that celebrated Black American bass singer and actor Paul Robeson had struggled to find accommodation during his visit to Glasgow.[40]

Spurred on by the mounting grievances against the expo, counteraction was led by the Independent Labour Party (ILP), a socialist group already supporting the emergence of dissenting voices and representatives from colonial workers' organisations. ILP opposition was led by activist Arthur Ballard, a carpenter by training whose London Socialist Book Centre was a gathering place for Leftist activists. Writing in 1938 in the *New Leader*, the newspaper of the ILP, Ballard described how: 'Walking amidst the wonderful buildings [at the *Empire Exhibition*] one thinks that the Empire is just a paradise on earth'. 'The average visitor, amidst this setting, may be carried away by this propaganda unless we are able to do something to present the real situation within the Empire'. The ILP group, Ballard explained, was planning an anti-imperialist exhibition 'to present the other side', appealing for funding to support this venture. 'We hope', he said, 'to make the Exhibition effective not only as propaganda but as a means of directly helping our comrades in the colonial countries to organise against the exploitation from which they suffer'. Any profits would go to develop working-class organisations in the colonies.[41]

The 1938 Glasgow *Empire Exhibition*'s counter-exhibition, organised by Ballard, was mounted across town at the Kingston Halls and named the *Workers' Empire Exhibition*.[42] The main evidence of this counter-exhibition comes in the form of newspaper reports, accompanying leaflets and

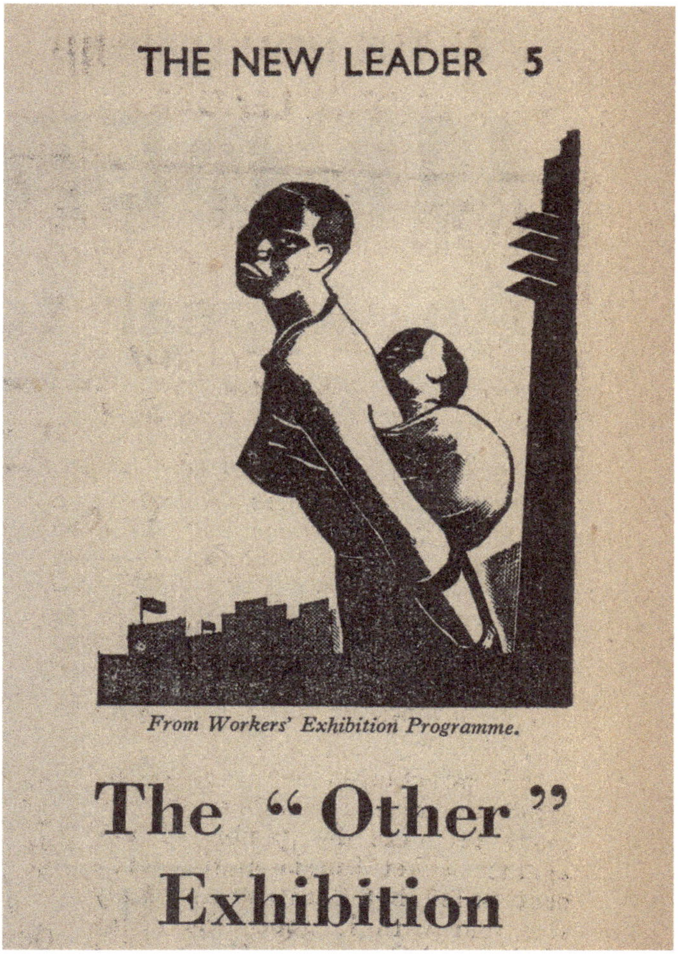

5.2 Illustration from the *Workers' Empire Exhibition* programme, reproduced in the *New Leader* (12 August 1938), p. 5. Image courtesy of the British Library. All rights reserved and permission to use the figure must be obtained from the copyright holder.

textual descriptions, rather than photographs. A small image of a subject of the British Empire with a baby strapped to her back standing by Tait's Tower headed an article in the *New Leader* (Figure 5.2). This described the blue and grey colour scheme of the *Workers' Empire Exhibition*, purposely designed to echo that of the *Empire Exhibition*, twelve panels clustered round a central column. The panels subverted messages in the main exhibition, showing 'that the real owners of the Empire are not the people of Britain or of the colonies, but the big financial and commercial interests centred in London', describing the main firms operating within the Empire, led by wealthy 'plutocrats' like Lord McGowan and the Duke of Montrose. The counter-exhibition had involved weeks of preparation, much research and the gathering of material from three continents. This research was translated into visual form by a group of London artists contributing voluntarily.[43] Sections were devoted to parts of the Empire, showing how India was being held 'in poverty and ignorance by brute force'; they gave the background to recent upheavals in the British West Indies including poor education and population malnutrition; low mining wages in South and East Africa; extermination of the native population in Australia; and oppression in Palestine, Ceylon and Ireland.[44]

The form of this fringe event – the *Workers' Empire Exhibition* – allowed for a full-blown counter-attack, with alternative narratives that exposed conspicuous power dynamics of class and race, allowing a developing anti-imperialist political agenda to reach a new audience. And the collective rage of the exhibitionary attack through which these injustices were highlighted was extremely forceful and affecting.

This political manifesto in three dimensions was accompanied by a plethora of communiqués on the same subject produced in other forms. Visitors received a satirical leaflet that adopted the style of a holiday brochure. With straight-faced irony very different from the more straightforwardly polemical approach of exhibitions of the same period, the leaflet invited visitors to: 'Come and See THE EMPIRE BY THE ALL RED ROUTE' (referring to the red route used on maps to demarcate the extent of the British Empire). It mocked: 'VISIT THE EMPIRE – It is the duty of every British citizen to see OUR GLORIOUS EMPIRE. We must take a proper pride in OUR POSSESSIONS, which cover nearly one-third of the earth's surface. Patriotic workers should make use of their holidays to visit the Empire' (original emphases) (Figure 5.3). After introducing the reader to the luxury they would travel in, the pamphlet laid bare the gross inequalities of workers employed on the route, the many aggressions played out in each context, enslaved peoples and shockingly low life expectancies, exposing visitors' lack of knowledge of the realities of 'their' empire.[45] The leaflet highlighted the two things visitors needed to do to help end 'the tyranny of British Imperialism': the first was to support workers' organisations in the Empire fighting for political and economic freedom and justice, such as the International African Service Bureau; the second was to support, in Britain, the ILP, which, they claimed, was the only political party fighting imperialism.

Scottish Labour politician James Carmichael opened the *Workers' Empire Exhibition* with celebrated working-class novelist and political activist Ethel Mannin (Figure 5.4).[46] Mannin's opening speech was echoed in her article for socialist newspaper *Forward* titled 'Empire with the Lid Off' in which she decried the racism of colonialists, British ignorance about the Empire and the 'Imperialist mentality'. She noted optimistically

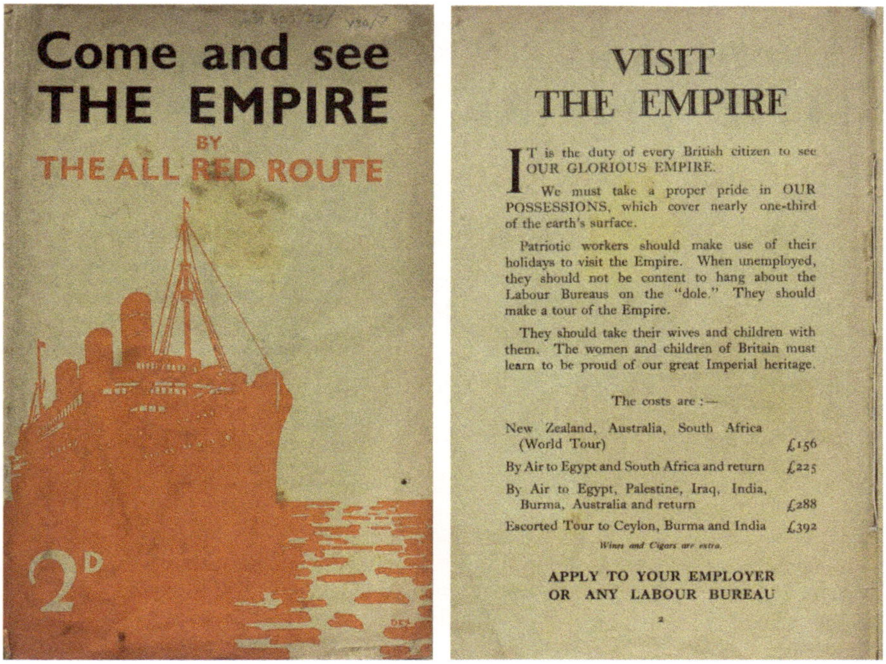

5.3 Brochure accompanying the *Workers' Empire Exhibition*, 1938, inviting visitors to 'Come and see the Empire by the all red route'. Archive of Jimmy Deane. Image courtesy of Warwick Digital Collections.

the increasing number of people, especially younger people, ceasing to take Empire for granted and pointed out that while the *Empire Exhibition* over at Bellahouston Park was being 'patronised by Royalty', every newspaper was full of news of 'street-fighting in Jamaica and the rushing of troops and a battle-cruiser to crush the revolt of workers struggling for the "privilege" of a bare living wage' – just one example, she said, of what could be found on looking below the outward pomp. The two exhibitions, Mannin said, were 'two sides of a medal': the *Empire Exhibition* showed the British Empire 'from the angle of the owning and governing classes … the capitalist side of the medal'; the *Workers' Exhibition* showed the Empire 'from the bottom … the workers' angle'. Mannin drew out further contrasts: at the *Empire Exhibition* you could see products of Empire while at the *Workers' Exhibition* you would see 'what it has cost in human blood and sweat and exploitation to turn out these products'.

The *Workers' Empire Exhibition*, Mannin said, was not spectacular; it did not have 'clever' or 'expensive floodlighting'. It was 'no amusement park'. Indeed, it was, she said wryly, 'as unspectacular as Hitler's quiet annexation of Austria [or] Mussolini's quiet extermination of the

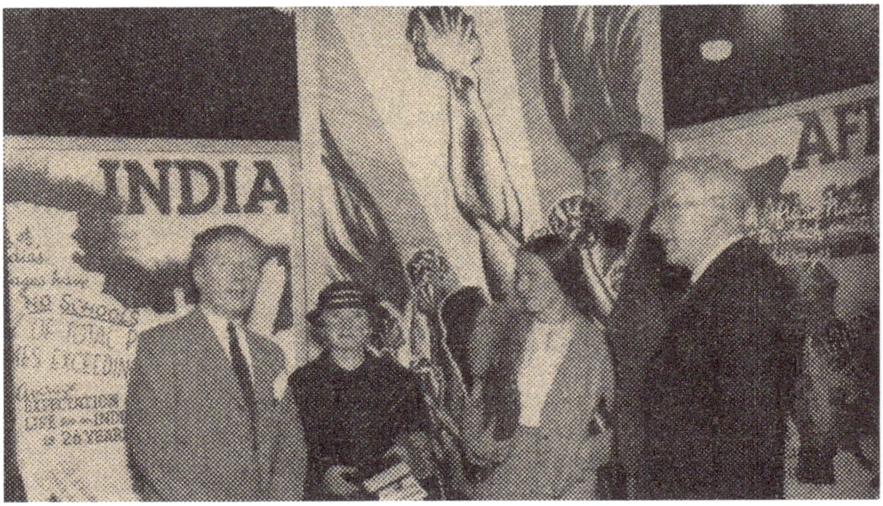

5.4 Opening of the *Workers' Empire Exhibition* by Scottish Labour politician James Carmichael with novelist and political activist Ethel Mannin, from the *New Leader* (19 August 1938), p. 5. Image courtesy of the British Library.

Abyssinians ...' It explained fascism within the British Empire, including the Colour Bar Act, and was a 'record of the living drama of the struggle for human liberty'.[47] Alongside the exhibition, the *New Leader* ran an eight-page 'Empire Special' amplifying the horrifying conditions in the Empire and the gross inequalities, where socialist activist Fenner Brockway again drew direct parallels between the fascist aggressions of Mussolini and Hitler and those of imperialism.[48]

It is hard to assess the impact this presentation had on its visitors because the only significant trace around the exhibition appears to have been left by those who supported it. But it is evident that the *Workers' Empire Exhibition* provided a focal point for a series of growing anti-imperialist and anti-fascist voices, being a perfect meeting point for such opposition, accompanied, as it was, by lectures to packed halls on colonial conditions, given by such prominent speakers as Kenyan anti-colonial activist Jomo Kenyatta and George Padmore, both of whom published articles attacking the Glasgow *Empire Exhibition*. Activist groups organised themselves round another fringe event, the Peace and Empire Congress held at Glasgow's McLellan Galleries in September 1938, where the Scottish Peace Council worked with the International Peace Campaign and the National Peace Council, to give delegates from Empire countries an opportunity for discussing common problems.[49]

Although modestly sized, the *Workers' Empire Exhibition* was notable enough to attract messages of congratulation from prominent socialist leaders, including independence activist Jawaharlal Nehru, who was to become India's first Prime Minister a few years later.[50] Soon after, it toured to central London, showing for eight days at Friends House on Euston Road alongside a colonial conference and public demonstration with speakers from Leftist groups, including the pacifist Peace Pledge Union, and from the colonies.[51] All these events provided an invaluable focus for the developing anti-colonial agenda in Britain. The *New Leader* reported the outcome of the colonial conference as to appoint a British Centre against Imperialism, and the election of a 'Council of Nineteen', with ten representatives of the Colonial Peoples and nine of British anti-Imperialists.

Under the heading 'Coloured Workers Speak Out', the *New Leader* noted that the keynote was Karl Marx's declaration that 'Labour with a white skin cannot emancipate itself while Labour with a black skin is branded'. This Marxist quotation was printed on a huge banner backing the platform 'showing two muscular workers, one white and the other black, linked by the symbol of their work'.[52] Following the conference a committee was appointed to co-ordinate the struggle against imperialism in the colonies and in Britain, linking the colonial workers' organisations and mobilising support in Britain.[53] These exhibitions in Glasgow created between them a representation of Empire that could be as much kicked against as supported, galvanising individual activists and groups. The *Workers' Empire Exhibition* – a small counter-exhibition – reinforced new anti-imperialist political identities and galvanised oppositional groupings at Britain's end of Empire. It succeeded in creating of the Glasgow *Empire Exhibition* a kind of anti-imperial manifesto, explicit proof of all the things that were manifestly wrong with the Empire.

Countering 'degenerate art' in Britain

I turn now to consider a very different example of a counter-exhibition in Britain, held in the same year: *Twentieth Century German Art* at London's New Burlington Galleries. Its focus was on countering Nazi propaganda and, in particular, that played out through the profusion of '*Schandausstellungen*' (exhibitions of shame) developed in Germany to demonstrate adherence to or rejection of certain people and ideas, to mock artworks, cultures, religions and races considered shameful. Despite the ever-growing knowledge in Britain of Nazi abominations, public opinion remained divided on the question of the Nazi impact up until the outbreak of the Second World War. The *Manchester Guardian*, for example, reported with admiration in April 1934 on the 'magnificent technical display' mounted by the Nazis at a *German People and German Work* exhibition in Berlin, despite this impressive form enveloping displays about 'the old Teutonic tribes', racial science and hygiene, displays against 'mixed marriages' and pro Nazi sterilisation laws, as the newspaper described them.[54] Hitler's adept, direct use of exhibitions as platforms extended to his attendance at events such as the *Schreckenskammer der Kunst* (*Chamber of Horrors of Art*) at Dresden in 1935, with art historian Herbert Read including a photograph of Hitler attending this exhibition in his article in *The Listener* 'Soviet Realism: Art in the USSR'.[55]

The *Entartete Kunst* (*Degenerate Art*) exhibition first staged at Munich's Archaeological Institute from July to November 1937 was the most powerful and chilling of the Nazi 'exhibitions of shame'.[56] Organised by the Reich Chamber of Fine Arts under director Adolf Zeigler and opened by Hitler, the exhibition, gathering 650 artworks considered to reflect symptoms of cultural decline, was part of a concerted campaign that had begun even before the National Socialists came to power in 1933.[57] Artists witnessed their work being systematically removed from German museums, paralleling their attempted erasure from German society, with many being forced to flee from the cities, several ending up in Britain.[58]

While *Degenerate Art* was focused towards creating erasure of the identities of artists and works considered a threat, it was a powerful example of the exhibition as argument. It was paired with another exhibition, the *Grosse Deutsche Kunstausstellung* (*Great German Art* exhibition), held nearby in Munich's new House of German Art, a building described by artist Robert Medley in British journal *Axis* as 'impressive rows of simple columns repetitively asserting "Noble Simplicity and Iron Discipline"'.[59] *Great German Art* opened in July 1937 but this time in order to celebrate art glorifying the Reich, the two exhibitions working as dialectical counterparts: 'good' art (at the House of German Art) versus 'bad' art (at the Archaeological Institute).[60] The contents of *Great German Art* had been solicited by newspaper advertisements calling on all German artist members of the Reich's Chamber of Culture, living at home or abroad, to submit works to provide 'as comprehensive and high quality a survey of contemporary German painting, sculpture and graphic arts as possible'. Although limited guidance was given on subject matter or style, most of the submitted work was figurative and glorified the German nation, Aryan people and its National Socialist leaders.

Aside from whole exhibitions mounted as arguments in Nazi Germany, sections of Nazi exhibitions were also presented in the form of picture books (variously called *Bildbucher*, *Bilderbucher* or *Anschauungsbucher*), recalling the visual-textual interplay of illustrated weeklies and allowing exhibitions to act as propaganda by pursuing particular arguments. Picture books were sometimes called *Anschauungsbucher*, a word choice creating slippage between visual perception (*Anschauung*) and ideology (*Weltanschauung*).[61] Exhibitions of this type, using a combination of text and photographs, were comparable to books in being reproducible and mass produced – a form for the machine age – and also to advertising, with its economical interplay of visual and textual elements. Under Stalin, denunciatory or defamatory exhibitions were also becoming explicit sites of defamation and expurgation.[62]

Twentieth Century German Art at New Burlington Galleries

While exhibitions in Germany had been absorbed into chief propagandist Joseph Goebbels' armoury of propaganda amplifying the Nazi programme, in Britain exhibitions were increasingly structured as rhetoric to present an opposing view. The exhibition of *Twentieth Century German Art*, held at New Burlington Galleries in July 1938, was mounted to counter the vituperative view of Modernist art and the reputational damage meted out by the German exhibitions of shame.[63] While the British government was still pursuing a policy of appeasement, attempting to avoid conflict with increasingly aggressive German and Italian fascist states, the exhibition allowed activists to offer an alternative view, a provocative counter-argument.

The exhibition's original title was due to be *Banned Art*. However, its committee, perhaps fearing reprisals, changed it to the more neutral and descriptive *Twentieth Century German Art*, claiming the show was 'not concerned with the political', rejecting the immediate context by suggesting it had been under discussion for 'ten or fifteen years'.[64] However, correspondence from the organisers shows it as being a direct response to Munich: 'You will remember that there was an Exhibition of so-called "degenerate art" in Germany about a year ago. We have collected works by the same artists, who are no longer allowed to paint and exhibit in their own country, and will be showing them at the New Burlington Galleries'.[65]

Its publicity was less moderate and more agitational in tone, however, inviting people to 'Go and see expelled and banned art. Visit and Support the Exhibition of German 20th Century Art'.[66] Meanwhile, the front cover of the book visitors received with their entrance fee linked it directly with recent exhibitions mounted by the Nazis, announcing it as 'the London opening of the famous Munich exhibition of "degenerate" German art'.[67] This 108-page book, *Modern German Art*, by 'very well-known German art critic' Peter Thoene, pseudonym of prominent Yugoslavian art historian Oto Bihalji-Merin, introduced by Herbert Read, was one of a series of anti-fascist studies published as Pelican Specials.[68]

Britain's first comprehensive survey of modern German art, the New Burlington Galleries exhibition showed sixty-four artists' work, all born or coming to prominence in Germany, almost all branded 'degenerate' by the German regime.[69] Many of the artists shown were, by then, living in England, with works including figurative oil paintings by German Modernists like Max Liebermann and Paula Modersohn-Becker,

Max Beckmann's triptych *Temptation*, Oskar Kokoschka's *Self-Portrait of a Degenerate Artist* and Franz Marc's *Blue Horses*. Organised in collaboration with the Freier Künstlerbund (Free Association of Artists), a federation of exiled German artists living in Paris, the exhibition fundraised for refugee relief, with Spanish painter Pablo Picasso as Patron and painter Augustus John as President.[70] It was the idea of Swiss painter Irmgard Burchard, previously a progressive gallerist in Germany before the rise of Hitler, working with Herbert Read and painter and patron Roland Penrose.[71]

Installed in a conventional fashion on New Burlington Galleries' fourth floor, wall-hung paintings were interspersed with sculptures on podiums. A first gallery was hung primarily with figurative oil paintings by German Modernists, a larger second gallery contained early twentieth-century Expressionist paintings, a small third gallery held abstract and semi-abstract works, a fourth hung with large oils.[72]

John Heartfield, founding member of the Berlin Dada group, was a notable exclusion from *Twentieth Century German Art*. Heartfield's work – overtly political, anti-Nazi photomontages – was well known amongst fellow refugees. It manipulated appropriated, mass-produced images to create satirical new meanings, putting him high on the Gestapo's death list. Heartfield's exclusion from *Twentieth Century German Art* was likely due to the blatantly political nature of his work. He arrived in England from exile in Prague a few months after the exhibition under the auspices of the Artists' Refugee Committee (ARC) formed the same year (as discussed in Chapter 6). The ARC was an idea forged in the network created at *Twentieth Century German Art* by German refugee painter Fred Uhlman, who had settled in London in 1936, and his English wife Diana Croft, who volunteered for and donated to the exhibition, forging contacts that would be crucial to their setting up of the committee.[73]

The New Burlington Galleries exhibition provided the focus for a cultural festival of talks and music: Max Beckmann spoke in German 'Über meine Malerei' ('On My Painting'), referring only obliquely to political events.[74] A music festival included a performance of work by Arnold Schoenberg and a staging of Kurt Weill's *Threepenny Opera*, with lead singing from German soprano Elisabeth Schumann and American actor and singer Paul Robeson, who was on his way home to the US from a pro-Republican tour of Spain.[75] Robeson was photographed in front of Beckmann's oil on canvas triptych, *Temptation* (1936–37), an image widely circulated in the contemporary press.

The exhibition was originally due to be open for three weeks but the run was extended several times, finally closing eight weeks later after twenty thousand visitors.[76] In a statement on 'Limitations and Possibilities' in the book accompanying the exhibition, critic Peter Thoene summarised the show's importance in allowing art to have freedom. Influential *New Statesman and Nation* critic Raymond Mortimer commented, however, that it would be tempting to acclaim the works in the show simply because Hitler had condemned them, but as a critic he said that his duty was to resist this, declaring the works 'extremely bad propaganda' and suggesting visitors might say 'If Hitler doesn't like these pictures, it's the best thing I've heard about Hitler' because 'the general impression made by the Show upon the ordinary public must be one of extraordinary ugliness'.[77]

Despite individual critics such as Mortimer's dislike of the work, the exhibition acted as a powerful anti-Nazi statement, a protest against the rise of Nazism and an act of defiance.[78] And, as hoped, the provocation worked and Hitler responded. Opening the second *Great German Art Exhibition* at Munich's House of German Art in summer 1938, Hitler contrasted the virtues of the semi-classical nudes and paintings of German peasantry with the work of 'Moscow and the Jews' on show at the New Burlington Galleries. He sneered, 'We rejoice that the democracies are opening their progressive doors to these degenerate elements for, after all, we are not vindictive. Let them live, we do not mind! For all we care, let them work – but not in Germany!'[79] Having succeeding in provoking Hitler, AIA members felt vindicated in responding further to his 1938 speech, by creating a poster headed 'Hitler Attacks London Art Exhibition', quoting from Hitler's response to the London exhibition as 'impertinence', attempting further provocation by asking 'Why does Hitler expel artists?' and proposing 'Because Fascism is afraid of those who think, of those who see truth, of those who speak the truth'.[80] AIA members distributed the poster and a call to visit *Twentieth Century German Art* at a demonstration in support of the Spanish Republic. These provocations from exhibitions at New Burlington Galleries were enough to have the gallery specifically singled out for castigation on grounds of its 1938 'anti-German exhibition of "degenerate art"' in the German Gestapo's *Informationsheft GB* (*Information Booklet GB*), compiled as an introductory handbook on Britain for the Nazi occupation troops.[81]

While exhibitions had long been used in Britain for presenting national trade and industry, they had, by the eve of the Second World War, become effective for countering prevailing political beliefs: for documenting and sharing political ideas on the Left in Britain, working in tandem with a series of other media including pamphlets, lectures and congresses. They showed themselves as a formidable focus for anti-imperialist discourses, with counter-exhibitions a potent strategy in Britain for opposing ideas and political positions, drawing inspiration from counter-exhibitionary strategies developed by the French Surrealists in the early 1930s. The next chapter discusses the way in which propaganda exhibitions proliferated as a way of displaying political solidarities in Britain.

Notes

1 Hitler speech at opening of Great Exhibition of German Art, Munich, 10 July 1938.
2 Adam Jolles, *The Curatorial Avant-Garde: Surrealism and Exhibition Practice in France, 1925–1941* (University Park, PA: Pennsylvania State University Press, 2013).
3 Jody Blake, 'The Truth about the Colonies, 1931: Art Indigene in Service of the Revolution', *Oxford Art Journal*, vol. 25, no. 1, 2002, p. 38.
4 Romy Golan, 'The World Fair: A Transmedial Theatre', in Jordan Mendelson et al. (eds), *Encounters with the 30s* (Madrid: Fabrica, 2012), pp. 183–4. Counter-exhibition held from September 1931 to February 1932, across the city at the former Soviet pavilion of the Paris 1925 Expo. See also Romy Golan, *Muralnomad: The Paradox of Wall Painting, Europe 1927–1957* (London: Yale University Press, 2009).
5 Blake, 'The Truth about the Colonies', p. 45.
6 Janine Mileaf, 'Body to Politics: Surrealist Exhibition of the Tribal and the Modern at the Anti-Imperialist Exhibition and the Galerie Charles Ratton', *Anthropology and Aesthetics*, no. 40, Autumn 2001, pp. 239–55.
7 Blake, 'The Truth about the Colonies', p. 45.
8 M. Kuijt, 'Exposing the Colonial Exhibition: Dutch Anti-Colonial Activism in a Transnational Context', *Reinvention: An International Journal of Undergraduate Research*, vol. 12, issue 2, 2019.
9 Alicia Foster, *Radical Women: Jessica Dismorr and Her Contemporaries* (London: Lund Humphries, 2019), p. 73.
10 Frances Haskell, 'Botticelli, Fascism and Burlington House – The "Italian Exhibition" of 1930', *The Burlington Magazine*, vol. 141, no. 1157, August 1999, pp. 462–72, p. 465.
11 Andree Hayum, 'Mussolini Exports the Renaissance: The Burlington House Exhibition of 1930 Revisited', *The Art Bulletin*, vol. 101, no. 2, 2019, pp. 83–108.
12 www.britishpathe.com/video/the-treasure-ship-1 (accessed 17 May 2023).
13 Letter of March 1930 quoted in Hayum, 'Mussolini Exports the Renaissance', p. 88; Clark, *Another Part of the Wood*, pp. 177–86; Haskell, 'Botticelli, Fascism and Burlington House', p. 470.
14 Deborah Sugg Ryan, 'Staging the Imperial City: The Pageant of London, 1911', in Felix Driver and David Gilbert (eds), *Imperial Cities: Landscape, Display and Identity* (Manchester: Manchester University Press, 2003), p. 117–35. Ryan explains how The Festival of Empire of 1911 at Crystal Palace in Sydenham advertised itself as 'a Social Gathering of the British Family', to encourage the 'firmer welding of those invisible bonds which hold together the greatest empire the world has ever known'.
15 Gopal, *Insurgent Empire*, p. 210.
16 Matera, *Black London*, pp. 14–15.
17 For example, see the handbook of the Wembley Exhibition, as quoted by Matera, *Black London*, p. 26.
18 Scott Cohen, 'The Empire from the Street: Virginia Woolf, Wembley, and Imperial Monuments', *Modern Fiction Studies*, vol. 50, no. 1, Spring 2004.
19 'The King Opens the Great Exhibition', *Manchester Guardian*, 23 January 1924.
20 Matera, *Black London*, p. 26.
21 'West Africa and the Empire Exhibition', *West Africa*, 12 April 1924, p. 322.
22 Matera, *Black London*, p. 28.
23 Matera, *Black London*, p. 25. Woodham and Britton discuss two key Wembley protests – one was West African Students' Protest which led to the closure of the African Village, the other was a Working Conditions Campaign. Woodham, 'Images of Africa'; S. Britton, '"Come and see the empire by the all red route!":

Anti-imperialism and Exhibitions in Inter-war Britain', *History Workshop Journal*, vol. 69, no. 1, 2010, pp. 68–89.

24 Virginia Woolf, 'Thunder at Wembley', quoted by Cohen, 'The Empire from the Street', p. 86.

25 According to Matera, but he gives no further details of this 'counter exhibition'; *Black London*, p. 27.

26 Gopal, *Insurgent Empire*, p. 370.

27 Matera, *Black London*, p. 16; Gopal, *Insurgent Empire*, p. 343.

28 Matera, *Black London*, pp. 15, 23, 65–6; Gopal, *Insurgent Empire*, p. 329.

29 An event I know from written records. WMRC MSS.292/963/2 Abyssinia/ Ethiopia: Italian invasion, 1935–1944 'Exhibition of original paintings, 1936'.

30 Alistair Borthwick, *The Empire Exhibition Fifty Years On* (Edinburgh: Mainstream Publishing, 1988), p. 8.

31 Barbara Bush, *Representing Empire* (Abingdon: Routledge, 2006), p. 183; Kinchin and Kinchin, *Glasgow's Great Exhibitions*, p. 129; MacKenzie and McAleer, *Exhibiting the Empire*, p. 9.

32 Bush, *Representing Empire*, p. 183.

33 Woodham, 'Images of Africa', p. 27; J. M. Richards, 'Glasgow 1938: A Critical Survey', *Architectural Review*, July 1938.

34 Editorial, *Forward*, 5 June 1938.

35 Gopal, *Insurgent Empire*, p. 376, quotes Eric Williams, *The Negro in the Caribbean* (Washington: The Associates in Negro Folk Education, 1942).

36 Gopal, *Insurgent Empire*, p. 333. George Padmore, 'Labour Unrest in Jamaica', *International African Opinion*, vol. 1, no. 1, July 1938; 'Anti-imperialist Exhibition in Glasgow', *International African Opinion*, vol. 1, no. 1, July 1938.

37 Bob Crampsey, *The Empire Exhibition of 1938: The Last Durbar* (Edinburgh: Mainstream, 1988); *Forward*, Saturday 17 September 1938; *The New Leader*, Friday 9 September 1938, p. 3.

38 *Scottish Trades Union Congress Souvenir* (Glasgow: Scottish TUC, 1938), p. 16.

39 Crampsey, *The Empire Exhibition of 1938*, p. 68 quoted in A. Peat, 'Scottish Internationalisms at the 1938 Empire Exhibition: Between Britain, Europe, and Empire', *Open Library of Humanities*, vol. 6, no. 1, 2020, p. 21.

40 This was not the first time Robeson's racist treatment in Britain had hit the headlines. In 1929, the *Times* had reported that London's Savoy Grill had refused to serve Robeson, which caused a major scandal. But for most of the city's Black residents this type of incident was all too familiar. Matera, *Black London*, p. 37, quoting Martin B. Duberman, *Paul Robeson: A Biography* (New York: Ballantine Books, 1989), pp. 123–4 and Ade Ademola, 'Colour Bar Notoriety in Britain', in Nancy Cunard (ed.), *Negro: Anthology Made by Nancy Cunard* (London: Wishart and Co., 1934), pp. 556–7.

41 Arthur Ballard, 'We Are Going to Run an Anti-Empire Exhibition!', *The New Leader*, Friday 3 June 1938, p. 8.

42 Held 13–27 August 1938.

43 'The "Other" Exhibition', *New Leader*, 12 August 1938.

44 *Forward*, 20 August 1938.

45 BL collection, YD.2011.b.1781, 'Come and See the Empire by the All Red Route'; Britton, '"Come and see the empire"', pp. 68–89.

46 University of Exeter Special Collections Correspondence between Ethel Mannin and Christopher Walker EUL MS 452, 'Success of Anti-Imperialist Exhibition', *The New Leader*, 19 August 1938. Also report in *New Leader*, 23 September 1938.

47 Ethel Mannin, 'Empire with the Lid Off: This Insidious Imperial Propaganda – Fascism Under the Union Jack', *Forward*, 13 August 1938, p. 6.

48 'Empire Special', *New Leader*, 29 April 1938, p. iv.

49 *Forward*, 1 October 1938.

50 *New Leader*, Friday 19 August 1938.
51 Groups included the ILP, LP, PPU; *New Leader*, 13 January 1939, p. 2. It was shown from Friday 20 January 1939.
52 *New Leader*, Friday 27 January 1939, p. 4.
53 *New Leader*, Friday 10 February 1939.
54 'A "German People and German Work" Exhibition: From Teutonic Times to Modern Days Display in Berlin of Nazi Ideas on Many Questions', *Manchester Guardian*, 23 April 1934, p. 13.
55 *The Listener*, 2 October 1935, p. 579.
56 After Munich (where it showed from 19 July to 30 November 1937), it toured to Berlin from February to May 1938, to Leipzig from May to June and to Düsseldorf from June to August.
57 Tymkiw, *Nazi Exhibition Design*; Bohm-Duchen, *Insiders Outsiders*, p. 24.
58 Stephanie Barron and Sabine Eckmann (eds), *Exiles & Emigrés: The Flight of European Artists from Hitler* (New York: Harry N. Abrams, 1997), pp. 13–14.
59 Robert Medley, 'Hitler's Art in Munich', *Axis*, Early Winter 1937, p. 28.
60 Barron and Eckmann, *Exiles & Emigrés*, p. 13.
61 Tymkiw, *Nazi Exhibition Design*, p. 135.
62 Adam Jolles in 'Stalin's Talking Museums', *Oxford Art Journal*, vol. 28, no. 3, 2005, pp. 429–55 and Jolles, *The Curatorial Avant-Garde*.
63 Held from 8 to 30 July 1938. Discussed in Anthony Penrose's essay in Bohm-Duchen, *Insiders Outsiders*, p. 177. The exhibition poster (reproduced in Bohm-Duchen, p. 221) states the patrons included: Augustus John (President), the Bishop of Birmingham, Sir Kenneth Clark, Lord Ivor Churchill, Pablo Picasso, Herbert Read and Sir Michael Sadler. Penrose states that the exhibition lost money but that two at Penrose's London Gallery, supplemented by generous donations, helped raise funds for the same cause. TGA Ewan Phillips collection has photographs of the interior of Twentieth Century German Art at the galleries, London 1938.
64 Cordula Frowein, 'German Artists in War-Time Britain', *Third Text*, vol. 15, Summer 1991, p. 50.
65 WMRC, TUC archives, MSS.292/943/3/57 Exhibition of Twentieth Century German Art: letter from Irmgard Burchard, Hon. Organiser to Sir Walter Citrine, General Secretary of the Trades Union Congress, with publicity leaflet and card, 9 Jul 1938.
66 Radford, *Art for a Purpose*, p. 104.
67 Peter Thoene, *Modern German Art* (London: Pelican, 1938), front cover.
68 These Penguin Specials included E. Mowrer's *Germany Puts the Clock Back*, G. T. Garratt's *Mussolini's Roman Empire* and the Duchess of Atholl's *Searchlight on Spain*.
69 Wasensteiner observes that 62 of the 64 appeared on the list of Entartete Kunst here: www.geschkult.fu-berlin.de/e/db_entart_kunst; Lucy Wasensteiner, *The Twentieth Century German Art Exhibition: Answering Degenerate Art in 1930s London* (Abingdon: Routledge, 2018). Although 271 works are listed in the catalogue, at least 315 were exhibited and more than 600 were brought to London. Wasensteiner has traced the origins and onward journey of many of the works.
70 Freier Künstlerbund (FKb), whose purpose was to inform the public about opposition to the German government in exhibitions among other forms, organised the exhibition *Five Years of Hitler's Dictatorship* the same year in Paris as a retort to *Degenerate Art* held at the Paris Maison de la Culture in November 1938, as Keith Holz states in Bohm-Duchen, *Insiders Outsiders* p. 214.
71 Helping Penrose and Read with *Twentieth Century German Art* was Ewan Phillips, who would later become the director of the ICA from 1948.
72 Wasensteiner, *The Twentieth Century German Art Exhibition*, pp. 26–7.
73 As Anna Muller-Harlin notes in Bohm-Duchen, *Insiders Outsiders*, p. 190.

74 According to Barron and Eckmann, *Exiles & Emigrés*, p. 17, they note that Beckmann had refused to participate in group exhibitions organised by German exiles in Paris but agreed to speak at the New Burlington Galleries exhibition opening.
75 Bohm-Duchen, *Insiders Outsiders*, p. x. 'Today's Arrangements ... Mr Paul Robeson Singing at the Exhibition of Twentieth Century Art, New Burlington Galleries', *Manchester Guardian*, 3 August 1938, p. 18.
76 Wasensteiner disputes that Read did much of the exhibition's legwork. The exhibition catalogue lists Read as the Chairman and organisers as Irmgard Burchard, Dunstan Curtis, John Harrison, Brian Howard, The Rt. Hon. The Earl of Listowel and Roland Penrose, p. 5.
77 *New Statesman and Nation*, 16 July 1938.
78 Wasensteiner quotes *Twentieth Century German Art*, p. 6. She claims that as a gesture it was ultimately futile, failing either to make an effective anti-Nazi statement or to convince its British audience of the merits of 'degenerate' art.
79 David Aaronovitch, *The Times*, 9 June 2018, p. 8.
80 Poster is TGA AIA collection 7043/17/2. Cited by Morris and Radford, *The Story of the Artists International Association*, p. 50 and reproduced by Frowein, 'German Artists', pp. 52, 49.
81 Sibyl Oldfield, *The Black Book* (London: Profile, 2020), p. 33, compiled from May to July 1940. *News Chronicle*, 11 July 1938: 'Herr Hitler complained of the impertinence of a London Exhibition of pre-Nazi German artists, which he likened to the staging of the opposition Reichstag Fire Trial'. *Scotsman*, 11 July 1938: 'Modern art had no place in Nationalist Socialist Germany. He intimated that modern artists who had not yet left Germany should do so now'.

6

Exhibitions as solidarities

'Artists', wrote critic Myfanwy Evans in 1937, 'were in the middle of a thousand battles: Hampstead, Bloomsbury, surrealist, abstract, social realist, Spain, Germany, heaven, hell, paradise, chaos, light, dark, round, square'.[1] Evans described artists' lives being shaped by a series of intersecting formal, stylistic and political clashes. Some of these imaginative 'battles' they fought collectively, through exhibitions. Increasingly these exhibitions were being taken out of spaces of art and mounted in public spaces in direct response to political developments.[2] What bonds and ideas did such battles encompass and how might exhibitions be a way of signalling particular alignments?

Philosopher Sally J. Scholz, in her work on political solidarity, discusses solidarity as denoting a relation or unity between people, either emphasising the cohesiveness or fellow feeling of a group, the shared project that informs the unity, or accentuating obligations to fellow citizens by virtue of membership in a state.[3] 'Political solidarity', Scholz states, unites individuals based on their shared commitment to a political cause in the name of liberation or justice and in opposition to oppression or injustice. Exhibitions operated as a focus for political organising, uniting individuals around political causes and issues.

This chapter traces how exhibitions mounted in Britain from the 1930s became the focus for solidarities across a series of conditions outlined by Scholz: acting as a shared project and denoting fellow membership of particular groupings. Exhibitions were a conduit for such solidarity, providing a voice to newly arriving refugees and becoming the focus for conviviality between individuals. They allowed those rebuilding their lives in new locations to form cohesive social contacts and they enabled artists to signal connections with causes and people near and far, across time. The mobilisation of exhibitions as a productive form by newly arriving refugees was directly related to their circumstances: their vilification by the Nazis, alienation from previous contacts and dislocation from home. In their coming together, exhibitions offered a space of solidarity and a platform for sharing messages of hope. Exhibitionary solidarities were articulated as a gesture of friendship across time and space, fulfilling social scientist Émile Durkheim's idea of solidarity as a necessary component of a functioning civilisation and of a fulfilling human life.[4]

In becoming tethered in this moment to the itinerant conditions in which refugees were creating, the form of the exhibition itself became displaced and nomadic: created outside of galleries and museums, moving between public sites, from shops to bombsites, in keeping with art historian T. J. Demos's proposition of 'modernity-as-exile'.[5] Being more dependent on arguments than on particular works, information exhibitions operated well for artists and designers in exile, being made up of reproducible elements, principally as photographs and text. Czech-born philosopher Vilém Flusser described coming to London at the start of the war as being 'overcome by that strange dizziness of liberation and freedom which everywhere characterises the free spirit' and exile as 'an ocean of chaotic information'. 'If he is not to perish, the expellee must be creative', Flusser observed, going on to describe 'creation as a dialogic process, in which either an internal or external dialogue takes place. The arrival of expellees in exile evokes external dialogues, and a beehive of creativity spontaneously surrounds the expellee. He becomes the catalyst for the synthesis of new information'.[6] This process of synthesising new information in exhibitionary form became a significant project for newly arriving artists and designers in Britain.

We can consider such exhibitions as solidarities as 'the infrapolitics of the powerless', to quote social anthropologist James C. Scott's phrase in *Domination and the Arts of Resistance* describing the way powerless people use cultural forms as vehicles of resistance and insubordination to create and defend social space.[7] In this sense, exhibitions provided a space from which newly arriving people and political activists could safely mount a riposte, by signalling solidarities in form, messaging and mode of production. The manner in which these exhibitionary alignments from the British Left could be played out was not singular or settled and there were multiple, fractured directions for solidarities. From the mid-1930s, the British Left looked in several directions, with alignments forming with the Spanish Republicans against Franco and with the anti-fascists against German Nazis and Italian Fascists, while the role of the Soviet Union in its relationship with Britain was also the subject of tense debate, as I signal in this chapter.

This chapter discusses the many exhibitions mounted in solidarity with Spain, several supporting the plight of refugees to Britain and others in solidarity with the Soviet Union by a series of groups including the

Artists International Association (AIA) and the Free German League of Culture (FGLC). These examples cover the period starting in the late 1930s with the Spanish Civil War and end in the midst of the Second World War, after the Soviet Union had joined the Allies. Exhibitions enabled new arrivals to build cultural capital and to re-assert themselves following their attempted erasure under the Nazis. They also offered opportunities for women in their continuing struggle for equality.

Exhibitions as solidarities with Spain

Exhibitions became a significant focus for British Leftist resistance against the nationalist position in Spain and an important form from which to demonstrate solidarity with the Republicans in the Spanish Civil War. This war, a nodal point of the 1930s in Europe, tested the alliances and allegiances of leaders across the continent and challenged the pacifist principles of Leftist anti-war groups.[8] Over 2,500 British men and women travelled to Spain to assist the Republicans by joining International Brigades or with paramilitary units established by the Communist International.[9]

The first British civilian to be killed fighting on the government side in Spain in August 1936 was young AIA member Felicia Browne.[10] Browne had attended the Slade School before joining the Communist Party of Great Britain in 1933. While on a drawing excursion to Spain, where she sketched soldiers and local people, Browne joined the Republican militia. Her death activated other artists and writers in Britain, who responded in a series of ways, through protests, posters, banners, fundraising campaigns, writings, artworks and exhibitions.[11] News of Browne's death inspired other artists to align themselves with the International Brigades.[12] Priscilla Thornycroft, studying at the Slade School in London at the outbreak of the Spanish war, was inspired to begin campaigning.[13] Alongside newspaper coverage, the AIA's retrospective exhibition *Drawings of Felicia Browne* at London's 46 Frith Street, shown only weeks after her death in October 1936, drew attention to the cause.[14] Proceeds from commemorative booklet sales amounted to £200 for the Spanish Medical Aid Committee.[15]

Browne's death became emblematic for many artist-activists, marked in historical pageants such as the Communist Party of Great Britain's 1930s series. The first pageant, referred to in the *Daily Worker* as an 'English History Demonstration', was staged in September 1936.[16] At the end of the procession, the final point in the narrative was a 'great portrait' of Felicia Browne, flanked by a Red Flag and the Spanish Flag. Browne's portrait fitted within a tradition, which the *Daily Worker* described as 'leading down the centuries, directly to the struggle which is going on even more intensified to-day'. Political activists recognised the potency of acknowledging historical continuities and of collapsing time in order to bring past struggles closer to the present.[17] Echoing past struggles enabled activists to reinforce and amplify their current engagements.

Artists Aid Spain

While news of British casualties in Spain caught artists' imaginations, attacks by British Union of Fascists members on work by Jewish artists in Britain made the situation more acute. Two sculptures by Jewish artist Jacob Epstein were daubed with anti-Semitic slogans in October 1936. Artists' support from Britain for the Republican faction continued to grow, with posters and exhibitions acting as a focus for fundraising. In December 1936, the AIA mounted one of its most successful exhibitions in support of the Spanish Civil War: *Artists Aid Spain*.[18] Organised by AIA sub-group the Women of the AIA, assembled in just two weeks and held in London, it aimed to raise medical funds for Spain. It featured works donated by leading artists working in a range of styles: Ben Nicholson, Paul Nash, Eric Gill, Duncan Grant, Vanessa Bell, Eric Ravilious, Edward Bawden and Moholy-Nagy, who was living in London. Exhibition proceeds went to buy an 'Artists' Ambulance', intended for use by the International Column established to defend Madrid.

After being forced to cut short a visit to Spain, Anglo-Jewish painter David Bomberg, born to Polish Jewish immigrant parents and raised in London's East End, joined the cause from February 1937. Bomberg submitted a resolution to the London Group proposing they affiliate with the AIA and Surrealist groups in their support of anti-fascism in politics and art; that funds be granted for Spanish Medical Aid; and that 'honorary membership of the London Group be extended to certain left-wing poets and writers'.[19] Several British organisations formed to fundraise for Spanish aid used exhibitions as vehicles. The Communist-aligned Spanish Medical Aid Committee, set up at 24 New Oxford Street, London, held events including a 'St Pancras and Holborn Spain' week on London's Euston Road, with a fundraising exhibition showing how the Spanish government was 'building a new Spain'. Accompanying talks were given by people billed as 'Just back from Spain and the Refugee Camps of France' and Labour MP Ellen Wilkinson.[20]

Exhibitions continued to be at the heart of British fundraising efforts for Spain. In March 1937 the National Joint Committee for Spanish Relief (NJCSR), an umbrella organisation set up to bring together a number of Spanish aid organisations, including Communist-aligned ones, held an exhibition and sale of rare first editions, autographed books and authors' manuscripts. Located at Foyle's Gallery on London's Charing Cross Road, proceeds went to the Spanish Medical Aid Committee.[21] Contributors included prominent novelists H. G. Wells and Rebecca West. The month after, the NJCSR mounted another fundraising exhibition (whose existence

I know about from descriptions not images). This was led by the AIA and organised by artist Peggy Angus, held at St George's Gallery in London's Hanover Square, showing work by Duncan Grant, E. McKnight Kauffer, Paul Nash and Bernard Leach.[22] Ex-Byam Shaw Art School student and AIA member Felicity Ashbee contributed her work to the NJCSR, making three posters for the organisation captioned 'They face famine in Spain'.[23]

Spanish Aid exhibitions proliferated. The Spanish Exhibition Committee, formed in a London Inn of Chancery to represent groups wanting to send medical aid to Spain, was yet another body using exhibitions for fundraising.[24] A leaflet announced a 'SPANISH EXHIBITION ... In support of anti-fascist Spain', held at London's 36 Ludgate Hill, EC4 from February to March 1937, to raise money for 'Relief Work in Spain' and opened by a representative from the Spanish Embassy. The Committee's headed paper named an impressive and influential group of members drawn from Leftist politics and the arts: artist Duncan Grant, poet Francis Meynell, political activist Fenner Brockway and Spanish Civil War correspondent John Langdon-Davis.[25] Evidence of how the Committee's exhibitions looked is scarce, pieced together through text flyers and modest written advertisements. Emphasising the documentary form of the exhibition, one leaflet promised visitors to 'see in posters, photographs and pictures the background and causes of the war, Franco's allies, the International Brigade, care of the refugees, the work of reconstruction in schools, hospitals and factories'.[26] The exhibition functioned in giving visual form and image to situations otherwise far removed from British life.

Spain: The Child and the War, organised by the Holborn and W. C. London Committee for Spanish Medical Aid and held in October 1937 at the showrooms of toymakers Paul and Marjorie Abbatt on Wimpole Street, showed Spanish drawings by children aged nine to sixteen previously shown in Valencia. The display was accompanied by an illustrated pamphlet describing help going to children in Republican Spain.[27] The Abbatts, well known as patrons of leading Modernist designers, were close friends of architect Ernö Goldfinger, who had designed their shop interior.[28] In the same year as *Spain: The Child*, the Abbatts organised a section on 'The Child' for the British Pavilion at the Paris International Exposition, 1937, with Goldfinger as designer.

Exhibitions as interventions: the AIA's *Exhibition for Peace, Democracy and Cultural Development*

Under Prime Minister Neville Chamberlain, the British government's policy towards Spain was one of sympathy but 'non-intervention', even if this risked the downfall of Spanish democracy. In the face of this, the British Surrealist Group developed an idea of exhibitions operating as interventions and demonstrations of their alignments; bringing poets, artists and intellectuals together, to be exposed to a set of ideas that would initiate an emotional response to change or subvert minds. This strength of feeling from the Surrealists challenged the AIA's own anti-war stance and assumption of pacifism.[29]

For three days in April 1937 the AIA, working with the Society of Industrial Artists (SIA) and the British Surrealist Group, held the First British Artists Congress at London's Conway Hall addressing the theme of 'Unity of artists for peace, democracy and cultural freedom', accompanied by an exhibition at 41 Grosvenor Square from April to May 1937, fundraising and showing support for 'Peace, Democracy and Cultural Development'. The vast exhibition, containing over one thousand artworks, brought together painters and sculptors of many styles, to prove 'the broad aesthetic basis of the membership'.[30] The intersection of activities, which acted principally as a form of solidarity, makes clear the sociability and conviviality of these events, despite their deadly serious focus.

Marking the same occasion, sculptor Henry Moore designed a manifesto-style pamphlet for the British Surrealist Group, printed by Farleigh Press, to draw Congress and Exhibition visitors' awareness to the British government's policy of non-intervention in Spain (Figure 6.1). 'On the occasion of the Artists' International Congress and Exhibition WE ASK YOUR ATTENTION', it announced, going on to set out a lengthy argument for the necessity of British government intervention: 'IF only in self-defence we must END ALL FORMS OF NON-INTERVENTION, INTERVENE IN THE FIELD OF POLITICS, INTERVENE IN THE FIELD OF IMAGINATION' (as per the original formatting). It is evident in this Surrealist rhetoric that exhibitions were being imagined as a materialisation of the end of the policy of non-intervention, themselves becoming interventions in debate. The leaflet continued: 'Economic justice is the first object of our intervention, but we demand also that vindication of the psychological rights of man, the liberation of intelligence and imagination. INTERVENE AS POETS, ARTISTS AND INTELLECTUALS BY VIOLENT OR SUBTLE SUBVERSION AND BY STIMULATING DESIRE' (original emphases).[31]

The foreword to the guide for the *Exhibition for Peace, Democracy and Cultural Development* stated that the AIA 'does not in any sense stand for a uniformity of expression: for the dragooning of artists or the sinking of individuality', suggesting their programme had a clear focus on solidarity around political questions but a lack of clarity or dogma around how to unite visually. Instead, the AIA's exhibition demonstrated, they said, 'the richness and variety of the work that can be produced in the vanguard, thereby holding out a promise which isolation and defeatism themselves must salute'.[32] This richness and variety was reflected in the range of styles and approaches taken by exhibiting artists, separated into an abstract room, including work by members of the Circle group, as well as Kandinsky and Léger.[33] In its ambition to act as a stylistic and thematic

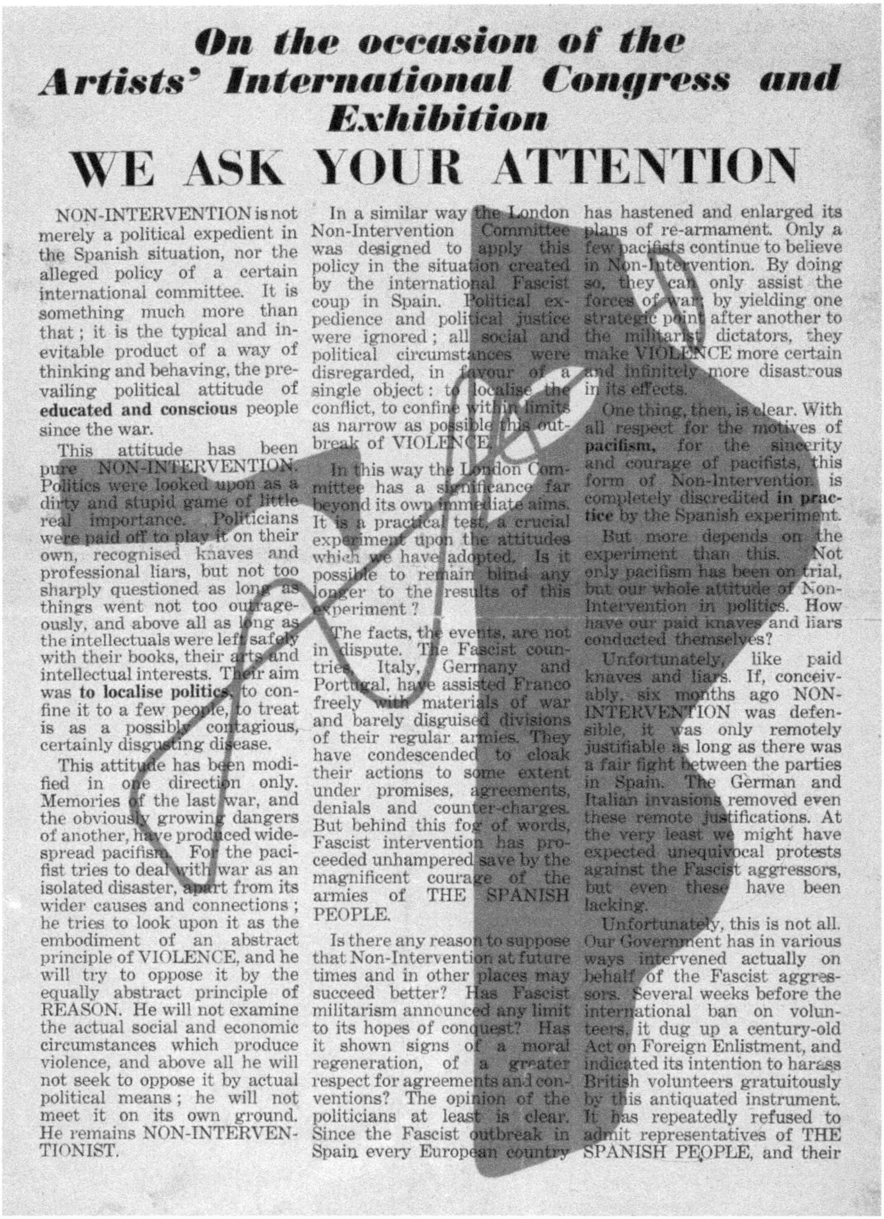

On the occasion of the Artists' International Congress and Exhibition

WE ASK YOUR ATTENTION

NON-INTERVENTION is not merely a political expedient in the Spanish situation, nor the alleged policy of a certain international committee. It is something much more than that; it is the typical and inevitable product of a way of thinking and behaving, the prevailing political attitude of **educated and conscious** people since the war.

This attitude has been pure NON-INTERVENTION. Politics were looked upon as a dirty and stupid game of little real importance. Politicians were paid off to play it on their own, recognised knaves and professional liars, but not too sharply questioned as long as things went not too outrageously, and above all as long as the intellectuals were left safely with their books, their arts and intellectual interests. Their aim was **to localise politics**, to confine it to a few people, to treat is as a possibly contagious, certainly disgusting disease.

This attitude has been modified in one direction only. Memories of the last war, and the obviously growing dangers of another, have produced widespread pacifism. For the pacifist tries to deal with war as an isolated disaster, apart from its wider causes and connections; he tries to look upon it as the embodiment of an abstract principle of VIOLENCE, and he will try to oppose it by the equally abstract principle of REASON. He will not examine the actual social and economic circumstances which produce violence, and above all he will not seek to oppose it by actual political means; he will not meet it on its own ground. He remains NON-INTERVENTIONIST.

In a similar way the London Non-Intervention Committee was designed to apply this policy in the situation created by the international Fascist coup in Spain. Political expedience and political justice were ignored; all social and political circumstances were disregarded, in favour of a single object: to localise the conflict, to confine within limits as narrow as possible this outbreak of VIOLENCE.

In this way the London Committee has a significance far beyond its own immediate aims. It is a practical test, a crucial experiment upon the attitudes which we have adopted. Is it possible to remain blind any longer to the results of this experiment?

The facts, the events, are not in dispute. The Fascist countries, Italy, Germany and Portugal, have assisted Franco freely with materials of war and barely disguised divisions of their regular armies. They have condescended to cloak their actions to some extent under promises, agreements, denials and counter-charges. But behind this fog of words, Fascist intervention has proceeded unhampered save by the magnificent courage of the armies of THE SPANISH PEOPLE.

Is there any reason to suppose that Non-Intervention at future times and in other places may succeed better? Has Fascist militarism announced any limit to its hopes of conquest? Has it shown signs of a moral regeneration, of a greater respect for agreements and conventions? The opinion of the politicians at least is clear. Since the Fascist outbreak in Spain every European country has hastened and enlarged its plans of re-armament. Only a few pacifists continue to believe in Non-Intervention. By doing so, they can only assist the forces of war; by yielding one strategic point after another to the militarist dictators, they make VIOLENCE more certain and infinitely more disastrous in its effects.

One thing, then, is clear. With all respect for the motives of **pacifism,** for the sincerity and courage of pacifists, this form of Non-Intervention is completely discredited **in practice** by the Spanish experiment.

But more depends on the experiment than this. Not only pacifism has been on trial, but our whole attitude of Non-Intervention in politics. How have our paid knaves and liars conducted themselves?

Unfortunately, like paid knaves and liars. If, conceivably, six months ago NON-INTERVENTION was defensible, it was only remotely justifiable as long as there was a fair fight between the parties in Spain. The German and Italian invasions removed even these remote justifications. At the very least we might have expected unequivocal protests against the Fascist aggressors, but even these have been lacking.

Unfortunately, this is not all. Our Government has in various ways intervened actually on behalf of the Fascist aggressors. Several weeks before the international ban on volunteers, it dug up a century-old Act on Foreign Enlistment, and indicated its intention to harass British volunteers gratuitously by this antiquated instrument. It has repeatedly refused to admit representatives of THE SPANISH PEOPLE, and their

6.1 'On the occasion of the Artists' International Congress and Exhibition WE ASK YOUR ATTENTION', pamphlet designed by artist Henry Moore for the British Surrealist Group to draw awareness to the British government's policy of non-intervention. Published by Farleigh Press. Image reproduced by permission of the Henry Moore Foundation.

broad church, the exhibition lacked a clear installation strategy. It was criticised as overcrowded, visually incoherent and downright poor. Sometime AIA member and painter William Townsend wrote in his private diary of the 'monstrous' and 'depressing' show, while *Left Review* dismissed Surrealist works as having 'the vagueness and chaos of anarchism' and the 'worker artists' as having poor style and poorly chosen subjects.[34] A packed 'Surrealist' room included work by prominent artists who were both directly and indirectly linked with the movement: Ernst, Tanguy, Miró, Klee, Picasso, Man Ray, Dalí, Delvaux and Giacometti. The archival records of the exhibition I have seen do not clarify how it was hung or captioned, or, indeed, the role contributing painters played in its coming together.

The AIA at the Paris International Exposition

While Germany and the Soviet Union were sizing each other up through pavilions set on either side of the central axis of the Exposition Internationale des Arts et des Techniques dans la Vie Moderne, held in Paris in 1937, Britain's national pavilion was critically disdained for being conservative, whimsical and old-fashioned (as discussed in Chapter 1). Elsewhere at Paris, British artists engaged more directly with increasingly volatile international politics. Members of the AIA were invited to create a demonstration in the form of a League of Nations room and an International Peace Campaign room at a small pavilion for 'Peace Democracy and Cultural Freedom'. These initiatives were related to the International Peace Campaign headed by British statesman Lord Cecil (who had spearheaded the formation of the League of Nations as a vehicle for peace). AIA artists worked on these peace exhibitions, with funding support from the French Popular Front peace organisation. AIA contributors included Misha Black, James Holland, Elizabeth Watson, Betty Rea and Nan Youngman.

The AIA artists created a series of painted slogans and photo-based murals combining peace statements such as '*Enthusiasme Force Efficacite*' ('*Enthusiasm Force Efficacy*') and '*Une Idee Fait Son Chemin*' ('*An Idea Makes Its Way*'). Maps and statistics explained the role of the League of Nations in maintaining world peace.[35] Dynamic shapes directed visitors' attention. This combination of media allowed them to work quickly, cheaply and on site, rather than having to find space elsewhere to pre-prepare. Photographs allowed them to represent their struggle more directly and with greater impact. These contributions allowed the AIA to express public solidarity with others with the same beliefs in a highly visible, international context. The AIA was appropriating space, in the Lefebvrian sense, in a context otherwise dominated by jostling national interests. Elsewhere at the Paris exhibition, totalitarian governments were moving towards more traditional means of representation including monumental paintings, tapestries and mosaics.[36]

Guernica in Britain

The AIA's painted murals and photomurals had formal and material qualities in common with the work shown at the Spanish Republican Pavilion, which had been formulated, in the increasingly tumultuous context of the Spanish Civil War, as international propaganda for the Republic, an invitation to join in solidarity with a government under attack. The building, devised by architects Josep Lluís Sert and Luis Lacasa, had extensive internal and external photographic murals by leading photomural artist Josep Renau, incorporating striking slogans and instructive pictograms; a systematic use of the photographic representation of Spain to serve the needs of a political programme, introducing visitors to the government's concerns.[37] Despite Renau's much-admired work, the most widely celebrated work of the Paris exhibition and of the Spanish Pavilion was Picasso's monumental black and white oil mural *Guernica*, evoking the brutal aerial bombing of the Basque city of Guernica by German aircraft at the request of Franco's Nationalist army. This painting, commissioned by the Spanish Republican government, arrested visitors' attention near the Spanish Pavilion's entrance. Works by Picasso and Renau had much in common, stretching floor to ceiling, being of equal width and using the same black and white tonal range, uniting them as structural forms of montage. Although not hung adjacent to the photomurals, Picasso's painting was experienced as part of the same visual continuum.[38]

After the closure of the International Exposition, Picasso sent *Guernica* on the road, believing in its power as political message and its capacity to move and activate people. Experienced directly in exhibitions as opposed to small, grainy reproductions in illustrated magazines or black and white newsreels, it could be a travelling focus for building international solidarities. Showing first in Norway, *Guernica* went on to visit twenty cities, before finally being installed at MOMA New York at the end of 1940.[39] Artist Roland Penrose toured *Guernica* round England, sponsored, as he explained, 'by a strong committee of left-wing politicians, scientists, artists and poets', under the auspices of the NJCSR. Penrose's intention was essentially pacifist: to 'draw attention to the horrors of war' and the way in which Britain's non-interventionist policy was allowing Franco a free rein to commit atrocities.[40] Details of the mode of *Guernica*'s installation are consistently eclipsed by the contemporary interest in the artwork's relative effectiveness as pacifist propaganda, but it is clear that Roland Penrose believed fervently not only in the message of *Guernica* but in its power and immediacy when witnessed at first hand. For some the painting was ineffective as a political statement, however. Philosopher Jean-Paul Sartre asked whether *Guernica* would win even one heart to the Republican cause, given its incomprehensibility, while art historian Anthony Blunt opined that the painting was 'not an act of public mourning but the expression of a private brainstorm which gives no evidence that Picasso had realised the political significance of *Guernica*'.[41] Herbert Read rejected Blunt's suggestion that Picasso was politically naive, describing *Guernica* as 'a monument of protestation', 'a monument to destruction, a cry of outrage and horror amplified by the spirit of genius'; 'it is only when the widest commonplace is infused with the intensest passion that a great work of art, transcending all schools and categories, is born; and being born, lives immortally', defending the work both as politically engaged and great art, conditions he felt were rarely satisfied together.[42]

The British venues chosen to host the mural in Britain give an insight into the way its tour fitted with established patterns for showing politically engaged artworks. In London, *Guernica* went to New Burlington Galleries and Whitechapel Art Gallery, while in Manchester it was mounted at a car showroom. At New Burlington Galleries, in October 1938, *Guernica* was displayed alongside sixty of Picasso's preparatory paintings, sketches and studies. These, Penrose believed, made 'a more direct appeal than did the picture itself, which demanded a greater effort to be understood' and attracted lots of attention and three thousand visitors but not many funds for the NJCSR.[43] Patrons offering financial support included *News Chronicle* editor Gerald Barry (who would become Director-General of the Festival of Britain a decade later), socialist politician Fenner Brockway, writers E. M. Forster and Virginia Woolf, designers Ashley Havinden and E. McKnight Kauffer.

In January 1939 *Guernica* travelled to Whitechapel Art Gallery, to be shown alongside Spanish War documentary films and workshops. Labour Party leader Clement Attlee was photographed in front of it, providing a succinct visual means of signalling his support for and solidarity with the International Brigades fighting with the Republicans in Spain (Figure 6.2). Roland Penrose, painter Julian Trevelyan and art critic Eric Newton gave tours of *Guernica* at Whitechapel.[44] Many visitors came: over a fortnight at Whitechapel the artwork attracted twelve thousand visitors.[45] From Whitechapel, *Guernica* travelled north to a Manchester motor showroom, under the aegis of fundraising group Manchester Foodship for Spain, who explained that 'its awful symbolism portrays the ruins of human intelligence and human kindness. It is a damning commentary on War'. The *Manchester Evening News* reported 'human faces and lifelike animals distorted with agony and exterminated by ruthlessness', showing 'the anguish of destruction'. They concluded, 'No-one could fail to be impressed by a tremendous work which, more than any words, condemns the crime of war'.[46] *Guernica*'s journey through Britain enabled efforts to raise money and awareness of the war in Spain.[47] Moreover, it

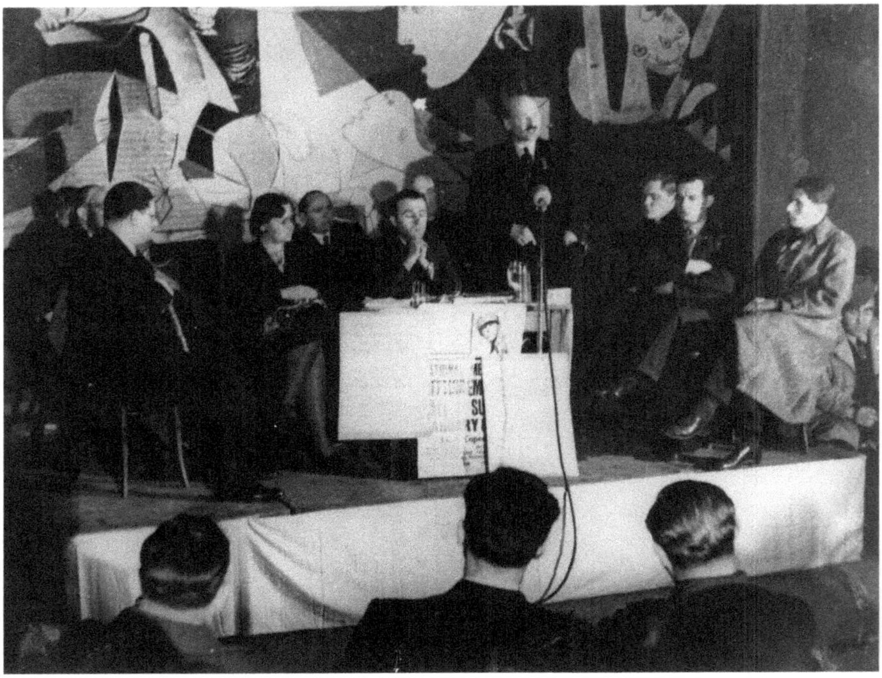

6.2 Labour Party leader Clement Attlee at Whitechapel Art Gallery, speaking in front of Picasso's *Guernica* in January 1939. Uncredited photographer.

demonstrated exhibitions as useful vehicles for communicating politics and enabling people to signal allegiances and solidarities.

Offering solidarity to artists under persecution: the Artists' Refugee Committee

Many German and Austrian artists had fled to Prague by 1938 but, as the Nazis became poised to take over Czechoslovakia, these fugitives needed a new sanctuary. Through international cultural networks, anti-Nazi German and Austrian artists who had fled to Prague but who now sought a passage to Britain made contact with Roland Penrose.[48] One such artist was Czech painter and cartoonist Josef Čapek, a left-wing political activist and part of the Oskar-Kokoschka-Bund (OKB) group named after the Austrian painter in exile in Prague. Čapek had written to political activist Margaret Gardiner appealing for help. Gardiner, who had earlier founded workers' group For Intellectual Liberty, contacted Penrose, her neighbour in

London's Hampstead. Penrose responded by inviting neighbours to lunch to discuss how they could join together to help.[49] They agreed to form the Artists' Refugee Committee (ARC) in November 1938, which invited art societies the London Group, the Royal Academy and the AIA to serve on the committee and to offer guarantors and hospitality for the OKB artists, a requirement in order to secure British entry visas.[50] Penrose represented the Surrealists on the ARC committee, while three of his Hampstead-based neighbours joined.[51] The group raised £1,700 to carry out their work and to give hospitality to refugees, succeeding in saving the lives of at least twenty people.[52]

Showing the work of Jewish artists in London: *Exhibition of German-Jewish Artists' Work*

While the AIA was creating exhibitions as anti-fascist statements, from 1933, London-based art dealers were also mounting exhibitions both as anti-fascism and in solidarity with artists of the Jewish faith. London-based art dealer Carl Braunschweig held *Exhibition of German-Jewish Artists' Work: Sculpture-Painting-Architecture* at Parsons Galleries, 315 Oxford Street in June 1934.[53] Braunschweig, born in 1886 to a German Jewish family, had fled to England in late 1933.[54] He worked on the show with the Central Association of German Citizens of Jewish Faith in Berlin, seeking to show how Jewish artists had contributed to recent German art in 'closest harmony' and doing 'honour' to it, the catalogue explained. Braunschweig's show contained 221 works by 86 contemporary German Jewish artists who had worked and exhibited in Germany before 1933 including painters Martin Bloch, Hans Feibusch, Lotte Laserstein, Max Liebermann and Adele Reifenberg. Most had not previously shown work in Britain, more than a third were women, many were still in Germany experiencing persecution by the Nazis and some had already scattered abroad to Britain, France, the Netherlands, Switzerland, Spain and Palestine, as the catalogue explained. The show placed a focus on artists 'no longer able to display their productions in public exhibition or public galleries in their native country'; 'their work cannot be discussed in the Press'. Many of the artists, the catalogue noted, had been reduced to poverty and lost touch with supporters.[55] In making this exhibitionary statement in 1934, Braunschweig aligned himself with artists later labelled 'degenerate'.[56]

Dutch pioneer of abstraction Piet Mondrian, accompanied by his friend the artist Winifred Nicholson, moved from Paris to London in September 1938 to escape the threat of Nazi invasion. Although not Jewish, Mondrian had immediate cause to fear as two of his works had been included in the 1937 *Degenerate Art* exhibition, effectively putting him on a blacklist. He was helped by his friends Russian Constructivist Naum Gabo, who had arrived in 1936, and abstract painter Ben Nicholson.[57] Expressionist artist

Ludwig Meidner was in exile in England for fourteen years from 1939. He had been highly regarded in Germany, the subject of two solo exhibitions in Berlin and Dresden before 1933, represented by four paintings at *Degenerate Art*. Meidner had no solo show in England although his work was included in two group exhibitions at London's Ben Uri Gallery.[58] In the absence of existing exhibiting opportunities and forced to develop new connections through which to find work and support, these refugees artists formed their own groups to organise exhibitions, as I go on to discuss in the next section.

Exhibitions as platforms for refugee artists: Free German League of Culture

Many artists of extraordinary talent, fleeing from Central and Eastern Europe and beyond, arrived in Britain during the 1930s, through a number of routes including the patronage of the ARC and the RIBA Refugee Committee, established in 1939.[59] Painter Oskar Kokoschka arrived in England from Prague in 1938, under the auspices of the ARC. While in Germany, Kokoschka, like Meidner, had seen his identity erased through the inclusion of twelve of his paintings in *Degenerate Art*. Kokoschka took British nationality and remained in the country until 1954.[60]

Another to arrive under the auspices of the ARC was German photomontage artist John Heartfield, who entered Britain with Labour MP Ellen Wilkinson as 'guarantor'.[61] Born Helmut Herzfelde, Heartfield had become prominent in German Communist circles, including the Rote Frontkämpferbund (Red Front League of Struggle) founded in 1924, whose clenched-fist symbol Heartfield designed. Heartfield had been compelled by Hitler's rise to power to flee to Czechoslovakia in 1933. In April 1934, Heartfield had been a major participant in an international exhibition of caricatures held at Prague, in which contributions by George Grosz and T. T. Heine, whose cartoons were admired by Trotsky, were prominently displayed. The tone of the exhibition, and especially of Heartfield's works, was so sharply anti-fascist that the German and Italian governments put pressure on Czechoslovakia, with Czech police responding by removing the offending pictures. Heartfield already had a British audience, through work such as his provocative photomontage *The Happy Elephants*, which satirised Prime Minister Chamberlain's 'fairy-tale', as he saw it, of the Munich Accord with Hitler, giving British people a false sense of security. The work was published in *Picture Post*, the popular magazine launched in the same year by Hungarian refugee Stefan Lorant. Further Heartfield works had been published in British illustrated magazine *Lilliput* within weeks of his arrival in London.[62]

Many refugee artists who had achieved critical acclaim in their home contexts failed to achieve the same level of recognition or success in

Britain where there was limited appreciation for German Expressionism and the New Objectivity. Limited sympathy in the British contemporary art world for the work of refugee artists made critic Herbert Read unusual in his interest in developments in German art.[63] Despite his fame in Germany, while in exile in Britain Oskar Kokoschka did not have a solo exhibition of his own work, a contradiction referred to by fellow refugee artist Fred Uhlman, who described 'Germany's most distinguished painter, a man famous all over Central Europe, but at that time almost completely unknown in England except to a few connoisseurs and dealers'.[64] While exiled in Britain exhibitions did offer opportunities for artists to connect with each other, however, and to assert their shared beliefs and solidarities, providing a platform and a voice.

In December 1938, Kokoschka and Heartfield were co-founders of London-based solidarity organisation the Freie Deutsche Kulturbund (Free German League of Culture in Great Britain, or FGLC) as a 'German, anti-Nazi, antifascist non-party refugee organisation'. Its multiple cultural support for those experiencing cultural, social and political exile in Britain included mounting exhibitions, which offered the chance to raise the plight of refugees and to provide mutual support for recent arrivals. An important model for the FGLC was the Assoziation revolutionärer bildender Künstler Deutschlands (Association of Revolutionary Visual Artists), known as 'Asso', which many refugees who reached Britain from Germany had been members of. The initiative for Asso's foundation had come from artists keen to mobilise the graphic talents of artists in the service of the Communist Party, including creating visual displays, leaflets, broadsheets, banners and placards and giving an organisational framework to political art but without imposing a common style; focusing on questions of solidarity and common concern, rather than formal or stylistic qualities of the work.[65]

The FGLC's stated aim was to create mutual understanding between refugees and British people: emphasising and strengthening the solidarity of the refugees with 'all democratic, freedom-loving, progressive forces'; looking after the social interests of refugees; cultivating and developing relations with 'other friendly organisations and personalities'. The emphasis was on cultural and social rather than political aims, reflecting the restrictions imposed on refugees' political activities by the British government. Despite this, the FGLC's activities attracted the attention of British Security Service MI5, which kept members under close surveillance, believing it 'a communist front organisation'. An MI5 Special Branch report in December 1939 described the FGLC as 'an organisation which aims to arrange lectures and other events on artistic and literary manners but is more or less recognised to foster communist sympathies among refugees and British subjects'.[66] At its peak, the FGLC had 1,500 members, with group exhibitions a major strand of its work, successfully providing newly arriving German artists a community with which to work and exhibit. These regular exhibitions were mounted in a Hampstead house made available by George Bell, Bishop of Chichester, with larger exhibitions at central London venues.[67]

FGLC exhibitions provided an opportunity to see works by artists not shown before in Britain, and to help support these artists. The *First Group Exhibition of German, Austrian, Czechoslovakian Painters and Sculptors* held in June 1939 at the Wertheim Gallery, run by Lucy Wertheim in London's Burlington Gardens (a short-lived venture running from 1930 to 1939), showed work by artists in exile from their countries.[68] The standard of sculpture was particularly praised by the press – works like Georg Ehrlich's *Fisherman's Son* and Siegfried Charoux's *Terracotta Group*, for example – while Kokoschka, who showed four paintings, was criticised by the *New Statesman* for 'his disregard for the subtleties of his art is too undisguised'.[69]

While in Britain, Kokoschka acted principally to promote other artists' work, including FGLC member and fellow German artist-in-exile John Heartfield. Unlike other refugee artist friends, Heartfield was achieving a reasonable amount of success and financial stability while in Britain, receiving regular commercial commissions, including designing book covers.[70] In December 1939 Kokoschka mounted a politically provocative one-person show of Heartfield's work called *One Man's War against Hitler: Exhibition of John Heartfield* at Paul Wengraf's Arcade Gallery with montages from Berlin and Prague, documenting Heartfield's fight against Nazism. The Arcade was one of very few London galleries to remain open throughout the war.[71] Wengraf, born in Vienna in 1894 to a family of art dealers, had set up a business at home but fled to London in 1938 after the German annexation of Austria, opening the Arcade Gallery on Old Bond Street in March 1939, supported by his friend Danish engineer Ove Arup.[72] The rather esoteric gallery programme ranged from Netherlandish Mannerism and early Baroque to Neoclassicism, as well as work by Jewish artists trapped in Nazi-occupied Europe to help them escape persecution. One such artist was former Bauhaus student and Communist Party member Friedl Dicker, whose work Wengraf exhibited in August 1940, in the hope that doing so would aid her emigration from Czechoslovakia, which tragically did not materialise. Wengraf received exemption from internment and continued to exhibit even at the height of the Blitz.

Heartfield was interned for six months in 1940.[73] The FGLC held another exhibition of his work in June 1941, to mark his fiftieth birthday: a display of photomontages published in Communist newspaper *Arbeiter-Illustrierte Zeitung* (*Workers' Illustrated News*). Alongside this exhibition, Heartfield gave a talk about the evolution of photomontage from photography. In the audience was MI5 informant Kurt Hiller, who was there to assess Heartfield's capacity to be politically radical and subversive in Britain. Hiller reported Heartfield demonstrating 'how he and his fellow workers had always used [photomontage] in the service of the truth (as the Communists understand it) and how great a part this could play in social and revolutionary struggles'.[74] Heartfield remained on the M15 watch list and his written applications to remain in Britain for work and health were denied, precipitating his return to Berlin in 1950.

Exhibitions supporting the plight of Jewish people continued to be mounted through the Second World War. In February 1943 *Artists Aid Jewry* was held at Whitechapel Art Gallery, with a poster carrying a woodblock by Erich Kahn, printed in English and Hebrew. The exhibition consisted of 137 works by Jewish artist members of a group of organisations: the Jewish Cultural Club, the Free Austrian Movement and the FGLC. A pamphlet explained the exhibition as 'All in Aid of Jews who fell victims to Nazi-barbarism'. Its intention was 'to show works relating to Jewish life and problems'. As a token of sympathy with 'their unfortunate brethren' the exhibitors had agreed to give half of the proceeds from sales to Mrs Churchill's Fund, for aid to Russia. Appealing to visitors' good will, it concluded: 'The Artists express their hope that the visitors for their part will support this effort in the same spirit'.[75]

Refugee exhibitions during internment

If exhibitions mounted in Britain created a focus for solidarities with people in other parts of the world currently experiencing war and oppression, this became amplified once Britain entered the war. Many artists who arrived in Britain in the years immediately preceding the Second World War had been prohibited from exhibiting their work. Perhaps worse, the Nazi regime had ridiculed these artists' work in exhibitions. Ironically, the British internment camps in which 22,000 Germans and Austrians, as well as 4,000 Italians, were imprisoned as 'enemy aliens' in 1940 offered a certain creative freedom, to exhibit and to show artistic work of any style without ridicule.[76]

Relationships between FGLC members were cemented by common experience of internment as enemy aliens from 1940 to 1942. During this period, in July 1941, Kokoschka inaugurated the exhibition *Refugee Artists and Their British Friends*: over a hundred works shown at FGLC's Hampstead clubhouse to show solidarity with and for the benefit of interned refugees.[77] Even internees without artistic training were drawn to pursuing creativity in the midst of internment. Businessman H. G. Gussefeld turned tent pegs into letter-openers in animal shapes and used cannibalised linoleum for linocuts.[78] Living in these camps for several months, on the Isle of Man and in about a dozen places around Britain including Glasgow, Liverpool and Manchester, allowed artists to make art – sometimes in dedicated studio spaces – and to mount camp exhibitions as productive and enjoyable diversionary activities.[79]

While interned in Hutchinson Camp, art historian Klaus Hinrichsen took on the role of mounting exhibitions. The first, for which no known photographs exist, was reviewed as 'astounding' in *The Camp*, the makeshift homemade newspaper which was circulated in the camp.[80] The exhibition was held in September 1940 and included work by German painters Fred Uhlman and Erich Kahn. German Dadaist artist Kurt Schwitters, also interned in the camp, had offered porridge sculptures, 'exuding a sickly and evil smell', which the organisers turned down in favour of more conventional portraits and landscapes.[81] Schwitters, whose work had been condemned as 'degenerate' by the Nazis, was exiled to Britain from Norway in 1940, remaining until his death in 1948.[82] In 1940, during internment at Hutchinson, Schwitters painted portraits of fellow internees (including Fred Uhlman), landscapes and 'Merzbilder' collages, incorporating scrap wood and ceiling materials, dismantled tea chests, toilet paper, discarded cigarette boxes and sweet wrappers and even left-over porridge, as Hinrichsen recalled.[83] One Sunday he decorated the staircase of the office building with 'Miro-like designs'.[84]

Hinrichsen's second Hutchinson camp exhibition was held in November 1940 on the large first floor of an empty building set aside specially by Hutchinson's Camp Commander, with all the trappings of professional galleries including private view invitations, a catalogue and speeches, concluding with a performance of chamber music.[85] Photographs show framed paintings nailed to walls covered in wooden boards or hung against rough breezeblocks and unframed works on paper attached loosely to walls (Figure 6.3). Many were portrait heads of fellow inmates, probably painted from life. There were painted landscapes, photographs, a sculpted head and tiny figurines. The works represented a mix of styles. Hinrichsen explained they ranged from 'accomplished oil paintings, sculptures and graphic work to untutored but moving "sketches from the German Anti-Nazi fight and the Spanish fight for Liberty", to Punch and Judy puppets'.[86] Amongst them were carved wooden pieces with the appearance of tribal objects, displayed symmetrically.[87] Most of the styles on show were suppressed in Germany so this represented an opportunity to work in a way that had been forbidden for years.[88]

Wartime exhibitions at Bilbo's Modern Art Gallery

Another Isle of Man internee who used exhibitions as a mechanism for giving voice to fellow refugees was German Jewish refugee Jack Bilbo. German-born Bilbo, originally Hugo Baruch, had co-founded the Kampfbund gegen den Faschismus (Fight against Fascism) in 1930 to oppose Nazi policies. Having fled first to France, he arrived in London

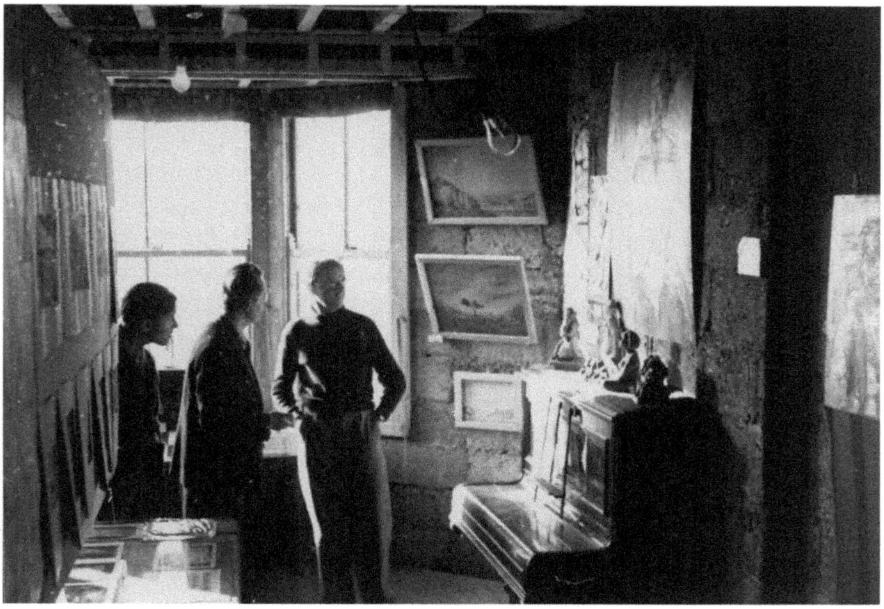

6.3 The second Hutchinson camp exhibition of November 1940. Daniel, Major H. O. Photo: Tate © The estate of Hubert Daniel. TGA 20052/2/7/13. All rights reserved and permission to use the figure must be obtained from the copyright holder.

in 1936. An anarchist and self-taught artist, Bilbo began painting and sculpting after arriving in Britain, using blackout paper for painting after the outbreak of war. Bilbo was interned at Onchan Camp on the Isle of Man (along with designer F. H. K. Henrion, who was released early to work on exhibition design at the MOI, as discussed in Chapter 7). The lively cultural community at the camp included regular lectures, amateur dramatics, musical performances, debating and newspaper the *Onchan Pioneer*, a vehicle for news and ideas.[89] At Onchan, Bilbo hosted two exhibitions of fellow internee artists in his camp cabin, deliberately including amateur artists in the exhibitions in order to eschew the conventional attachment to art world training.[90] The first exhibition he presented with the motto 'The world is a cage, forged by human stupidity. Art will break this cage', pointing to his belief in art's emancipatory qualities.[91]

On release from internment, Bilbo served in the Pioneer Corps but was invalided out in 1941, soon after opening the Modern Art Gallery on London's Baker Street in the midst of war, 'with the sole aim of giving the modern artist … an unbiased platform, and of creating for the people an oasis of sanity … believing in the necessity also for an intellectual fight against Hitlerism and all it stands for'.[92] This was an act of courage and defiance at a time when most London galleries had closed in fear of bombing and, for Bilbo, it was a way of performing solidarity with fellow refugee artists.[93] The gallery showed works by well-known avant-garde artists including Picasso and Surrealist Eileen Agar, as well as lesser-known refugee artists including Austrian Anna Mayerson, German Samson Schames and Czech Jacob Bornfriend. *Anti-Fascist Exhibition* held at the Modern Art Gallery in July 1943 was a display of cartoons by Vicky (Viktor Weisz), who contributed regular cartoons to *News Chronicle*. Proceeds from the exhibition went to the Stalingrad Hospital Fund, one of many fundraising efforts in solidarity with the Soviet Union, as I will discuss later in this chapter.

Bilbo held Kurt Schwitters' first solo exhibition in Britain at the gallery in December 1944, by now in a space on the Haymarket. Schwitters – like many other incoming artists of the time – had found it difficult to integrate into British artistic circles and this was a clear act of support and solidarity. Through his friend Naum Gabo, the Russian Constructivist sculptor and painter who lived in Britain from 1935 to 1946, Schwitters had been introduced to people within the British art establishment, such as Barbara Hepworth and Ben Nicholson.[94] But institutional support for Schwitters' work was limited, so his 1944 exhibition at the Modern Art Gallery was significant. It included thirty-nine works in collage, oil painting and sculpture; the catalogue, written by Herbert Read, hailed Schwitters as 'the supreme master of collage' and a Modernist poet whose work paralleled that of James Joyce, pointing out that despite not having been boosted by critics and art dealers to the same extent as some of his contemporaries he was 'one of the most genuine artists of the modern movement'.[95] A flyer for the exhibition advertised 'Paintings and Sculpture by Kurt Schwitters (The Founder of Dadaism and "Merz")'.

At the Munich *Degenerate Art Exhibition* in 1937 one of Schwitters' large assemblages had been hung at an angle under the heading '*Total Verrueckt*' ('*Completely Crazy*'). Hitler was photographed smirking in front of it. Schwitters had continued to court disapproval for his Merz work while in Britain, hiding it while in internment and only showing his portraits and landscapes publicly.[96] In the programme concurrent with the Modern Art Gallery show, Schwitters performed his abstract sound poem *Ursonate*, which he had written twenty years earlier in Germany and had also performed while interned. Read referred to these poems as integral to his work and as an art of 'abstract incantation'.[97] Bilbo recalled 'Kurt … being quite pathological, enjoyed himself immensely. So did my guests, because outside the bombing went on, which seemed to be logical, and therefore wasn't so amusing, and inside the house Kurt Schwitters went on with his illogicality, which was amusing'.[98] Two BBC representatives, invited to record his performance, left before he had finished.[99]

Showing British people the underground struggle against Hitler: *Allies inside Germany* on Regent Street

Exhibitions of solidarity were mounted in unlikely places during the Second World War, and shops, temporarily ceasing to function as places of trade in the midst of war, offered excellent sites for the proliferation of exhibitions mounted by political activists. The blitzed John Lewis department store site became used for many ambitious exhibitions, including the AIA's major 1943 exhibition *For Liberty*, which I introduce in Chapter 7. Meanwhile, in an empty shop on Regent Street the group around German photomontage artist John Heartfield mounted *Allies inside Germany* and Ernö Goldfinger took the empty Boots the Chemists at Piccadilly Circus for an exhibition called *The Two Mrs Britains*, each driven by an anti-fascist agenda.

The Leftist messages of these exhibitions increasingly converged with and were supported and amplified by the British establishment. This was the case with *Allies inside Germany*, 1942, an exhibition showing refugees playing an important part in anti-fascist resistance from within Germany, by documenting the opposition to Hitler inside Nazi Germany – 'the underground struggle of the German Anti-Nazis against Hitler', as its flyer explained – and raising money to produce propaganda material to be dropped over Germany.[100] The exhibition flyer's cover image, drawn in red by artist René Graetz, showed a cloth-capped worker standing with a clenched fist (Figure 6.4). Berlin-born Graetz had arrived in Britain in 1939, was interned on the Isle of Man in 1940 and then deported to Canada for a year. He had previously been frustrated by the large number of still lives and portraits in the FGLC and AIA joint exhibition of 1941, believing the isolation from the masses no longer justified the artists adopting a quietist position: 'In the gigantic struggle now taking place', he argued, 'the artist must take a decisive stand'.[101]

Allies inside Germany was held from July to August 1942 in an empty, two-storey shop at 149 Regent Street, in London's West End.[102] The exhibition was organised by the FGLC Chairman and Communist politician Hans Johann Fladung, with the English Anti-Nazi Committee backed by British Labour Party member and peer Lord Wedgwood. This was a multidisciplinary information exhibition, a first for the FGLC, whose previous exhibitions had all centred on artworks. *Allies* was instead focused around twenty-seven panels referring to a number of global, anti-fascist struggles, connecting locations and reinforcing political alliances. The design was a collaboration between Heartfield, who worked on some photomontages of photographic material, with drawn and painted images accompanied by texts by Heinz Worner and Hans Fladung, who were responsible for tracking down the exhibition's documentation, and René Graetz, who created the exhibition's graphic layouts.[103]

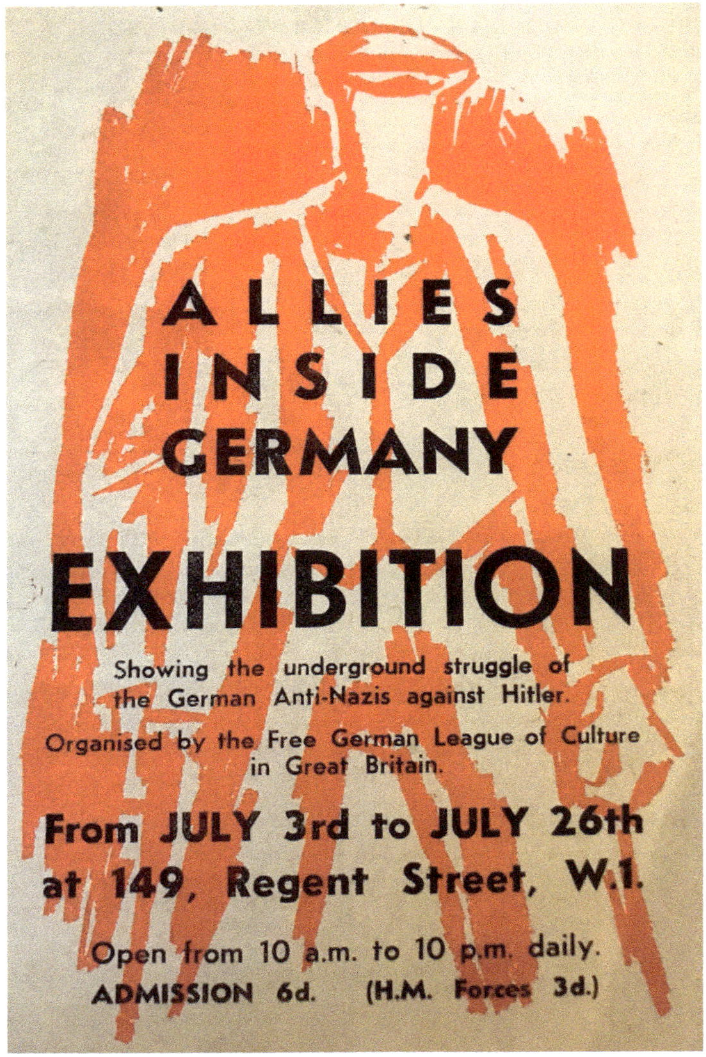

6.4 Flyer for *Allies inside Germany*, with a cover drawing by René Graetz, held in an empty shop at 149 Regent Street, London, to show 'the underground struggle of the German Anti-Nazis against Hitler'. René Graetz. © DACS 2023. All rights reserved and permission to use the figure must be obtained from the copyright holder.

German economist and, it was later revealed, Soviet spy Jurgen Kuczynzki provided statistical information. The panels in *Allies* had a homemade feel, with photographs collaged from German and exile newspapers and documents, some smuggled out of Germany, thematising resistance to Hitler inside Germany, the role of refugees in such efforts, the way that

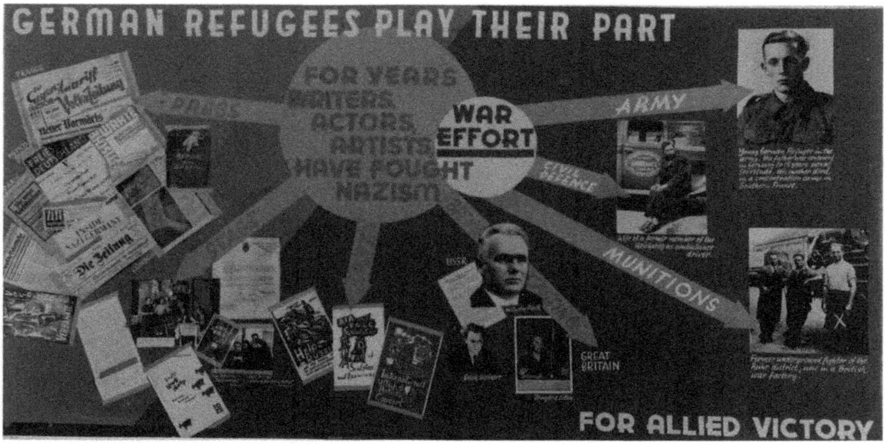

6.5 'German refugees play their part for Allied victory', panel 27 from *Allies inside Germany*. Uncredited photographer. Image courtesy of Akademie der Künst Berlin, Heinz Worner archive. © DACS 2023. All rights reserved and permission to use the figure must be obtained from the copyright holder.

literary and cultural life was being censored and silenced and international responses.[104] Slogans and labels were painted, cut out and printed on top.

Introductory panel 'Hitler Comes to Power' portrayed this as a class struggle in which the working classes were fighting back against a middle class duped into supporting the Nazi regime. It showed photographs of antagonism, with anti-Nazi protests by the working classes, and examples of underground working-class newspapers. Another panel illustrated the Nazi's progressive occupation of Europe – Austria, Czechoslovakia, Poland and onwards – contrasting the public demonstrations of military power with photographs of horrifying atrocities. Nazi leaders – from Hitler to Göring – were pictured, their 'crimes' listed. As counterpart, several panels showed pockets of resistance: peasants and Christians, as well as highlighting German intellectuals whose work stood against the current regime and those who had been exiled or silenced, illustrated by photographs, quotes and publications (Figure 6.5).

Pamphlets from resistance groups in Britain offering international solidarity were displayed. One panel, entitled 'Five Minutes Before Twelve', showed the cut-out heads of Hitler alongside Stalin quoted as saying 'Hitlers come and go, but the German people and the German state remain' while 'German Refugees Play Their Part for Allied Victory' stated that 'For years writers, actors, artists, have fought Nazism', showing how press, stage, art, broadcasting, civil defence and army had been pioneered by refugees.

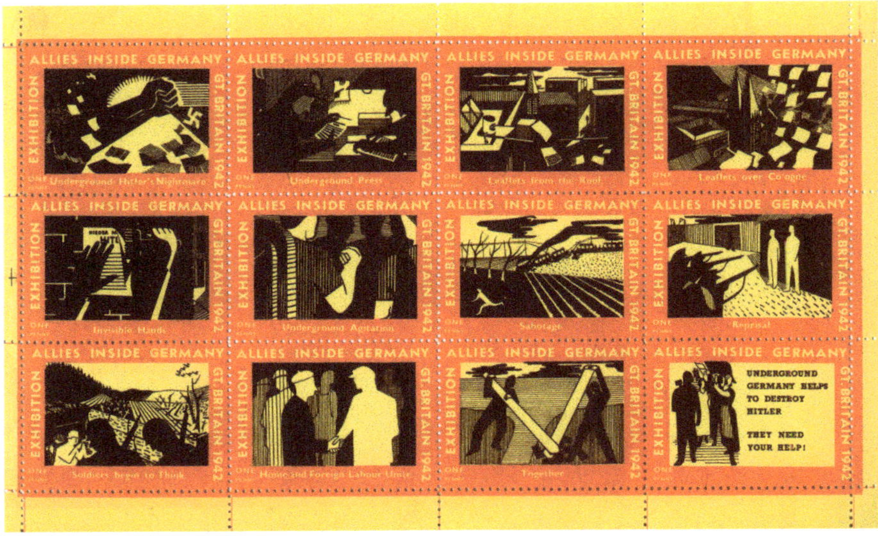

6.6 One-penny stamps collection designed by German refugee H. A. Rothholz issued for the *Allies inside Germany* exhibition. H. A. Rothholz Archive, University of Brighton Design Archives. (Uncatalogued). Image courtesy of the H. A. Rothholz estate.

Showing the level to which *Allies* had the establishment seal of approval, Royal Mail issued a series of twelve one-penny stamps on the occasion of the exhibition. These were designed by German refugee H. A. Rothholz, who had arrived in Britain in 1933 and was interned in Liverpool, the Isle of Man and Canada. Each postage stamp had a black and white linocut set in a striking red frame with yellow font, dramatising the role and impact of underground resistance in Germany; titles including 'Leaflets from the Roof', 'Sabotage', 'Home and Foreign Labour Unite', 'Underground Press' and 'Soldiers begin to Think' (Figure 6.6).[105]

Allies was the most high profile of all FGLC exhibitions, attracting thirty thousand visitors but a mixed press response.[106] The *Daily Mirror* said 'perhaps the most moving "exhibits" are some of the men who have organised the show. Several of them have escaped from concentration camps', referring to the exhibition organisers' status as refugees (without being strictly accurate).[107] The *Manchester Guardian* reported approvingly that the exhibition showed how German people were fighting Nazism, while *New Statesman and Nation* was impressed by 'the visual proof of the growth and boldness of the Opposition', including photographs of 'Catholic processions and public demonstrations'.[108] The *Spectator* published a letter from German historian in Britain Peter F. Wiener, who described *Allies* as 'in rather bad taste', weakening the wider effort by making the British public believe the 'dangerous myth' that there were allies within Germany that could bring the collapse of the regime. 'It is not for me to decide whether this untimely exhibition is the result of ignorance, homesickness or definite political activities'.[109] Press attention came from as far afield as *Australian Quarterly*, which reported the exhibition a success that had travelled beyond London to six cities including Manchester and Bristol.[110] In its travelling guise the exhibition was renamed *We Accuse – 10 Years of Hitler Fascism*, drawing more attention to the dangers of the Hitler regime and to the need to form a second front.[111]

The Resistance Exhibition, Warrington

One tantalising archival find that has yet to bear further fruit, but is suggestive of further solidarity exhibitions held across Britain during wartime, is the transcript of a speech made by French conservative Resistance leader Louis Marin at the opening of *The Resistance Exhibition*, in July 1944 in the northern town of Warrington (near Manchester).[112] Marin had sought refuge in Britain in April 1944, after the German Gestapo had issued a warrant for his arrest. The only hints of what the exhibition comprised are through Marin's words. A search of the local Warrington press may well bring further information. Marin's speech described the exhibition as about 'armed', 'organised' resistance by ordinary people in France. He used the occasion to suggest that it would only be through remembering current atrocities and oppression that the world would 'do better next time'. His concluding comment affirmed exhibitions as a useful medium through which governments could form public opinion and the public could judge the work of politicians: 'With a government depending on public opinion, exhibitions like this are an excellent thing, because governments based on public opinion and free discussion are responsive to the views of the people'. Marin ultimately affirmed exhibitions as playing an important role in supporting democracies.

Exhibitions of refugee children's work: *The War as Seen by Children*

The impact of the war on children was of great interest to the public and the focus of scrutiny by politicians and journalists. In an article for *Picture Post*, Tom Harrisson of Mass Observation discussed 'A Child's View of the War'. This was accompanied by images of children preparing for the exhibition *'Five to Ten' Exhibition of Children's War Pictures*, including one of Ursula and Ernö Goldfinger's daughter Elizabeth Ann, preparing her contribution at home at 2 Willow Road in Hampstead, north London. The exhibition was held first locally, at 10 Park Crescent in Hampstead, before touring onwards to New York.[113]

Wartime exhibitions were providing a platform for newly arriving refugee artists. They also gave refugee children a voice. Child refugees' work was shown at *The War as Seen by Children* organised by Kokoschka in 1942 to fundraise for the Refugee Children's Evacuation Fund. The exhibition shared the FGLC's Hampstead base to collect for Theydon Bois School for German Refugee Children, set up to teach children in German, with the aspiration that they could return home after the war.[114] Initially modest and co-organised with Johann Fladung, the exhibition was eventually opened by Czech Foreign Minister Jan Masaryk (part of the Czech government-in-exile in London) and evolutionary biologist Julian Huxley, later UNESCO's first Director-General. It showed work by refugee children of twelve nationalities, intending to educate adults about their experiences.

Kokoschka's opening speech showed the extent to which the exhibition was focused towards creating solidarities between the children. It would, he hoped, create 'a friendly and comradely spirit amongst [the children] themselves', 'a most important asset for future international co-operation'.[115] These children, 'such little victims of the Fascists' barbarity as were lucky enough to find hospitality in England', had experienced 'the Fascist menace of death and destruction' and were 'now struggling together with their elders for the victory of a life of happiness and beauty' to arouse in its viewers greater determination to fight the 'inhuman forces'. 'Art', Kokoschka continued, 'is one of the means through which man expresses his participation in life'. The children offered a model for adults, Kokoschka suggested; through learning to understand and respect each other they had become 'a working model for an ideal human society of the future'. Although, he cautioned, that despite this some children had still somehow grown to be soldiers of 'the atrocious Nazi regime', questioning how the system of education had produced current conditions.

Based on its popularity and perceived educational value, *The War as Seen by Children* was enlarged and reopened in a ceremony led by Lady Clark at the Cooling Galleries on New Bond Street, central London, in January 1943. This time it was more ambitious, including works by children of all the United Nations, touring to forty-five British towns, to be shown in schools and galleries, as well as touring across the US, with paintings selected by a panel chaired by Herbert Read. Exhibits were themed: 'How the child sees the present struggle', 'How refugee and evacuated children re-discover and re-build life and their national culture', 'How the refugee or evacuated child sees its new surroundings', 'How the children of the United Nations feel united in a common purpose' and 'How they see their future'.[116] The exhibition reminded the public of the plight of vulnerable refugee children who were liable to be forgotten in the fraught context.

'Comrades in arms exhibitions': *Aid to Russia* at 2 Willow Road

Britain's newly close relationship with the Soviet Union, as they joined the Allies midway through the Second World War, became the subject of a series of exhibitions, extending the hand of friendship, offering a materialised form of alliance and providing an occasion for fundraising. The Anglo-Soviet relationship was profoundly unstable: for twenty years after the Bolshevik Revolution of 1917, the British Left had been broadly supportive of the Soviet experiment, with many British artists such as illustrators Pearl Binder and Clifford Rowe spending time living and working in Russia. But, as historian Paul Corthorn notes, the British Left's relationship with Russia had become increasingly fractured by the late 1930s, as knowledge became widespread of Stalinist show trials and the purges, with increasing condemnation in Britain of Stalin as a ruthless dictator.[117] This opposition changed once again after the Nazi invasion of the Soviet Union in 1941, when Stalin entered the Second World War on the side of the Allies and Britain was thrown into a temporary alliance with the Soviets, culminating in the Anglo-Soviet Alliance of May 1942.[118] British Labour and the Left expressed renewed respect for the Soviet Union, which was seen as a key partner in the fight against Hitler. This resurgence of solidarity with the Soviet Union was expressed through many exhibitions and events in Britain, despite the potential conflicts involved in that support (Figure 6.7).

Some wartime Anglo-Soviet exhibitions were officially sanctioned and mounted by the government through the British Ministry of Information (discussed at length in Chapter 7), including *Comrades in Arms (A Picture of Russia at War)* held at Charing Cross Station in April 1942 and featuring portraits of Churchill and Stalin, as well as cartoons given as 'the gift' of propaganda from Stalin to Lord Beaverbrook in September 1941.[119] Keen to express support for the Soviet Union, artists and designers developed exhibitions outside government to show solidarity and to fundraise. One of the first was *Aid to Russia*, mounted by architect Ernö Goldfinger and his artist wife Ursula Goldfinger at their recently built home at 2 Willow Road, Hampstead in June 1942 to fundraise for the Aid to Russia Fund. Ernö Goldfinger, born into a Hungarian Jewish family in Budapest, had studied architecture in Paris before moving to London in 1934. His wife, painter Ursula Goldfinger, had trained as an artist in Paris under Purist painter Amédée Ozenfant.

By the time they mounted *Aid to Russia*, Ernö Goldfinger had gained almost two decades of experience as a highly productive exhibition designer. It is evident that exhibitions offered him the outlet for his architectural talents that was, at the time, unavailable elsewhere.[120] Goldfinger had mounted an exhibition on the life of Karl Marx at the 1925 USSR Exhibition Pavilion, a structure designed by Russian architect Konstantin

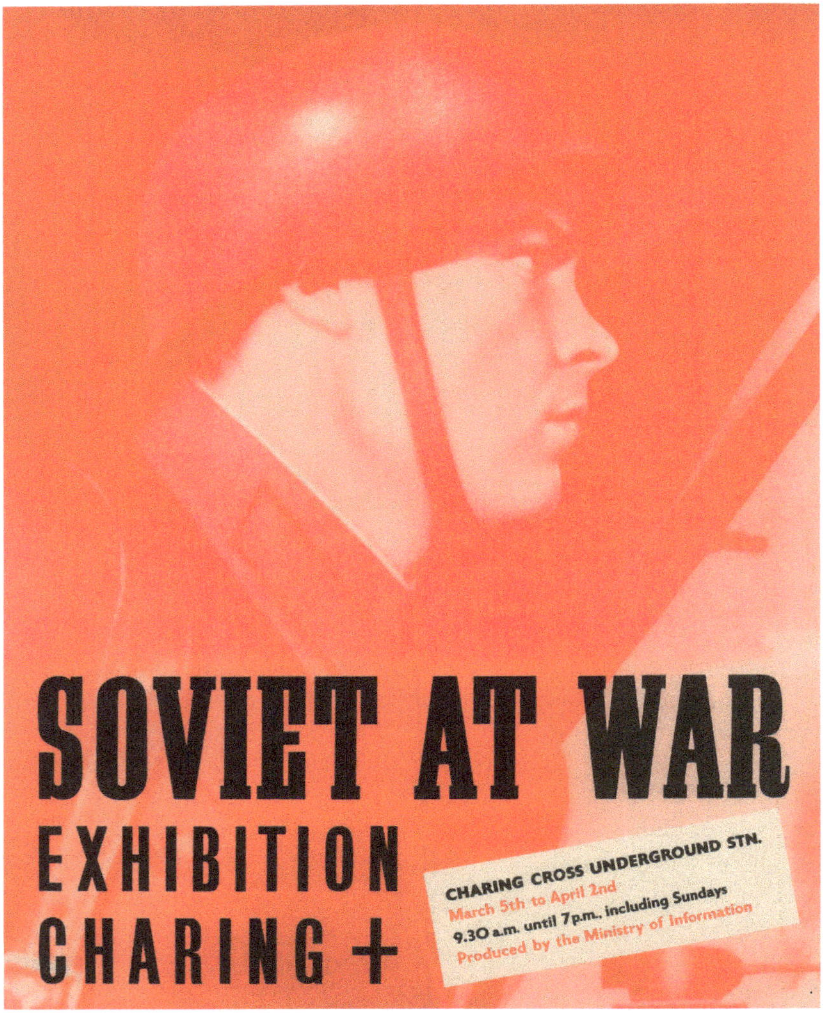

6.7 Poster for *Soviet at War*, mounted by the Ministry of Information and held at Charing Cross Station in April 1942. Uncredited artist. 1983/4/5522 © TfL from the London Transport Museum collection.

Melnikov and re-erected in Paris in 1933.[121] After moving from Paris to London in 1934, Goldfinger, a MARS Group member, designed the children's section of the 1938 *New Architecture* exhibition at New Burlington Galleries (discussed in Chapter 3). He had designed various toy and exhibition displays for toymakers Paul and Marjorie Abbatt, including the Children's Section of the British Pavilion at Paris 1937, as well as the ICI stand at Olympia in 1938.[122]

For *Aid to Russia*, the Goldfingers worked with Virginia Penn as exhibition secretary and her architect husband Colin Penn, President of the Association of Architects, Surveyors and Technical Assistants (AASTA).[123] Pivotal to the selection process was well-known art critic and collector Peter Watson, who had co-founded *Horizon* magazine with writer Cyril Connolly.[124] Ernö Goldfinger acted as exhibition treasurer with Ursula Goldfinger as registrar; their strong social and cultural networks and influence meant they succeeded in gathering works from sixty eminent contemporary artists including Pablo Picasso, Kurt Schwitters, Fernand Léger, Jacob Epstein, Paul Klee, Paul Nash, Roland Penrose, Eileen Agar, Henry Moore and Rita Kernn-Larsen, as well as lending works from their own collection.[125] The exhibition poster carried a simple pencil line drawing by Rolf Brandt: a pair of legs carrying a canvas inscribed with the signatures of artists contributing to the show.

An exhibition opening party was a convivial vehicle for gestures of solidarity, offering an occasion of levity on which to connect with others, to generate publicity, to gather donations and to inspire purchases for the cause from art collectors. The Goldfingers' home was a highly suitable space, its middle floors – with connecting living and dining rooms – having movable walls that offered extended wall space as a gallery hang. *Aid to Russia* opened on 4 June 1942 as a gathering of the great and the good, with Madame Maisky, wife of the Russian Ambassador, present and zoologist Sir Peter Chalmers Mitchell in the chair.[126] Nancy Cunard, heiress and anti-fascist activist, attended. Cunard's booklet *Salvo for Russia*, containing four radical poems (including one by Cunard) and ten engravings critiquing war by Julian Trevelyan, John Piper, Mary Wykeham and Surrealist artist and occultist Ithell Colquhoun, was on sale, produced in aid of the Comforts Fund for Women and Children of Soviet Russia.[127] By the time the exhibition closed three weeks later, it had been favourably reviewed in the *New Statesman*, the *Observer* and *The Times*.[128]

The same month as *Aid to Russia*, Goldfinger designed *Eastern Front*, another solidarity exhibition for the Society for Cultural Relations with the USSR, a body for Anglo-Soviet friendship and to aid cultural understanding of the USSR. In this case the exhibition was held at the showrooms of the Rootes car manufacturers on Piccadilly in June 1942.[129] The Rootes showrooms were a regular exhibition space during the war, hosting MOI exhibitions such as *Colonial Life* of 1943, developed for the Colonial Office, which included sections on colonies at war and education in the colonies (Figure 6.8).[130]

Artists Aid Russia at the Wallace Collection, London

Soon after *Aid to Russia* closed, the AIA's exhibition in solidarity with the Soviet Union, *Artists Aid Russia*, opened at Hertford House, central London base of the Wallace Collection. Ernö Goldfinger was at the helm

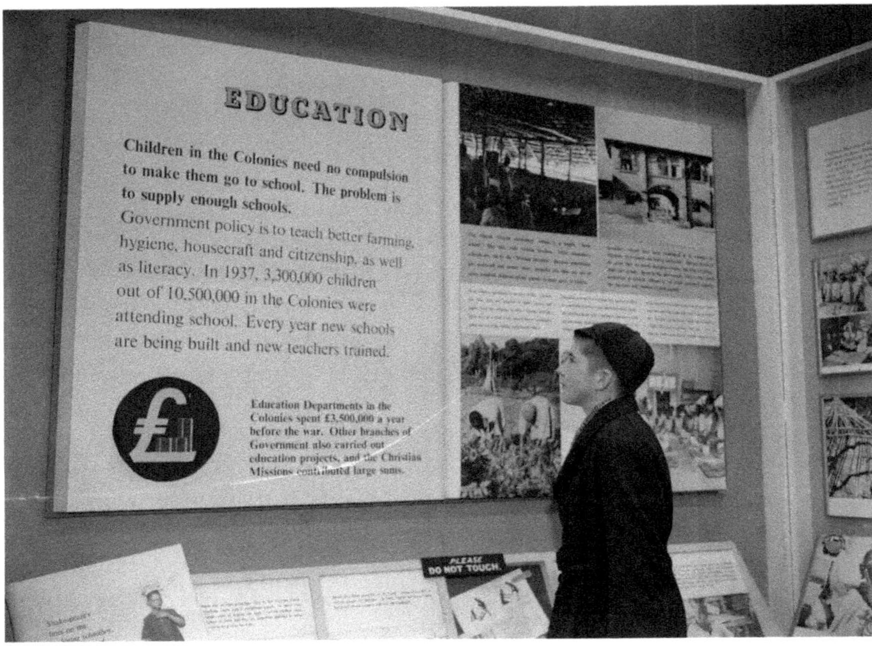

6.8 A young boy looks at a display panel about education in the colonies, at the *Colonial Life* exhibition, produced by the Ministry of Information and held at the Rootes car manufacturers on Piccadilly in June 1943. Richard Stone © IWM D 17418. All rights reserved and permission to use the figure must be obtained from the copyright holder.

once again. Collection objects evacuated from Hertford House had left space for 900 paintings, shown from July to August 1942, raising £2,000 in aid of Mrs Churchill's Aid to Russia Fund.[131] Graphic designer F. H. K. Henrion created the exhibition's catalogue and poster, recalling the poster was 'printed on a page of newspaper, which had to be passed by the censor to decide whether the words were acceptable to be made public. Airbrush was used over the palette, which was faded out, three colours used' (Figure. 6.9).[132] The catalogue cover carried a painter's palette and a hammer and sickle, with a quote from English Romantic poet William Blake's 1793 poem 'The Sword and the Sickle':

> The sword sang on the barren heath,
> The sickle in the fruitful field;
> The sword he sung a song of death,
> But could not make the sickle yield.[133]

This poem connected the English Romantic tradition with the Communist sickle symbol to celebrate the Anglo-Soviet relationship.

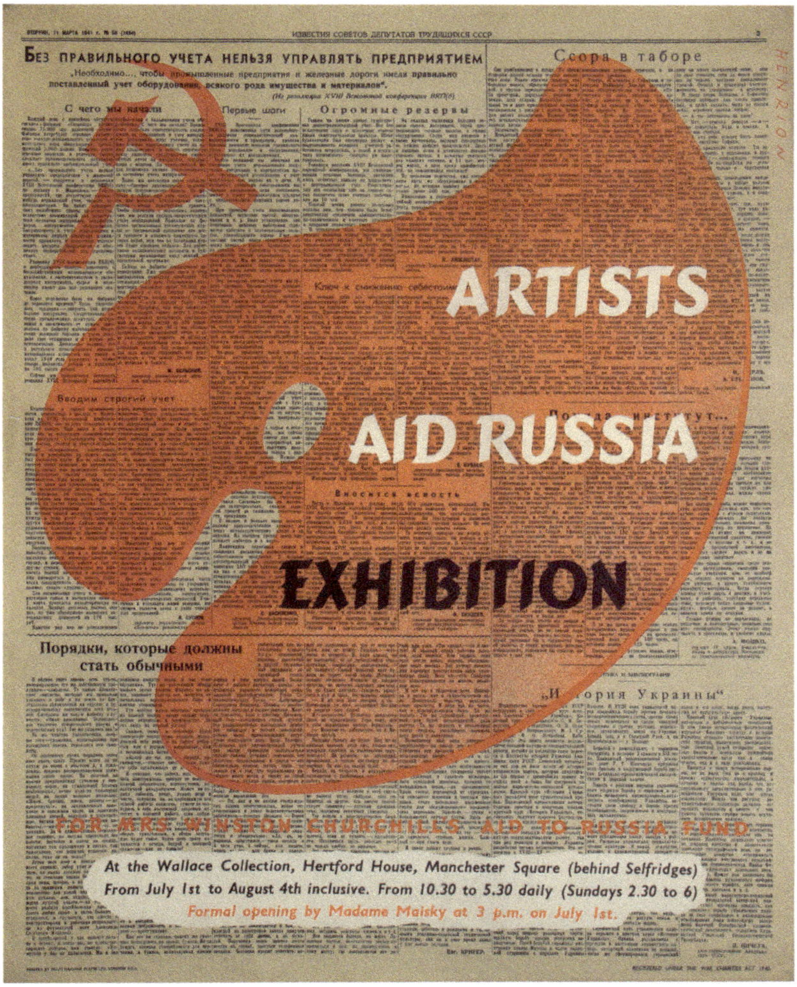

6.9 *Artists Aid Russia*, Hertford House, 1942. F. H. K. Henrion designed the exhibition's catalogue and poster. F. H. K. Henrion Archive, University of Brighton Design Archives. © Estate of F. H. K. Henrion. All rights reserved and permission to use the figure must be obtained from the copyright holder.

Ernö Goldfinger designed another solidarity exhibition: *Twenty-Five Years of Soviet Progress*, held at Hertford House in November 1942, the committee chaired by eminent architect Sir Edwin Lutyens. This time, the exhibition was a group initiative by the National Council for British-Soviet Unity, the Society for Cultural Relations with the USSR, the Anglo-Soviet Public Relations Association and the Russia To-Day Society.[134] Unlike the two earlier art exhibitions Goldfinger had co-organised, *Twenty-Five Years* was a documentary exhibition, aimed at educating the public about life in the USSR since the socialist state was founded, structured around information boards. Hertford House's ornate, chandeliered rooms were banked with documentary photographs sourced from books, government departments and news outlets, including images of Churchill, Stalin and Roosevelt, with hammers and sickles aplenty, advocating for Soviet politics, geographical resources and peoples of the USSR, its leaders, the sixteen states, health care and education and recent war action. Anti-fascist posters and cartoons were included in the agglomeration of information.

A strongly pro-Soviet introductory board presenting the overarching theme of *Twenty-Five Years* explained the founding of the Soviet state in November 1917 in the unmistakable tone of Soviet propaganda, giving an account of the 'sacrifice' and 'hard work' of the Soviet people in founding 'a new civilisation'. It glowed: 'They planned production for the well-being of the common man. Today the Soviet People win the admiration of the world for their heroic resistance to the Fascist aggressor. In the name of the British People we greet them as the defenders of civilization'.[135] This text abandoned any tone of reticence or scepticism about the Soviet record of the past decades and instead trumpeted their achievements.

Goldfinger designed charts further championing Soviet systems of healthcare. While working on official propaganda for the Ministry of Information and Office of War Information (as I discuss further in Chapter 7), F. H. K. Henrion designed exhibition publicity for *Twenty-Five Years*: a poster and flyers with striking red and green abstracted hammer, sickle and tank silhouetted against each other (Figure 6.10). Alongside the pictures, maps and charts, flyers promised films, lectures (on subjects as diverse as Soviet medicine, women and agriculture), stamps, songs, music and costumes. The Russian Ambassador was called upon, once again, to open the exhibition, alongside Royal Academy President, architect Edwin Lutyens. Elements of *Twenty-Five Years* then travelled on to be shown at Leicester City Museum & Art Gallery, before touring onwards alongside a programme of British-Soviet cultural events.[136]

Goldfinger continued creating exhibitions that celebrated these links, from his own home to shops and galleries around London. In March 1943, Goldfinger followed up *Aid to Russia* with *Red Army Week Exhibition* at his home at 2 Willow Road, selling art in support of the Hampstead Anglo-Soviet Committee.[137] At the same time, Goldfinger mounted *25 Years of the USSR and the Red Army* to raise money for a new Stalingrad hospital, held from February to March 1943 in an empty shop on Regent Street (Figure 6.11).[138] The exhibition, structured through 'chapters', combined documentary photographs with posters designed to drum up support for the USSR. F. H. K. Henrion designed an 'Aid the Wounded' poster to mark Red Army Day, 23 February 1944, encouraging donations to the Joint Committee for Soviet Aid 'under the Patronage of H.M. The Soviet Ambassador'.[139]

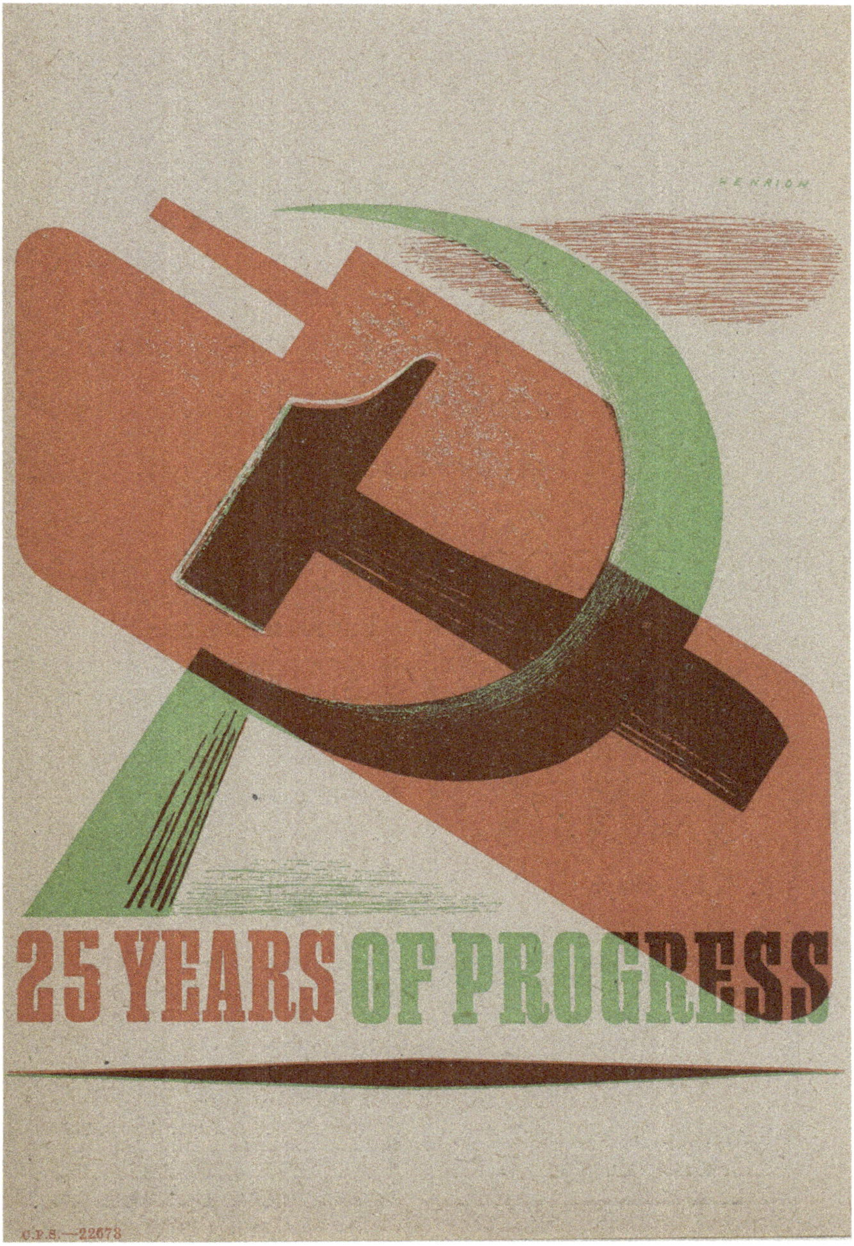

6.10 *Twenty-Five Years of Progress: National Exhibition of Soviet Life in War and Peace*, held at Hertford House, poster designed by F. H. K. Henrion.

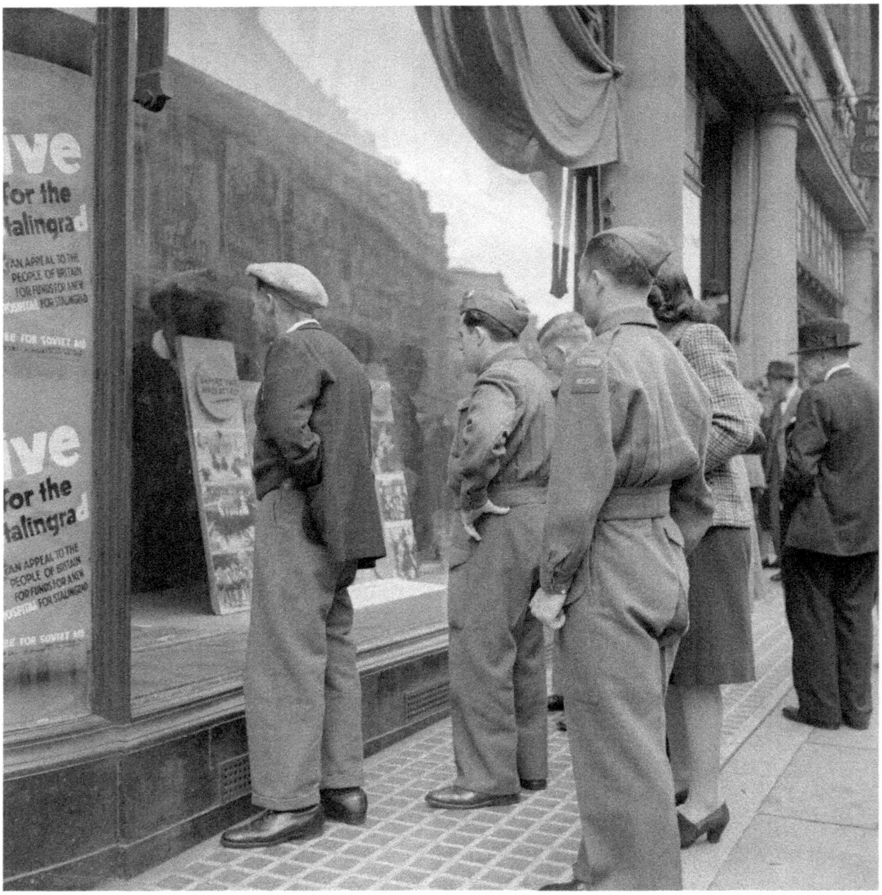

6.11 *25 Years of the USSR and the Red Army*, held in an empty shop on Regent Street, from February to March 1943. Ministry of Information official photographer © IWM D 016084.

Russian aid was not the only cause on British minds. In 1943 Hertford House hosted another fundraising exhibition, this time organised by the AIA: the *Artists Aid China Exhibition* was held from 31 March to 25 May 1943 in support of the Lady Cripps United Aid to China Fund.[140]

Ursula and Ernö Goldfinger's *Two Mrs Britains* exhibition for London Women's Parliament

Aside from the collaborations for aid to Russia, the Goldfingers lent their skills to organisations for other political causes. In 1943, Ernö and

Ursula Goldfinger were appointed by the Communist-aligned London Women's Parliament (LWP) to design an exhibition about women's role in fighting fascism, a solidarity exhibition connecting and supporting women. The LWP, chaired by Mary Morse, had been set up to stimulate discussion of women's problems and to lobby to achieve a better deal for women through parliamentary action. The Goldfingers worked with organiser Freda Grimble, a stalwart of women's organisations. The Goldfingers mounted their anti-fascist LWP exhibition *The Two Mrs Britains* at the Piccadilly Circus branch of Boots the Chemists in September 1943, another exhibition to find a useful base in a disused West End shop. It told the story of two women – 'the Mrs Britains' – from the First World War to the present day, through a series of images setting the two women's progress against images of world events (Figure 6.12).[141]

Photographs of the two central characters were taken by Viennese-born photographer Edith Tudor-Hart. The child of Jewish socialists, Tudor-Hart had trained as a photographer at the Bauhaus in Dessau and fled persecution to Britain in 1933. After arriving in Britain, she had been commissioned

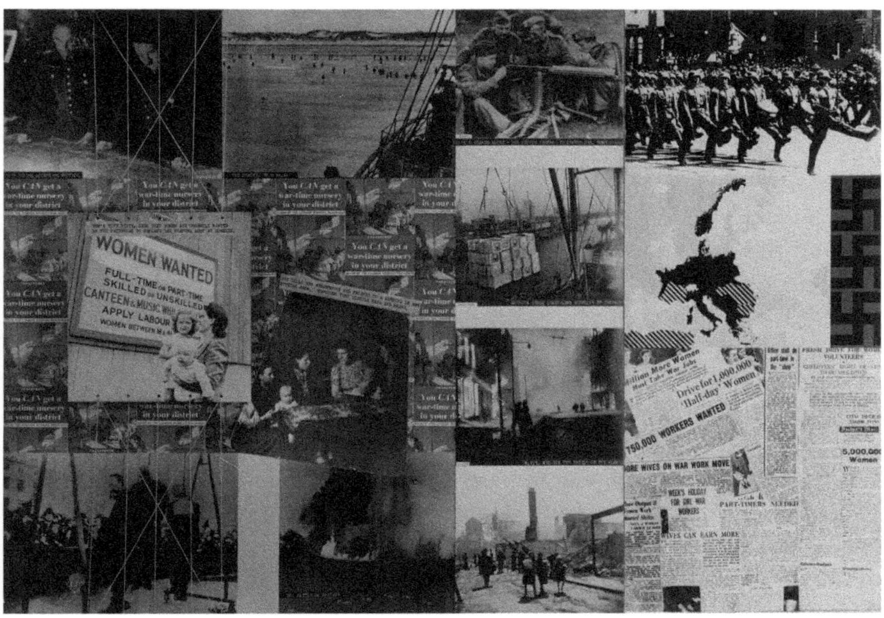

6.12 Display board for *The Two Mrs Britains* at the Piccadilly Circus branch of Boots the Chemists in September 1943, with photographs of the Mrs Britains by Edith Tudor-Hart. Uncredited photographer. Image courtesy of Architectural Press Archive / RIBA Collections. All rights reserved and permission to use the figure must be obtained from the copyright holder.

to take photographs for several magazines such as *The Listener* during the 1930s. A life-long Communist Party member, it was later revealed that Tudor-Hart had acted as a courier and spy recruiter for the Soviet Union.[142] Tudor-Hart's photographic models for the Mrs Britains were Muriel Smith, who Ernö Goldfinger had spotted working at Chalk Farm Bus Depot, and a Mrs Fiorentini from Tottenham, who he had met on a bus.

In the exhibition's narrative, shown through a photographic visual series, one of the Mrs Britains' hopes for a better future after the First World War had been dashed by unemployment and the rise of fascism. The other, her politically active daughter-in-law, was ensuring through vigorous campaigning and greater involvement in all aspects of life that the fruits of victory would not again be squandered.[143] The Goldfingers' striking exhibition boards consisted of a thread of visual narrative about the two central women, progressing from 1914 onwards. Edith Tudor-Hart's images followed the main protagonists through everyday life – working in a factory, cooking for their children or looking for work with their children in tow. This thread enabled viewers to follow and identify with the individuals through a narrative that was set against contextual images of current affairs, which the Goldfingers sourced from a range of agencies and publishers.[144] The Goldfingers' visual skill even in collaging stock images is shown in the lacing of the Tudor-Hart image, suspended midway on the panel, echoing a ship's rigging.

Poet Miles Tomalin wrote the exhibition's script, with emotive texts urging women to engage with politics: 'WOMEN! – you can be proud of the great part you are playing in this mighty war against the evils of Fascism ... WOMEN! – fight now for victory as you have seen Betty Britain doing'. Such labelling was intended to activate women through photographs, drawings and graphs, making them 'more conscious in their responsibilities and opportunities', as organisers explained, and inspiring pride in the part they were playing to fight fascism. The pictorial boards culminated in the question: 'WOMEN OF BRITAIN ARE YOU WITH HER IN THE FIGHT?' How they responded I do not know, as the archive does not reveal.

Exhibitions of solidarity were focused towards a limited number of alignments: the principal ones being support for Spain and Russia and acknowledging struggles for resistance against fascist regimes. In focusing collectively on these causes, artists and designers themselves created conditions of alignment and solidarity with each other in Britain, a new or temporary home for a large number of those who were central to developing this form. The Ministry of Information's wartime exhibitions programme – the focus of the next chapter – saw exhibitions being used to convey a wide range of agendas and ideas, with the mode of communication honed as 'story-telling'.

Notes

1 Myfanwy Evans, *The Painter's Object*, p. 5.
2 Andrew Stephenson, '"Strategies of Situation": British Modernism and the Slump c.1929–1934', *Oxford Art Journal*, vol. 14, no. 2, 1991, pp. 30–51.
3 Sally J. Scholz, 'Political Solidarity and Violent Resistance', *Journal of Social Philosophy*, vol. 38, issue 1, March 2007, pp. 38–52.
4 E. Durkheim, 'From Mechanical to Organic Solidarity', *Sociology: Introductory Readings*, vol. 2, no. 1, 2010.
5 Demos, *The Migrant Image*.
6 Vilém Flusser, *The Freedom of the Migrant: Objections to Nationalism*, ed. Anke K. Finger, trans. Kenneth Kronenberg (Urbana, IL: University of Illinois Press, 2003), pp. 3, 81, 86.
7 James C. Scott, *Domination and the Arts of Resistance* (New Haven, CT and London: Yale University Press, 1990), p. xiii.
8 As Paul Preston asserts in *The Spanish Civil War* (London: Harper Perennial, 2006).
9 National Archives ref KVS/112 is a list of the people who volunteered with the International Brigades in Spain 1936–9.
10 Felicia Browne digitised collection, TGA 201023, includes letters Browne wrote from Spain (EG reporting on bombing raids happening nearby). Tom Buchanan, 'The Lost Art of Felicia Browne', *History Workshop Journal*, issue 54, 2002, pp. 181–201.
11 Preston, *The Spanish Civil War*, p. 6.
12 Morris and Radford, *The Story of the Artists International Association*, p. 31.
13 Priscilla Thornycroft (Oral History, IWM catalogue number 31966), later known by the name Priscilla Siebert; Jackson, *British Women*, p. 156.
14 The exhibition was accompanied by a commemorative booklet, the text written by former pacifist artist member of the Bloomsbury Group, Duncan Grant. A few months later, in July 1937, Grant's partner Vanessa Bell's 29-year-old son Julian Bell would die in Spain whilst volunteering as an ambulance driver. Radford, *Art for a Purpose*, p. 49.
15 Browne was not the only AIA member to be killed in Spain: *Daily Worker* cartoonist Caro was a casualty. In summer 1936, based on the principle of collective security, the AIA affiliated with the newly launched International Peace Campaign (IPC), which held a Congress in Brussels in September 1936, the largest gathering of the World for Peace to date, to which AIA sent a delegate. TGA 7043/20 'AIA: The First Five Years, 1933–1938'; Tate, 'Felicia Browne: Unofficial War Artist', *YouTube*, 1 March 2016: www.youtube.com/watch?v=Y-QXhHpbV_k.
16 Photograph caption for Woolf's piece for *Daily Worker*, 14 December 1936.
17 Twenty thousand marchers, a long procession and carnival of flags, banners and colours, wound its way from the Embankment via Hyde Park to its eventual termination point in Shoreditch in London's East End, displaying hundreds of banners, many prepared by AIA artists. A large number of banners depicted heroic episodes and characters from Britain's radical past, the *Daily Worker*'s reporter describing the 'ghosts of England's fighters for freedom, ghosts of stalwarts, dead and gone' that must have marched alongside the marchers who appeared to the pageant spectators in a chronological sequence to evoke a sense of an unfolding radical tradition, with banners dedicated to Lilburn's Levellers, the Tolpuddle Martyrs, Robert Owen, the Chartists, Keir Hardie and William Morris. Prominent banners were dedicated to individuals not normally associated with radical labour history such as Simon de Montfort and Thomas More. Reported in *The Daily Worker*, 21 September 1936, pp. 1, 5. As discussed by Thomas

Linehan, 'Communist Culture and Anti-Fascism', in Nigel Copsey (ed.), *Varieties of Anti-Fascism: Britain in the Inter-war Period* (London: Palgrave Macmillan, 2010), p. 45.

18 Jim Fyrth, *The Signal Was Spain: The Aid Spain Movement 1936–1939* (London: Lawrence and Wishart, 1986).

19 W. Lipke, *David Bomberg: A Critical Study of His Life and Work* (London: Evelyn Adams & Mackay, 1967), p. 81. Bomberg was not the only artist forced to leave Spain upon the outbreak of war: painter Edward Burra was obliged to leave, making a series of work reflecting his experience of trauma and violence.

20 Warwick Digital Collections (WDC), Library Special Collections, JD 10.P6 PPC 2074, Medical Aid for Spain: The work of the Spanish Medical Aid Committee [prepared for the Committee by George Jeger and W. Arthur Peacock]. Also, 292/946/42/10; 292/946/18a/43 and 292/946/42/29 'Bulletins St Pancras and Holborn Spain Week'.

21 WDC, TUC archives, 292/946/18b/23 Newsletter of 5 March 1937, Spanish Situation – Pamphlets, Leaflets, etc. 1936–1938, 292/946/18b/23.

22 WDC, TUC archives, 292/946/16a/47 Spanish Rebellion.

23 Jackson, *British Women*, p. 154.

24 Formed at 82 Cliffords Inn, London EC4.

25 WDC, TUC archives, 292/946/16a/47 'Spanish Exhibition Committee'. Correspondence from 21 January 1937 discusses the Spanish Exhibition Committee with Adler at the Labour History Archive and Study Centre (People's History Museum) LP/WG/SPA/197–198. See also Fenner Brockway, *Inside the Left* (London: Allen & Unwin, 1942), p. 298.

26 WDC, TUC archives, 292/946/18b/36 'Spanish Exhibition', leaflet published by Farleigh Press.

27 WDC, TUC archives, 292/946/42/110, 'The Child and the War'.

28 Alan Powers, *Abbatt Toys: Modern Toys for Modern Children* (London: Design for Today, 2020). The Abbatts' gallery was used for regular exhibitions with a social focus: in March 1940 for showing projects for young children and their mothers, as Jean-Louis Cohen describes in *Architecture in Uniform: Designing and Building for the Second World War* (Paris: Canadian Centre for Architecture, 2011), p. 159. See also Erin McKellar, 'Designing the Child's World: Ernö Goldfinger and the Role of the Architect, 1933–1946', *Journal of Design History*, vol. 33, no. 1, 2020, pp. 50–65.

29 Radford, *Art for a Purpose*, p. 58 (formative principles of AIA put to the test) and p. 87 (about conflict between AIA and Surrealists).

30 Alongside this, a meeting was held at Conway Hall to discuss 'The Relation of Art to the State and Public'. Radford outlines the concerns discussed at the meeting in detail in *Art for a Purpose*, pp. 54–5.

31 Henry Moore Institute. Moore continued to be vociferous in his support for the Spanish Republicans, making a lithograph called *The Spanish Prisoner* in 1939, depicting a Spanish refugee behind the barbed wire of a French detention camp, intended to be sold to raise funds for Spanish refugees fleeing to France but never, in fact, published due to the outbreak of the Second World War. While fellow sculptor Barbara Hepworth showed her solidarity with the Republicans through her wooden sculpture *Project: Monument to the Spanish Civil War* (1938–9), a casualty of the Second World War when it was destroyed.

32 TGA AIA collection 7043.2.5.

33 Radford, *Art for a Purpose*, p. 57 points out that an installation view is illustrated in the 1937 *Circle* book.

34 UCL Townsend Journals (unpublished entry, 23 April 1937).

35 'Display for Peace', *Display*, 20, September 1937, p. 282; Radford, *Art for a Purpose*, p. 59; Morris and Radford, *The Story of the Artists International Association*, p. 37; Robin Stemp, 'Nan Youngman: Part 1', *The Artist*, November 1987, p. 22.
36 Golan, 'The World Fair', pp. 180–1.
37 Jordana Mendelson, 'Josep Renau and the 1937 Spanish Pavilion in Paris', in Jorge Ribalta (ed.), *Public Photographic Spaces: Exhibitions of Propaganda, from Pressa to The Family of Man 1928–55* (Barcelona: Museu d'Art Contemporani de Barcelona, 2008), pp. 313–49, p. 314.
38 James Holland, 'The Thirties', *The Designer*, November 1979, p. 4; Holland sketched Guernica for *Left Review*, August 1937. See also Lynda Morris and Christoph Grunenberg (eds), *Picasso: Peace and Freedom* (London: Tate, 2010).
39 Golan, 'The World Fair', p. 183.
40 Flyer in WDC, TUC archives, 292/946/18b/66(ii), mentions Wilfrid Roberts MP as Chair of organising committee working with Penrose, with Herbert Read as Deputy Chair, Earl of Listowel, E.L.T. Mesens and Mrs Sybil Stephenson. Patrons included Gerald Barry, E. McKnight Kauffer and Virginia Woolf. Penrose suggests in *Picasso: His Life and Work* (London: Gollancz, 1958) that *Guernica* venues included Liverpool but mentions only Leeds and Oxford in the end; Penrose, *Scrapbook* (London: Thames & Hudson, 1981), p. 87 and Penrose, *Picasso*, p. 286. See also Anthony Penrose in Bohm-Duchen, *Insiders Outsiders*, p. 176.
41 Jean-Paul Sartre, *The Age of Reason* (London: Penguin, 1967); Blunt, *The Spectator*, August 1937.
42 Herbert Read 'Picasso's Guernica', *London Bulletin*, no. 6, October 1938, p. 6; reprinted in G. Schiff (ed.), *Picasso in Perspective* (Hoboken, NJ: Prentice Hall, 1976), p. 105. Stephen Spender also wrote about it in *New Statesman and Nation*, 16 October 1938.
43 WDC, TUC archives, 292/946/18b/66, Catalogue of *Guernica* at New Burlington Galleries. Roland Penrose, *Picasso*, p. 286.
44 Photograph in Bohm-Duchen, *Insiders Outsiders*, p. 177. Elizabeth Cowling, *Visiting Picasso: The Notebooks and Letters of Roland Penrose* (London: Thames & Hudson, 2006), p. 42.
45 According to Remy, *Dictionary of Surrealism*, p. 20.
46 A press release headed 'Manchester Foodship for Spain' announced it would be at 32 Victoria St, Manchester, in February 1939; *Manchester Evening News*, 31 January 1939.
47 Nicola Ashmore's project 'Guernica Remakings' interrogates the history of Guernica, mapping its changing display contexts and interpretations, and its potency as the focus for contemporary activism: http://guernicaremakings.com/about/about-nicola-ashmore/.
48 Egbert, *Social Radicalism*, p. 522.
49 Egbert, *Social Radicalism*, p. 798 (footnote 51).
50 AIA leaflet 'Activities Since 1938' (updated but published in July 1942), quoted in Egbert, *Social Radicalism*, p. 798 (footnote 510). See also Fred Uhlman, *The Making of an Englishman* (London: Gollancz, 1960), pp. 212–15.
51 Egbert, *Social Radicalism*, p. 505.
52 Anthony Penrose puts this figure higher at £4,000 in Bohm-Duchen, *Insiders Outsiders*, pp. 177–8. See also Egbert, *Social Radicalism*, p. 798 (footnote 51); David Brown, 'Refugee Artists in Great Britain', *Art and Artists*, April 1984; Frowein, 'German Artists', p. 52.
53 Parsons Gallery was part of the Oxford Street showroom of Thos. Parsons and Sons, a paint manufacturing business.
54 Sue Grayson Ford, *Brave New Visions: The Émigrés Who Transformed the British Art World* (London: Sotheby's, 2019), p. 14.

55 Reproduced in *Brave New Visions*, p. 14. Details of the artists whose works were shown are listed here: www.artist-info.com/exhibition/Parsons-Gallerie s-Id385722 (accessed 21 October 2021).

56 The Six Point Group was a non-party feminist organisation founded by Lady Margaret Rhondda in 1921 to agitate for changes in British law. The Group co-operated with other feminist organisations in 1935 for the *Anti-Nazi Exhibition*, according to Julie Gottlieb, 'Varieties of Feminist Responses to Fascism', in Nigel Copsey (ed.), *Varieties of Anti-Fascism: Britain in the Inter-War Period* (Basingstoke: Palgrave Macmillan, 2010).

57 In 1940 Mondrian fled to New York to escape the escalating war.

58 Janet Wolff, *The Aesthetics of Uncertainty* (New York: Columbia University Press, 2008), pp. 104–6 discusses the arrival of Polish Jewish artists Jankel Adler and Josef Herman.

59 Valeria Carullo, 'Refugee Architects on the Brink of World War Two', talk for *Insiders Outsiders* festival, September 2021: www.youtube.com/watch?v=imKG skNPC-s (accessed 1 October 2021).

60 Wolff, *The Aesthetics of Uncertainty*, p. 103.

61 Egbert, *Social Radicalism*, pp. 523–6. See also Frowein, 'German Artists', p. 52.

62 Heartfield's photomontage 'The Happy Elephants' appeared in *Picture Post*, vol. 1, no. 3, 15 October 1938, p. 9; and Richard Carline, 'John Heartfield in England', in *John Heartfield 1891–1968: Photomontages* (London: Arts Council of Great Britain, 1969). Barron suggests his overtly political work ruled him out in 'John Heartfield in London', in Stephanie Barron and Sabine Eckmann (eds), *Exiles and Emigres: The Flight of European Artists from Hitler* (Los Angeles: Los Angeles County Museum of Art, 1997), p. 74. Heartfield stayed in Britain until 1950 when he returned to East Germany.

63 Wolff, *The Aesthetics of Uncertainty*, pp. 8–9, 101–2; Monica Bohm-Duchen, *Art in Exile in Great Britain 1933–45* (London: Belmont Press, 1986); Radford, *Art for a Purpose*, p. 39; Frowein, 'German Artists', p. 47.

64 Uhlman, *The Making of an Englishman*, p. 213.

65 Willi Guttsman, 'The Influence and Failure of Weimar Radicalism in Emigré Art in Britain', *Third Text*, vol. 5, no. 15, 1991, pp. 43–4.

66 Charmian Brinson and Richard Dove, *A Matter of Intelligence: M15 and the Surveillance of Anti-Nazi Refugees, 1933–50* (Manchester: Manchester University Press, 2015), p. 115.

67 The house was at 36 Upper Park Road, London NW3. The FGLC lasted for seven years.

68 For more information about Lucy Wertheim see: https://townereastbourne.org.uk/whats-on/exhibitions/lucy-wertheim.

69 *Jewish Chronicle*, July 1939, p. 43. Sculpture praised in *The New Statesman and Nation*, 24 June 1939.

70 Remaining in Britain until 1950, Heartfield took on non-political freelance commissions for book covers for publishers such as Secker & Warburg and Lindsay Drummond. Pamphlet covers designed by Heartfield in 1942–3 are held in the Wiener Holocaust Library collection.

71 Carline, 'John Heartfield in England'.

72 Ford, *Brave New Visions*, pp. 7, 10.

73 After release, in April 1941 Heartfield gave a talk on 'What Peasant Breugel [sic] Has to Say to Us'; letter of 1941 reproduced in Peter Pachnicke and Klaus Honnef (eds), *John Heartfield* (New York: Harry N. Abrams, 1992), p. 311.

74 Brinson and Dove, *A Matter of Intelligence*, p. 119 quoting from TNA KV2/1010.

75 Held in February 1943. Pamphlet in Whitechapel Archive carried an image by Erich Kahn with Secretary Worner's address at 36 Upper Park Rd NW3 at the bottom, reproduced Bohm-Duchen, *Insiders Outsiders*, p. 219.

76 Wolff, *The Aesthetics of Uncertainty*, p. 111.
77 Charmian Brinson and Richard Dove (eds), *Politics by Other Means: The Free German League of Culture in London, 1939–1946* (London: Vallentine Mitchell, 2010), p. 63.
78 Nicholas J. Saunders documents such processes in his book *Trench Art: Materialities and Memories of War* (Abingdon: Routledge, 2020). A wartime exhibition at the Knoedler Gallery in Bond Street showed art works done by British prisoners of war in enemy prison camps (documented in IWM D series 15002–15010), including portrait heads and landscapes based on postcard views (given the limited vista from inside camps).
79 Klaus Hinrichsen, 'Visual Art behind the Wire', in David Cesarani and Tony Kushner (eds), *The Internment of Aliens in Twentieth Century Britain* (London: Frank Cass, 1993), p. 195.
80 Leo Baeck Institution, 'Art Exhibition', *The Camp*, 21 September 1940. Dave Hannigan, *Barbed Wire University: The Untold Story of the Interned Jewish Intellectuals Who Turned an Island Prison Into the Most Remarkable School in the World* (Guildford: Lyons Press, 2021) and Simon Parkin, *The Island of Extraordinary Captives: A True Story of an Artist, a Spy and a Wartime Scandal* (London: Sceptre, 2022).
81 TGA 20052/1/5/19, Klaus Hinrichsen, typescript for Radio 3, 'Interned with Kurt Schwitters', p. 5.
82 Megan Luke, *Kurt Schwitters: Space, Image, Exile* (Chicago, IL: University of Chicago Press, 2014).
83 TGA 19043A, in a letter from Diana Uhlman to Kenneth Clark she reports on the exhibition Fred Uhlman has contributed to at Hutchinson camp; Hinrichsen, 'Visual Art behind the Wire', pp. 192–3, 202.
84 Luke, *Kurt Schwitters*, p. 162.
85 Klaus Hinrichsen collection: photograph of internees at the second art exhibition at Hutchinson Internment Camp November 1940, TGA 20052/2/11; TGA 20052/2/7/13; 20052/2/7/17; TGA 20052/2/7/18; Hinrichsen, 'Visual Art behind the Wire', p. 200.
86 Hinrichsen, 'Visual Art behind the wire', p. 203.
87 TGA 20052/2/7/13 Major H. O. Daniel, Photograph of internees at the second art exhibition at Hutchinson Internment Camp, November 1940.
88 Hinrichsen recalled his fellow internee, celebrated Expressionist painter Ludwig Meidner, chose not to join in with exhibitions; Hinrichsen, 'Visual Art behind the Wire', p. 201; TGA 20052/1/5/19, Klaus Hinrichsen, typescript for Radio 3, 'Interned with Kurt Schwitters'.
89 Leo Baeck Institution Archives.
90 Rachel Dickson, Sarah MacDougall and Ulrike Smalley, '"Astounding and Encouraging": High and Low Art Produced in Internment on the Isle of Man during the Second World War', in Gilly Carr and Harold Mytum (eds), *Cultural Heritage and Prisoners of War: Creativity behind Barbed War* (Abingdon: Routledge, 2012), p. 193. Quoted in Ford, *Brave New Visions*, p. 22. Hinrichsen, 'Visual Art behind the Wire', p. 199.
91 Rachel Dickson, 'Our Horizon is the Barbed Wire', in Bohm-Duchen, *Insiders Outsiders*, p. 153.
92 Jane England, *Jack Bilbo & the Moderns*, exh. cat. (London: England & Co. Gallery, 1990), p. 3; Ford, *Brave New Visions*.
93 Michael Paraskos, 'Herbert Read: The Eye of the Storm', in Bohm-Duchen, *Insiders Outsiders*, pp. 184–5. TGA 200410/10/3/8 and TGA 200410/10/3/9. Bilbo was interviewed by British Pathé in 1947: www.youtube.com/watch?v=FD1Oy7J1JXs. He self-published his autobiography, *Jack Bilbo: An Autobiography (The first forty years of the complete and intimate life-story of an Artist, Author,*

Sculptor, Art Dealer, Philosopher, Psychologist, Traveller and a Modernist Fighter for Humanity), in 1948, closing Modern Art Gallery the same year.

94 Luke, *Kurt Schwitters*, p. 171.

95 According to Paraskos, this delighted Schwitters.

96 Hinrichsen, 'Visual Art behind the Wire', p. 201; typescript for Radio 3, 'Interned with Schwitters', p. 8.

97 Schwitters TGA 7212/21.

98 On Monday 4 December, as noted by Ford, *Brave New Visions*, p. 34. England, *Jack Bilbo & the Moderns*, p. 3 quotes from Bilbo.

99 Luke, *Kurt Schwitters*, p. 171.

100 Keith Holz in Bohm-Duchen, *Insiders Outsiders*, p. 218.

101 René Graetz quoted in 'Widerstand statt Anpassung, Deutsche Kunst im Widerstand Gegen den Faschismus, 1933–1945' (exhibition catalogue), Berlin, 1980, quoted in Guttsman, 'The Influence and Failure', p. 45.

102 As the poster held at Marx Memorial Library noted, 'owing to great success the exhibition was extended from 26 July to 16 August'.

103 Holz, in Bohm-Duchen, *Insiders Outsiders*, p. 218; Jutta Vinzent, 'Muteness as Utterance of a Forced Reality Jack Bilbo's Modern Art Gallery (1941–1948)', in Shulamith Behr (ed.), *Arts in Exile in Britain 1933–1945: Politics and Cultural Identity* (Leiden: Brill, 2005), p. 330. René Graetz and Heinz Worner archives are in the Akademie der Künste, Berlin. Graetz was born in Berlin; his father was of Polish Russian extraction, as Elizabeth Shaw explains in 'Radical Émigré Artists and Their Return to DDR', *Third Text*, vol. 5, no. 15, 1991, pp. 57–61.

104 Holz, in Bohm-Duchen, *Insiders Outsiders*, p. 218.

105 Rothholz designed these fundraising stamps, depicting scenes of underground and resistance activity in Germany; UoBDA, H. A. Rothholz Archive DES/RHZ/1/1/13.

106 Holz, in Bohm-Duchen, *Insiders Outsiders*, p. 219 observes lots of similarities between these panels and those the DKK had created for the NY World's Fair, 1939 – 'Germany of Yesterday – Germany of Tomorrow, 1938–9' – which were never shown.

107 Hilde Marchant, *Daily Mirror*, 29 June 1942 – I have found no evidence that this statement is true for any of the contributors.

108 *Manchester Guardian*, 25 June 1942; *New Statesman*, 11 July 1942.

109 'Allies inside Germany: Letters to the Editor', *Spectator*, 17 July 1942, p. 12.

110 A. W. Stargardt, 'Allies inside Germany: The German Resistance Movement against Nazi-Fascism', *The Australian Quarterly*, vol. 16, no. 3, September 1944, pp. 23–9.

111 Frowein, 'German Artists', p. 54.

112 WMRC MSS.292/946/1/26, 'Text of speech by Louis Marin at the opening of the Resistance Exhibition, Warrington, 14 Jul 1944'.

113 Tom Harrisson, 'A Child's View of the War', *Picture Post*, 11 May 1940, pp. 17–19.

114 S. Roberts, 'Education, Art and Exile: Cultural Activists and Exhibitions of Refugee Children's Art in the UK during the Second World War', *Paedagogica Historica*, vol. 53, no. 3, 2017, pp. 300–17.

115 *The War as Seen by Children* catalogue published by Refugee Children's Evacuation Fund, with a foreword by J. G. Siebert and reproducing a speech by Kokoschka.

116 *The War as Seen by Children*, p. 6.

117 Paul Corthorn, 'Labour, the Left and the Stalinist Purges of the Late 1930s', *The Historical Journal*, vol. 48, no. 1, March 2005, pp. 179–207.

118 Lothar Kettenacker, 'The Anglo-Soviet Alliance and the Problem of Germany, 1941–5', *Journal of Contemporary History*, vol. 17, no. 3, July 1982, pp. 435–58.

119 *Comrade in Arms* was available in 52 sets to all regions, with 20 sets each to Army Bureau of Information and Ministry of Supply, appearing in 1,100 sites across Britain (Aulich, *War Posters*, p. 348 taken from National Archives INF1/676). The MOI's 'Soviet exhibition' shown at Sunderland Museum was reviewed in *Museums Journal*, vol. 42, September 1942, p. 150.
120 As noted by Cohen, *Architecture in Uniform*, p. 332.
121 James Dunnett and Gavin Stamp, *Ernö Goldfinger: Works 1* (London: Architectural Association, 1983), p. 19.
122 Dunnett in Dunnett and Stamp, *Ernö Goldfinger*, p. 13 [image p. 69] and p. 58 [image 68]. *Architects Journal*, 24 February 1938, 10 March 1938; *Architects Review*, September 1938; *Shelf Appeal*, March 1938.
123 Colin Penn was an active Communist Party member who, together with Goldfinger, built the Daily Worker offices in 1946, according to Robert Elwall, *Ernö Goldfinger* (London: Academy Editions, 1996), p. 69.
124 Peter Watson went on to found the ICA with Roland Penrose a few years later.
125 Barbara Pezzini '"Aid to Russia" A Wartime Modern Art Exhibition in a Modernist Setting', *The British Art Journal*, vol. 5, no. 3, p. 68. See also The National Trust Willow Road Archives.
126 Pezzini, 'Aid to Russia', p. 68. Kenneth Clark had been approached but had refused, citing a prior engagement.
127 Image of Goldfinger and Cunard in Dunnett and Stamp, *Ernö Goldfinger*, p. 46, which states incorrectly that the Willow Road exhibition was held in 1943.
128 The exhibition closed on 21 June 1942, having welcomed 1,778 and raising £200; Pezzini, 'Aid to Russia', p. 67.
129 Reports on Eastern Front in *Architects Journal*, 16 July 1942 and *The Times*, 23 June 1942, in Dunnett and Stamp, *Ernö Goldfinger*, p. 61.
130 Opened by the Duke of Devonshire, it then toured to Manchester. Image is at IWM D17866. 'An Empire Exhibition', *Manchester Guardian*, 29 February 1944, p. 2.
131 The Aid to Russia Fund of the National Council of Labour, initiated in 1941 by the Joint War Organisation, with Clementine Churchill chairing. The national Aid to Russia campaign was taken over by the Red Cross and focused on medical supplies.
132 Richard Hollis interview with F. H. K. Henrion in 1986 Imperial War Museum Catalogue 9592/ 1986–11–12: www.iwm.org.uk/collections/item/object/80009378.
133 William Blake, 'The Sword and the Sickle', in Geoffrey Keynes (ed.), *Blake Complete Writings* (Oxford: Oxford University Press, 1966), p. 178.
134 *Architects Journal*, 19 November 1942.
135 RIBA GolEr/400/1.
136 RIBA GolEr/281/1, letter from Trevor Thomas, Director to Ernö Goldfinger, 29 October 1942.
137 Dunnett and Stamp, *Ernö Goldfinger*, p. 61 – they reference *Hampstead News*, 4 March 1943.
138 IWM D photograph series D12681 and D12690.
139 Robin Kinross, *Modern Typography: An Essay in Critical History* (London: Hyphen Press, 1992), p. 107.
140 Tate Archives LON/ AIA 1943. Held in London at Hertford House, 31 March to 25 May 1943.
141 RIBA GolEr/400/2 LWP Exhibition.
142 Refugees from National Socialism in Wales: Learning from the Past for the Future: https://wp-research.aber.ac.uk/nsrefugeeswales/history/background-of-refugees/refugee-profiles/edith-tudor-hart/. See also https://discovery.nationalarchives.gov.uk/details/r/C14895030.

143 Elwall, *Ernö Goldfinger*, p. 59 includes Women's Parliament – image at RIBA. The zenith of Goldfinger's exhibition work would be his role years later in designing *This is Tomorrow* at the Whitechapel Art Gallery in 1956.

144 Goldfinger papers 'LWP exhibition', GolEr/400/2. Correspondence in RIBA archive (5 July 1943) shows that Miles Tomalin and Edith Tudor-Hart were amongst those organising the exhibition.

7

Exhibitions as weapons of war

Stressing the distinction between techniques used to create 'democratic' propaganda by 'factual information' and the 'hysteria-stimulating' of totalitarian regimes, exhibition designer Misha Black noted that 'in the field of the purely documentary or descriptive exhibition', 'a remarkable technique' had been developed in Britain during the Second World War.[1] This, he said, was the use of the 'informative and story-telling type of exhibition', differentiated from 'the simple display of commodities', in which Britain is 'superior to the most efficient foreign competition'. British designers had, Black observed, 'fully embraced the communicative potential of exhibitions', pioneering 'a new approach to contemporary exhibition design'.[2] Such communicative exhibitions were, he thought, particularly effective in 'creating the core of informed opinion on specific subjects' and their influence could 'spread as widely as a contagious disease', showing his interest in the comparative efficiencies of this medium when placed alongside others.[3] Black was at the forefront of developing exhibitions as weapons of the British war effort while employed at the Ministry of Information (MOI) and is pivotal to the focus of this chapter about exhibitions' use as weapons by the British government and allied groups during the Second World War.

This chapter explores the way that exhibitions became weapons of war through the MOI's programme, intended to teach practical skills and to create patriotic and responsible citizens. These affective exhibitions were formed to inspire particular actions and emotions in audiences at home and in ally countries where they were sent, including pride, optimism, reassurance and, at times, a sense of horror at atrocities. These were all forms of 'propaganda of integration', to use sociologist Jacques Ellul's phrase, creating awareness and acceptance of certain behaviours and conditions of war. Wartime government exhibitions were a development of the art of projection used across earlier public sites and spaces. Their designers were recruited to create 'good' citizens on the home front. MOI exhibitions were a form of argumentation aimed at a mass public. All of these exhibitions had photographs at their core, sometimes in the mode of photo-stories, at others as documentary and collage, helping people visualise wartime abstractions and revealing the mechanics of war to the British population.[4]

Designing Britain's war exhibitions

Many of those designers who had led the development of exhibitions as communication or information across pre-war contexts reappear in this chapter. Industrial designer Milner Gray was appointed to lead the MOI's new Exhibitions Branch. Gray had led several design consultancies by the outbreak of the Second World War, as well as being founder member of the Society of Industrial Artists in 1930. London-born Gray studied commercial art at Goldsmiths College London, leaving during the First World War to join the army, later transferring to the Royal Engineers camouflage school. He co-founded the Bassett Gray Group of Artists and Writers in 1921: a group of artists, writers and designers (including his friend painter Graham Sutherland and long-term collaborator Misha Black) who took on design commissions for exhibition stands, packaging, china and textiles. Gray reorganised Bassett Gray to create Industrial Design Partnership, successor to Bassett Gray, from 1934 to 1940.

On arrival at the MOI, Gray requested they appoint Black as, he explained in a memo, 'a constructive architect on the arrangement and layout of Exhibitions, with special regard to his knowledge of propaganda requirements'.[5] Black started at the MOI in January 1941, helping on technical aspects of mounting exhibitions, with Norbert Dutton assisting on organisation and design.[6] Other designers were also recalled from their wartime service to join the MOI's Exhibitions Branch. Many were drawn from the established networks of Black and Gray, including James Holland (who had co-founded the AIA with Misha Black in 1933 and was central to its work through its first decade), prominent architect Frederick Gibberd (a MARS member, who had contributed to the 1938 MARS Group exhibition at New Burlington Galleries) and commercial artist Richard Levin (introduced in Chapter 1).

As across many of the pre-war contexts in which exhibition design was being pioneered, several exhibition branch employees had only recently arrived in Britain. Architect Bronek Katz was born in Warsaw, German-born graphic designer F. H. K. Henrion joined the MOI's Exhibitions Branch on release from internment as an enemy alien on the Isle of Man and German-born architect Peter Moro joined the MOI's Exhibitions Branch after being interned for six months at Kempton Racecourse and then on the Isle of Man.[7] Hungarian-born architect Ernö Goldfinger (introduced in earlier chapters) also carried out occasional work on MOI exhibitions. Each of these brought extensive previous experience of mounting exhibitions, designed for different contexts and causes, to their work at the heart of government. Although having diverging professional backgrounds – some, like Henrion and Levin, coming from commercial art backgrounds while others, like Moro and Goldfinger, had trained as architects – all were united in finding forms for information design in a shifting ecology of communication. Many of those employed on exhibitions work had trained and practised in numerous locations across continental Europe but became central to shaping the visual representation of Britain's war effort. Working at the heart of government was a pragmatic decision for Left-aligned designers, who took commissions across many contexts at once. Despite experience at the centre, these designers continued to experience prejudice. Black had been turned down for naturalisation as a British citizen (despite living in Britain almost his entire life).[8] At the outbreak of war, Richard Levin had found it difficult to get war work because of his Russian father (despite having been born in Britain).[9]

Morale-boosting exhibitions in shop windows

At the outbreak of war, exhibitions were principally used by the British government as promotion for trade and industry and vehicles of soft power and national projection. War would change this, convincing politicians and civil servants that exhibitions could add an important new element to Britain's armoury of propaganda, by showing clearly elements that were otherwise distant or abstract; giving the public a more vivid and absorbing picture of how Britain was fighting, through three-dimensional environments that could envelope people in information.[10]

The MOI, re-formed to co-ordinate propaganda after the outbreak of war, did not immediately establish a team to create exhibitions and displays, being almost exclusively focused on planning for news.[11] Shops and department stores were first to identify the potential of commercial window displays to amplify government messages to the passing public. Early in the war, *Display* spotlighted shops that were already supporting topical aspects of the war effort: London department store Whiteleys' display manager had mounted a campaign appealing to the public to save waste paper; London department store Selfridges had joined the waste paper salvage campaign and, elsewhere, gave over their windows to Fougasse 'Careless Talk Costs Lives' posters, an engaging series visualising the potentially catastrophic consequences of being drawn into seemingly trivial chatter about the war effort.[12] Shops around the country followed suit. *Display*'s July 1940 cover showed the window of Southsea shop Handley's Ltd, devoted to War Savings Certificates. Such windows, the magazine enthused, had propaganda potential 'next only to the press and radio', drawing a direct comparison across media.[13]

Aware of the impactful and impressive way Nazis used window displays as propaganda, early on in the war the British display industry advocated for shop displays as a way to amplify key messages to the public. *Display* editor Richard Harman declared 'One of the biggest mediums of compelling public attention is display', observing that Goebbels had seized not only newspapers, radio and advertising but 'the whole of the shop windows of Germany too, for Germany is the most display-conscious nation in the world' with 'every display man and window dresser a member of a state display organisation, with a Nazi official at the head'.[14] Hitler's election campaigns had been fought through window displays, the giant word 'Ja' ('yes') repeated by the thousand in windows in every German town, co-ordinated by Goebbels' propaganda department. The British MOI, Harman concluded, had a long way to go in recognising the importance of window displays to the home effort.

Seeing the importance of speaking to people on the home front in places where they lived and worked, the MOI finally announced a non-commercial shop window display scheme in July 1940. Themes would be issued fortnightly, the intention 'to bring pictorially to the public mind some important national fact – sometimes it may be an instruction – that hitherto may have been only an impression gained from radio or newspaper, thus driving home the point concerned and fixing it indelibly on the public mind'. The first theme, 'Hold Fast and We Win', was an appeal for the public to show determination while the country was passing through such a testing time.[15] The MOI helped shops by providing photo and map enlargements and reproductions of newspaper cuttings. *Display* commented, 'Publicity divorced from reason and truth can play no part in the British scheme', but 'windows which echo the whole country's sentiments at the present time will now – not under the compulsion of a pistol or truncheon, but by common consent – appear in windows from end to end of a nation which is destined to become a fortress for freedom'. Here was a statement about the power of consensual and integrational propaganda, helping to build the wartime citizenry, with *Display* at the helm of the invisible government, ready to reinforce messages.

Launching the government's exhibitions programme

By late 1940, the MOI, under Director-General Frank Pick, had added exhibitions to their expanding communications environment. Pick had pioneered the use of exhibitions as publicity in his previous roles at the London Passenger Transport Board, Design and Industries Association and Council for Art and Industry, as discussed earlier in this book.[16] At the MOI, exhibitions became an element in the series of media deployed in tandem for each campaign, including radio, press, film, posters and booklets. A substantial report by civil servant A. G. Highet, who had a background in publicity for the General Post Office, justified the addition of exhibitions in wartime.[17] 'There are exhibitions and exhibitions, just as there are newspapers and newspapers', Highet wrote. 'Generally speaking, the standard of exhibition technique in this country is not comparable with that of the continental nations. This fact increases the importance of Government Departments being ahead of exhibition design here'. Highet's ambition was for the government to pioneer British use of exhibitions as propaganda.

The major shortcoming of propaganda exhibition design in Britain, Highet believed, was that they were not realising their three-dimensional character. The flat should be combined with the three-dimensional, using developments of 'the perspective photograph' and 'photo-montage'. Properly used in this way, exhibitions would, Highet thought, permit the designer to tell a story 'chapter by chapter as the visitor walks into or out of the display', using the metaphor of story-telling to describe their potential impact. He assessed how far exhibitions were a sensible use of public money for this new context, calculating the potential cost of exhibitions on a cost per head basis as roughly one penny per visitor to film displays and that, 'while the visitor to the cinema may see six films, the exhibition may be designed to explain one problem only', indicating the way the comparative merits of exhibition to the wartime context were being thought about in relation to other media.[18] Highet reflected that exhibitions had much in common with posters and leaflets, reflecting 'the mind of those responsible for it', concluding that displays in the windows of vacant shops were the best value for money.

Art historian and National Gallery Director Kenneth Clark advised on the development of an MOI exhibitions programme.[19] Clark, who had joined the MOI at its inception, had set up the War Artists' Advisory Committee (WAAC) in November 1939 with the very different aim of appointing artists to record and document – 'to draw up a list of artists qualified to record the (Second World) war at home and abroad ... to advise on the selection of artists from the list for War purposes and [to] advise on such questions as copyright, disposal and exhibition of works and the publication of reproductions'.[20] Under Clark, the National Gallery was hosting a series of wartime exhibitions of war artists, including one in 1941 organised by art historian and art dealer Lillian Browse.[21] The project of the MOI's Exhibitions Branch was different from the archival instinct of the WAAC, being focused towards providing useful information that could impact immediately on people's emotions and behaviours.

Sending exhibitions to ordinary working places

MOI officials calculated that exhibitions developed as part of a defined strategy could supplement wider campaigns. Architect G. S. Kallmann echoed this, using a musical metaphor to describe MOI exhibitions providing the 'opening bars' or 'the climax', rather than carrying the full extent of a government information campaign.[22] Site was key to reaching audiences; the most suitable identified by officials as 'where the public normally meets': 'stations, cinemas, the large suburban and provincial shops'.[23] Six exhibition subjects were to be planned each year for circulation as 'photographic documentaries, linked together by captions, arranged and numbered so as to tell a consecutive story … the captions to be devised by first-class caption writers'.[24] According to MOI guidance, suitable exhibition sites were ordinary working places that were commonly found across the country such as the works canteens of factories, underground shelters (particularly when exhibitions were dealing with the subject of infection), marquees in parks, fairgrounds and markets during the summer and, during the winter, empty shops and stations – in effect, taking the material to people where they gathered either for work or for leisure, an aspiration in the spirit of pre-war exhibition groups such as the Artists International Association (AIA). Museums, galleries and other familiar or established sites for visual culture were notably absent from this wartime list. Larger exhibitions, inviting mass gatherings, were discouraged in the guidance as too dangerous, at least initially.

Exhibitions of different sizes, to fit town halls and shop windows, made a visible and urgent response to war and allowed campaigns to be visually and materially represented, to 'show' the war as well as to 'tell' it, to justify to the public the rightness of policy directions and to bring war closer, becoming more than a distant abstraction. Such exhibitions stood to persuade the viewer of the veracity of a version of the world as it stood at that point; presenting not a constructed past but the world at that moment in a universal present, lacking in specificity or historicity in order to have a broad appeal and to remove human subjects from class conditions and structures of difference and to foreground a 'quality of usualness'.[25] They stood to connect with the viewer, making an urgent appeal, seeking empathy and engagement. A separate scheme, sponsored by MOI's Photographs Division, issued groups of photographs in sets of eight or ten to three hundred sites: libraries, art galleries, museums, information centres, all mounted on screens provided by the Ministry. Meanwhile, the Window Display Scheme sent out a set of posters to over a thousand sites for display every three weeks.

The way exhibitions should engage the public was described in internal memos and reports. They must grab visitors' attention, compelling examination of a theme 'even by the most disinterested spectator' through 'simple statements almost in the form of slogans', enabling visitors to leave 'with a few fundamental ideas stamped on their memory'. For a 'really serious student', those with a deeper interest, detailed information would be available in the form of 'elaborated statistics' and 'graphs'. 'Informatory matter' must be linked with 'a human appeal', making it directly relevant to the viewer; 'the use of mechanical movement' meaning displays considered visitors' manoeuvrings through them; the use of 'personal demonstration', with people employed to show visitors how to carry out relevant tasks; and 'the placing of common objects in unusual theatrical settings' to heighten engagement.[26] MOI exhibition designer Black shared this interest in understanding how to reach viewers, comparing exhibitions with other media 'automatically excluding all distracting elements' – in not having to compete with other media, 'the exhibition takes equal place with the film in completely encompassing the spectator and allowing only those distractions which are deliberately planned to accentuate the effect', he wrote, suggesting the carefully choreographed and controlled routes through which they envisaged visitors moving.[27] Whether visitors conformed to this behaviour is less certain.

In government statements, exhibitions had been identified as offering a means of engaging and informing British people, as an integral element of the multimedia environment of the information war, with media and politics becoming entangled in Britain's armoury of propaganda. Thinking about media theorist Fred Turner's major claim for exhibitions in the parallel period in the US, we might ask how orchestrated the messages were in shaping and directing not only wartime knowledge but the British 'democratic personality'. In his book *The Democratic Surround*, Turner argues that US authorities, seeking to counter the impact of fascism's mastery over the public through mass media, had built a 'democratic personality' during the 1940s and 1950s, through creating alternative communications environments, which Turner calls 'surrounds', where US citizens could be developed as rational and empathetic individuals.[28] Turner charts how anthropologists and psychologists developed this concept, drawing on the input of refugee artists of the Bauhaus who had newly arrived in the US, with highly developed theories of multi-screen display and immersive theatre. These artists helped visualise and build these 'surrounds', enacted across various environments, with major museum exhibitions key. The legacy of these 'surrounds' Turner sees continuing today.

The context that Turner analyses has clear parallels with the British setting of this book. Some of the ex-Bauhaus faculty Turner describes as architects of the US 'democratic surround' had previously spent time living and working in exile in London in the 1930s, most prominently architect Walter Gropius (living in London from 1934 to 1937) and artist László Moholy-Nagy (living in London from 1935 to 1937). Both took roles in shaping exhibitions culture in Britain (as discussed earlier in this book); neither was central to one institution but survived from precarious and piecemeal employment during their temporary stays.

There are key differences between the US and British wartime environments, however, which make Turner's 'democratic surround' less applicable to the British context. The US government exhibitions programme was more centralised, with much significance attached to major landmark exhibitions mounted at MOMA New York (and a few other cultural contexts such as music venues). A more extensive and de-centralised programme of wartime propaganda exhibitions was developed in the British context, proliferating across the country, in venues of varying sizes and scale, for various campaigns and causes, designed and led by dozens of people (as this chapter will show). Although the MOI were evidently looking to understand the psychology of their audiences, this was much less advanced science than in the US context. It is clear from accounts of the British information war that there was limited knowledge of using mass psychology, behaviourism or psychoanalysis to shape a mass individualism grounded in the democratic rhetoric of choice that Turner describes in the US.[29] There was also far less clarity in the British context about the formal or rhetorical distinctions between 'democratic' and 'fascist' exhibitions than that described by Turner in the US context and more willingness to borrow exhibition devices from across ideological lines. The impact of the British official wartime exhibitions programme is far less easy to evaluate than its transatlantic contemporary and its legacies harder to trace.

Creating exhibitions for the machine age

The type of exhibitions favoured by the MOI had evolved from the reproducible form pioneered at 1930s exhibitions at Charing Cross and exemplified in the 1938 MARS Group show, structured around photographs and photocopies. These were exhibitions for the machine age: portable and infinitely reproducible. Given small budgets, restricted materials (including wood, metal and material) and short lead-in times, makers of these exhibitions needed to innovate, by using a combination of practices and reproductive processes including photo printing, typesetting, blockmaking, stencils, silk-screen printing, colour spraying, wet and dry mounting, lacquering and punching.[30] F. H. K. Henrion described MOI exhibition designers using Photostats to enlarge photographs and type, with type often hand-drawn if fonts were unavailable.[31]

Copies of exhibitions were circulated, using folding and collapsible screens, on which exhibition material could be directly painted or fixed, made a minimal size for transporting and easy to unpack and mount.[32] The use of photographic techniques and of images from MOI's Photograph Division was key to their success, allowing the exhibitions to be topical and to make a direct appeal. Photographs were used, documenting everyday life in Britain, drawn from an extensive home front collection taken by MOI photographers depicting work, domestic life and landscapes under wartime conditions. Alongside this, the displays were built from pictorial charts, models, statistical figures and symbols, diagrams, incidental murals and decorations commissioned from artists. Guidance for exhibition-makers instructed that alongside central visual elements, 'especial importance' should be attached to 'the scenario and captions', with 'intelligent use of contemporary display technique'.[33]

MOI exhibitions sat alongside themes conveyed through posters, booklets, radio campaigns, sometimes being shown alongside films.[34] They were made closely in tandem with other forms publicising the same campaign, not as a 'slavish copy' but as a 'counterpart', 'carefully synchronised so that each plays its part in building up a recognisable and memorable character'.[35] They needed to be varied enough to communicate a mix of messages and to fit in venues of varying scales across the country. To manage this, exhibitions should fall into two main categories. The first – 'Inspirational or Prestige' exhibitions – focused on 'stabilising home morale', covering 'civilian morale', 'comparison of our air achievement with that of the enemy, our convoy system, resources of the Empire, the causes of the war, our war aims and the reasons and need for bringing the war to a just and permanent conclusion'. Inspirational or prestige exhibitions were calculated to demand or provoke a strong emotional reaction that would be productive in building a community in support of war actions.[36] The second category – 'Instructional or Utility' exhibitions – dealt with advice and instruction to the public, for example on first aid, evacuation, rationing and 'the necessity for conserving essential services' such as water, light and heat, action to be taken in the case of gas attack (reducing fear by explaining remedial measures) and salvage. These exhibitions would be created in collaboration with the Ministries leading on the relevant areas.[37]

Citizenship exhibitions: the Ministry of Information's *London Pride*

Exhibitions developed by the MOI to inspire or bring prestige were intended to connect with the public through making an urgent, emotional appeal. What mechanisms did they use to appeal to the emotions of their viewers? A detailed discussion of the first exhibition mounted by the MOI's Exhibitions Branch – *London Pride* – allows insights into how this affective form was conceived and created.

London Pride, a 'civilian morale exhibition' opened in December 1940 at Charing Cross Station's ticket hall, set out to 'establish the fact that "London can take it"' and celebrated the endurance of Londoners in the face of the adversity of the Blitz.[38] 'London Pride', as the exhibition was called, had a double-meaning: the first literal pride in London and the second a flower of that name, which appeared to take seed almost anywhere, inspiring hope of renewal when it was found growing on bomb-sites.[39] A stylised image of the flower appeared on the right wall flanking the exhibition's entrance with the words:

LONDON PRIDE. SMALL EMBLEM OF A GREAT DETERMINATION[40]

London Pride's materials and materiality, its particular combination of information, political messaging and visual structuring working together, were key to its use and impact.

Showing the perceived significance to communicating the government's programme of such an apparently small event, held in an Underground station, the Home Secretary Herbert Morrison opened *London Pride*. Clad in a heavy overcoat to deliver his opening speech, Morrison reassured the assembled audience that the government's shelter-building programme was well underway. This followed the start of the Blitz a few months earlier.[41] Morrison attended the opening with Ellen Wilkinson MP, the first woman to serve in a Labour government. The government's shelter and post-raid welfare programme had come in for extensive criticism and the exhibition opening offered an occasion at which Morrison could demonstrate to press and public that he intended to make good on the deficiency. His opening of *London Pride*, an exhibition about a city with which he was closely associated as former leader of London County Council (LCC), was reassuring and calculated to increase public confidence in the government's ongoing civil defence response to war. Morrison would later become a prominent cheerleader for the multiple government-funded exhibitions of the 1951 Festival of Britain, but his regular support for earlier exhibitions – as will be apparent from his regular appearances across this book – showed his long-running belief in the importance of exhibitions as suitable platforms for government messaging.

Milner Gray was *London Pride*'s lead designer. Its core material was an extensive series of photographs, carefully selected from the stock of photo-agencies such as Topical Press, one of a growing industry of press agencies developing to support the burgeoning illustrated magazine market.[42] Charing Cross Station, by then an established exhibitions venue (as I discuss earlier in this book), was taken over by the MOI for exclusive use for a year.[43] Being on a bustling thoroughfare, passed by thousands of travellers each day, in the liminal space between work and home, the exhibition site inhabited an ambiguous place for its viewers between labour and whatever leisure the war context allowed, between daily work and time outside, an ambiguity mirrored in its subject matter, positioned, as it was, somewhere between affecting private domestic actions and public life.

London Pride's displays were focused around a central island site, panels hung in a cross shape and photographs placed on one side of the angle, with descriptive captions facing them (Figure 7.1). Round the

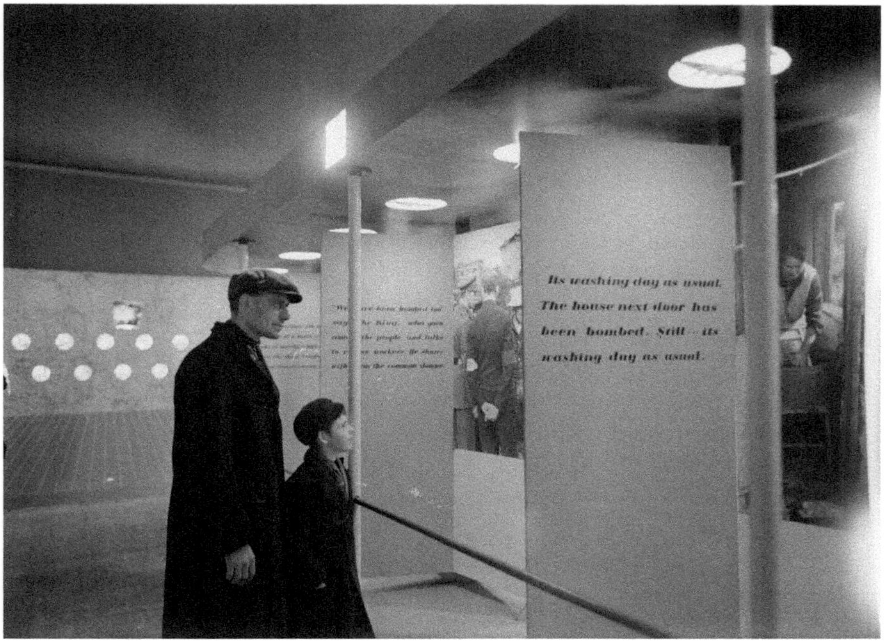

7.1 *London Pride* exhibition mounted by the Ministry of Information at Charing Cross Underground Station, 1940. Ministry of Information Photo Division photographer © IWM D 1756.

sidewalls were montages giving 'a panoramic composite view of life in the blitz', according to *Display*. A giant, enlarged photograph of Prime Minister Winston Churchill with the caption 'CARRY ON LONDON' in capitals amplified its impact and loomed over the right side of *London Pride*'s entrance, while a crowd scene with the capitalised caption 'LONDON CARRIES ON' filled the left side. While the first caption – 'CARRY ON LONDON' – took the form of a direct instruction to London, the second – 'LONDON CARRIES ON' – played with the same words to provide a commentary on London's hardiness. The crowd, pictured in a bombed area, were 'a study of London's Pride – her triumphant citizenship'.[44] Above the entrance, the title fascia 'London Pride' was mounted in a raised Playbill slab serif, while the typeface chosen for the central storyboards was a friendly sans serif, giving the text an informal, conversational appeal. Lettering reinforced the atmosphere of the exhibition, a slogan by the exhibition's entrance declaring with playful alliteration, 'Citizens of no mean City: they stand to their posts that liberty may live'.[45]

The sense of simultaneity, of many activities happening at once, was crucial to the impact of the displays. Seen from the entrance, the collage of black and white images crowded into the exhibition, the Prime Minister at one end waving across to the crowd on the other. The communal act of viewing the exhibition reinforced this sense of simultaneity, bringing passersby into a collaborative act of reinforcing and inspiring their sense of citizenry, in concert. This was a 'story' about London and Londoners' resilience but not played out over time; instead this was an urgent story played out in the present. London was shown 'now', not only as embodied through its people and buildings but as an abstraction through the presence of a map mounted on the wall to the left of Churchill, as if Churchill was gesturing towards it, parallel perspectival lines leading the eye from his enlarged photograph into the map's detail, giving visitors a complex sense of scale: both of being on the ground – in and with London – and of having an overview of it, being forced to have a multi-perspectival vision.

The urgency and immediacy of this appeal was the most striking quality of *London Pride.*[46] This came through the expanded photographic elements and was conveyed by the unfaltering focus on what was happening in that moment, combining topical news photographs with textual use of the present tense. Announced on the advertising poster, the exhibition's subtitle was 'a photographic record of how London carries on through the blitz', the present tense – 'carries' – crucial to its direct appeal.[47] The exhibition sought to democratise its subjects by showing a broad and authentic social mix, diverse and yet in unity.[48] *Display* referred to this as 'a complete cross-section of London life', from 'fire watcher down to the shelterer', 'from the King and Queen to the humblest resident of Stepney'.[49] Londoners pictured created recognisable types that viewers could identify with – photographs of 'typical Londoners who have been bombed', as *The Times* described them, ARP workers, firemen, nurses, police, postal workers pictured in around fifty or sixty photographs, to create a direct sense of relevance.[50] The displays addressed members of the public as individuals within the mass and the displays also presented individuals as having a kind of synecdochic character, whereby one person stood for the whole of society.[51] This idea of showing the 'typical', which *Display* echoed in its admiration for the 'typical character studies of Londoners', was repeated many times in the exhibition's text: 'here are typical Londoners ...'[52] This was a way of suggesting their proximity to the viewer.

Life-sized photographs showed London 'characters' going about ordinary tasks to create a sense of usualness: a woman doing her washing at home pictured scrubbing her clothes using a washboard with accompanying text in the present tense.[53]

> Its [*sic*] washing day as usual.
> The house next door has been bombed.
> Still ... its [*sic*] washing day as usual.

Beyond promoting the urgency of the present, text created impact through a looping quality: reassuringly humdrum normality despite the abnormality of the Blitz, emphasised by repetition. Such integration of text and image created a filmic quality to the exhibition, with its clamour of characters united within narrative, text working to advance the action given the images were not in themselves expressive enough of the intended message of Londoners' resilience, determination and normality in the face of the upheaval and catastrophe of war. What was in view was deeply affecting, calculated to shift attitudes to those around them: to feel connection and to deepen feeling. While the exhibition was consistently upbeat and focused on the congenial, what was happening out of sight, beyond the exhibition, was crucial to viewers' engagement: the very real horror of a potentially catastrophic war, in the process of wreaking havoc on the lives of ordinary people, mainly hidden from view except for glimpses of bombsites and armaments.[54]

Photographs conveyed immediacy, the state of things now, showing people settling down for the night in bomb shelters, buildings still smoking after recent bombings, assaulting the viewer with a sense that despite disruption people's lives were continuing and the sense of war as being common to all, even the monarchy. Displays showed the royal family leaving their protected spaces to engage with the wider populace. A photograph of the King visiting a bombsite was shown with the words '"We have been bombed too", says the King, who goes amongst the people', an image and caption singled out by *The Times* who observed that a 'companion picture showed "Arry [*sic*] and Bill"', a pair of rescue workers, 'telling her Majesty all about it'.[55] The mocked accent of Harry and Bill pointed once again to the deep class divide conspicuous through the images, reflecting this difference with seeming affection.

Developed through *London Pride* was the myth of a universal human condition, drawing all people together, ignoring differences of race and institution, a people happy and engaged despite immediate privations and evident discrepancies of class and situation. In *London Pride*, women were naturalised as carrying out caring and domestic duties as nurses and housewives, while men carried out more physically challenging duties as police and fireman. Viewers were entreated to relate these scenes to their own daily experiences, to identify with them. This was reinforced by photographs displayed flat at the bottom and angled down at the top – reminiscent of Bayer's earlier immersive installation work – to meet the gaze of the observer; by the enlargement of images which, being near to human scale, gave them a greater intensity; by strong reflector, angled lighting illuminating every element made possible within Charing Cross Station's ticket hall despite blackout specifications; and by now familiar display techniques such as port-hole windows allowing a playful way of glimpsing information (Figure 7.2).[56]

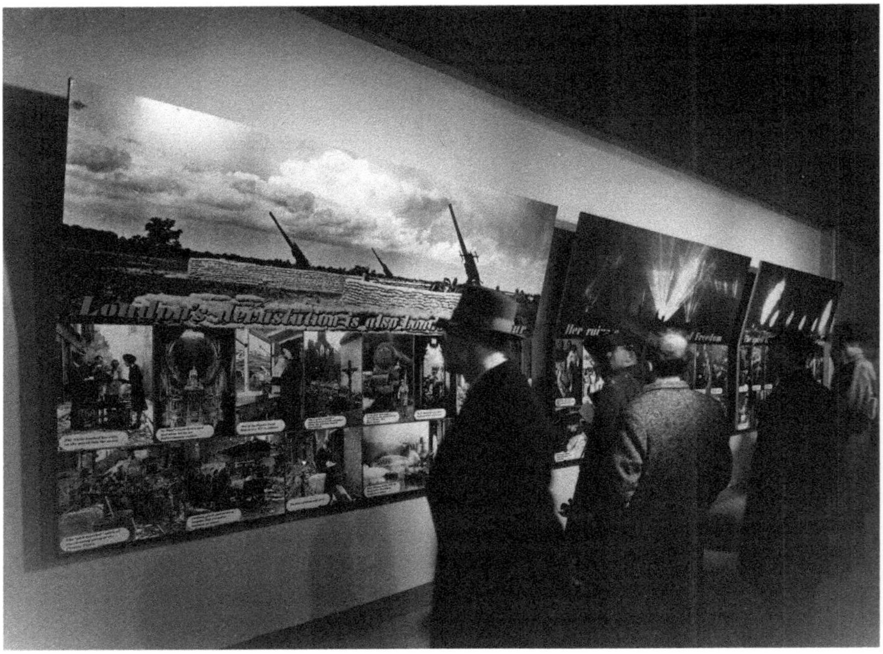

7.2 Visitors looking at *London Pride* with photographs displayed flat at the bottom and angled down at the top – reminiscent of Bayer's earlier immersive installation work. Topical Press. 1998–4092. © TfL from the London Transport Museum collection. All rights reserved and permission to use the figure must be obtained from the copyright holder.

London Pride as 'photographic essay'

Picking up on its literary form, with its combination of engaging enlarged photographs 'of heroic size' and lively captions, *Shelf Appeal* described *London Pride* as the MOI Exhibition Branch's 'first essay', enthusing that 'no picture was used which was not in itself interesting'.[57] *Display* praised it for 'simple dignity, but 'with sufficient unusual angles to arouse … interest'. The exhibition appealed first and foremost to visitors' feelings: 'Never has a display or exhibition made so big an appeal to the emotions or held so much topical and local interest. London Pride is outstanding'.[58] Above a series of vignettes of people carrying on in the midst of wartime damage ran the slogan: 'London's devastation is also London's splendour' and, further on, 'Her ruins are the ramparts of freedom', declarations suggesting London's destruction might just be the making of her, echoing the belief of contemporary architects in the picturesque possibilities of ruins and the potential that bombing offered for building back better.[59]

London Pride's success was built on the propaganda power of photography, the sequential viewing of the exhibition's storyboards in its photo-story format realised as people moved around the site.[60] The exhibition created a blurring of its subject and its viewer: who *felt* pride in London and who or what *was* the pride of London, with its long and important history; its buildings, bricks and mortar, largely immutable despite being under siege and subject to further destruction at any moment; its significance for the whole 'family' of the United Kingdom – 'a brave and cheerful family party', as the exhibition's text put it. London's resilience, even in the face of destruction, was part of this serious and sincere narrative. The exhibition spoke direct to a local audience, unlike the GPO film *London Can Take It*, made a couple of months earlier for a US audience, which described London for outsiders, with a US narrator.[61]

London Pride resembled weekly magazines' innovative fusion of typography with photography and its viewers were schooled in seeing such content through their familiarity with the weeklies.[62] *London Pride*, like other Ministry exhibitions, was structured around images drawn from the in-house Photograph Library, which collected and organised photographs from agencies and MOI photographers, for use in official propaganda and in the press.[63] The exhibition's look and feel, of images overlaid with narrative, echoed popular photo-weeklies such as *Life*, published in the US since 1936, and *Picture Post*, published in the UK since 1938, with wartime special issues subsidised by the MOI.[64] *Picture Post* fused the English tradition of social comment and reportage with developments in layout, typography and photography that flowered on the Continent in commercial, political and avant-garde circles in the interwar years.[65] *London Pride* shared *Picture Post*'s intimacy, with a focus in the exhibition on ordinary acts like the private, domestic task of doing washing, from sparsely populated photographs and from the descriptive accounts and personal testimonies that accompanied them. This allowed viewers to connect and identify with the exhibition's subjects, to create a 'structure of feeling', to use literary critic Raymond Williams's phrase, to think with and to feel through, to allow the material to appeal to them as personal, rather than merely held at a distance.[66]

Scripting wartime exhibitions

While mirroring illustrated magazines in conveying information through the hybridity of image and text, *London Pride*'s impact was achieved through its strong accompanying script. It is unclear who wrote *London Pride*, but the many professional writers from a range of literary backgrounds employed to work across MOI campaigns included poet Cecil Day-Lewis, poet Dylan Thomas (who wrote a commentary for the documentary *New Towns for Old*, 1942), poet Louis MacNeice (who wrote the script for an Albert Hall pageant in 1943), Eric Knight (who had collaborated on a script for the documentary *World of Plenty*), novelist Arthur Koestler (who wrote the script for documentary *Lift Up Your Head, Comrade*) and Lewis Mumford (who wrote the film *The Cities*).[67] Writer Robert Sinclair contributed text to exhibitions *How to Fight the Fire-Bomb* (on behalf of the Ministry of Home Security) and *The March of the Nation* (the story of the growth of American aid to Britain). George Orwell wrote text for the exhibition *Free Europe's Forces*, 'the story of the men of our allies who are fighting with us for freedom' and writer Gavin Starey scripted the *Women at War* exhibition. *Shelf Appeal* magazine enthused, 'All these writers and artists prepare their stories in close touch with the Government Departments and MOI officers concerned, and MOI Exhibitions Branch keeps a firm hand on the preparation of script and design at every stage'.[68] *Architectural Review* described the resulting exhibitions as 'photogenic' story-telling, 'mainly literary in character', signalling the importance of their text–image combination, which activated all 'forms of expression' in 'synchronisation'.[69]

Exhibitions' relative merits as propaganda were carefully considered. So, too, were the materials they were formed from. Photographs, as an immediate, easily reproducible and highly expressive medium, were central to the MOI's display strategy. Above any other form, photographs were considered the best way of evoking feeling in *London Pride*, for showing a collective response to a common enemy and for indicating the appropriateness of mass mobilisation in the face of a common threat. The MOI's Francis Bird elaborated on the sense of photography's impact: 'Photographs were one of the most potent instruments of war-time information ... The really superb picture ... could have the same effect upon public opinion abroad as a great victory'.[70] While reading about a remote victory in a news bulletin might resonate, photographs could bring the war 'closer'. Exhibitions like *London Pride* used photographs to present seemingly factual accounts of wartime, their constituent elements being library photographs assembled to tell new stories.[71] But while photographs could speak generically to this wartime context, they also reverberated, with powerful affect.

Photographs showing *London Pride* visitors, such as an MOI photograph of a man in flat-cap and overcoat with a child in cap and tie, themselves London 'types', involved a complex double-mediation: inviting viewers to identify with these exhibition-goers who were, themselves, in the process of reading and interpreting to identify with the exhibition's subjects, pictured in the same mode of brave and cheery endurance (Figure 7.1).[72] Photographing people interacting with the exhibition became an element in the Ministry's wider propaganda effort, to be shown elsewhere as evidence of the effectiveness of such media.

Mass Observation reported a steady stream of visitors to *London Pride*, which 'seemed to be received as well, or indeed better than most exhibitions at ... Charing Cross'.[73] *Display* was fulsome in its praise, describing the exhibition as 'one of the most attractive propaganda displays we have seen', 'the best presentation of British spirit that we have seen' and 'the best piece of propaganda display yet seen since the war started'.[74] The magazine criticised the MOI's decision not to tour it to 'all parts of this country and the neutral countries of the world'. If exhibited 'in New York, Buenos Ayres [*sic*], Montreal or Melbourne', 'it would be the rage of the town', *Display* opined.[75] The exhibition did tour but only to a dozen London department stores including Selfridges, Whiteleys and Kennards, while a modest version was created as four sets and toured through the US. *Architectural Review* said that despite its small scale, *London Pride* had 'the full orchestra of Corbusier-MARS effects', noting its being a 'semi-portable exhibition', transferrable to other sites.[76] *Modern Publicity in War*, a survey of 1941 publicity, reproduced images of *London Pride* to show the best of wartime display techniques.[77]

London Pride was only the first of the MOI's extensive series of story-telling exhibitions. Alongside the design of visual elements – photographs, illustrations, pictorial charts, statistics and diagrams – each exhibition's 'story' was told in a sequence of pictures and text, making the textual contribution as central to the experience as the visual one, akin to being physically immersed in an illustrated magazine. Travelling exhibition *Life Line*, telling the 'story' of the Merchant Navy, was a collaboration between designers F. H. K. Henrion and Charles Hasler, working with scriptwriter, folklorist and folk singer A. L. Lloyd. Resulting displays were created from standardised wooden panels, three-dimensional illuminated displays, models and pictograms offering succinct explanations of technical aspects, with a few small exhibits and printed pamphlets giving further information. *Life Line* had many hallmarks of other exhibitions designed by this team: it centred on expanded headshots of named individuals. The text explained these people as 'Willem Trotzenbergh – Fireman' and 'Saidi Ali – Fireman', their photographs accompanied in the exhibition by testimonial texts, which described Trotzenbergh as 'the son of a Rotterdam docker, Willem worked as a lorry driver for a paper factory, but he likes the sea better', drawing the viewer further into sympathising with and identifying with the workers while portraying the Navy as diverse and egalitarian (Figure 7.3).[78]

'Instructional or utility' exhibitions at Charing Cross

Numerous small government exhibitions at Charing Cross amplified the focus of wartime campaigns, offered practical skills and addressed matters of general war interest. These included *Private Scrap Builds a Bomber* (encouraging salvage), *Gangway Please* (explaining war transport), *The Story of Lin* (a picture of China at war), *The Unconquerable Soul* (the story

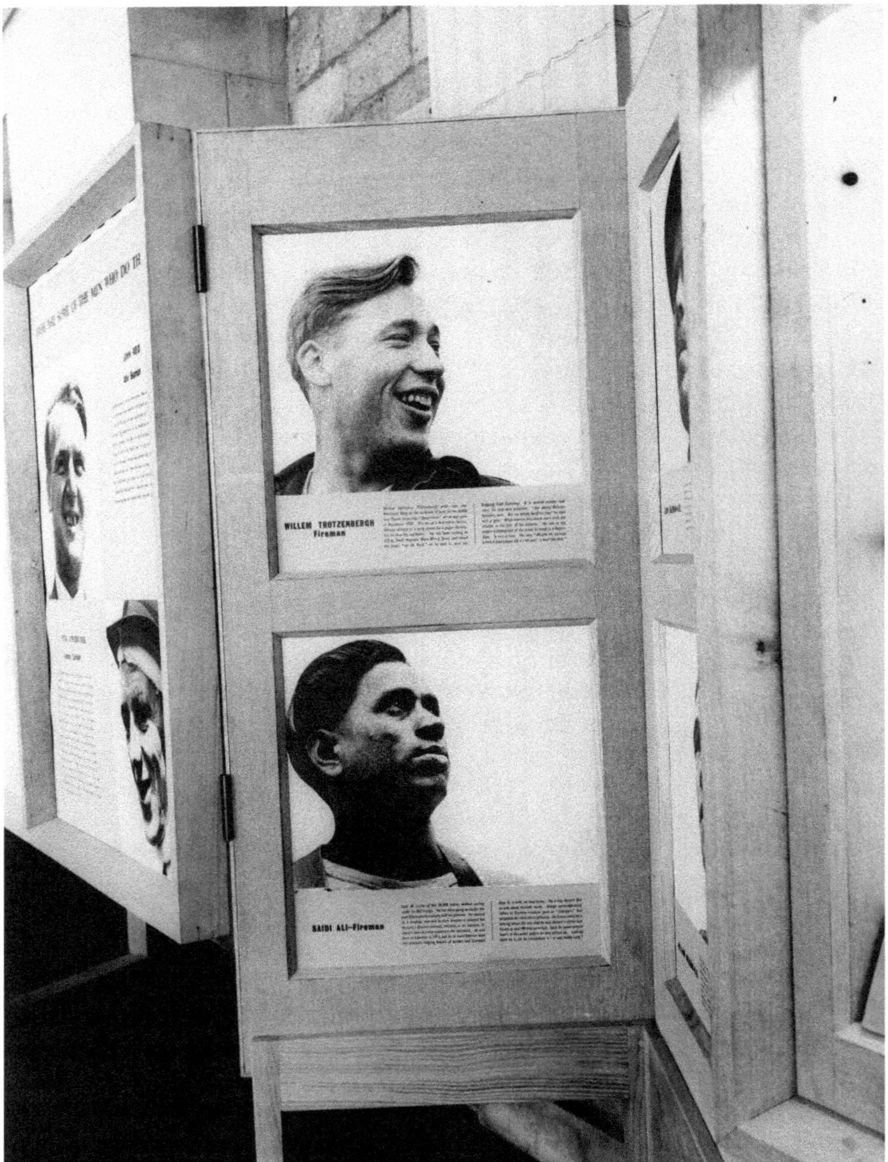

7.3 *Life Line* exhibition, mounted by the Ministry of Information to tell the 'story' of the Merchant Navy, introducing 'Willem Trotzenbergh – Fireman' and 'Saidi Ali – Fireman'. F. H. K. Henrion Archive, University of Brighton Design Archives.

of resistance in occupied countries), *Ocean Front* (the air-sea war against Japan) and *Jungle Front* (war in the South Pacific). *The Navy at Work* at Charing Cross allowed visitors to imagine spending the day with men of an HM Destroyer: Bill, 'thirty-eight, married, an Active Service man', and Fred 'twenty, single, a Hostilities Only man', with documentary photographs showing Bill going about his day-to-day tasks on board.[79]

Many exhibitions communicated information focused on keeping civilians safe by teaching practical skills and explaining how to prepare for Nazi aggressions, such as gas attacks. *Poison Gas*, produced by the MOI on behalf of the Ministry of Security, was 'a gas mask exhibition', mounted at Charing Cross Station from August 1941.[80] A flyer designed by F. H. K. Henrion, showed the uncanny spectre of a gas-masked figure alongside a striking Playbill font (Figure 7.4). Among the questions the displays addressed were 'How many types of war gases are there? Is a civilian gas mask different from civilian duty respirators? Can you see a gas cloud? What would happen after a gas attack? Can animals be protected against gas?'[81] The show's main message was that, unlike bombs, gas would not kill if civilians used proper protection. Its central theme was trust: who and what equipment they could rely on. Peter A. Ray worked as collaborating designer on the exhibition's graphic presentation, with text written

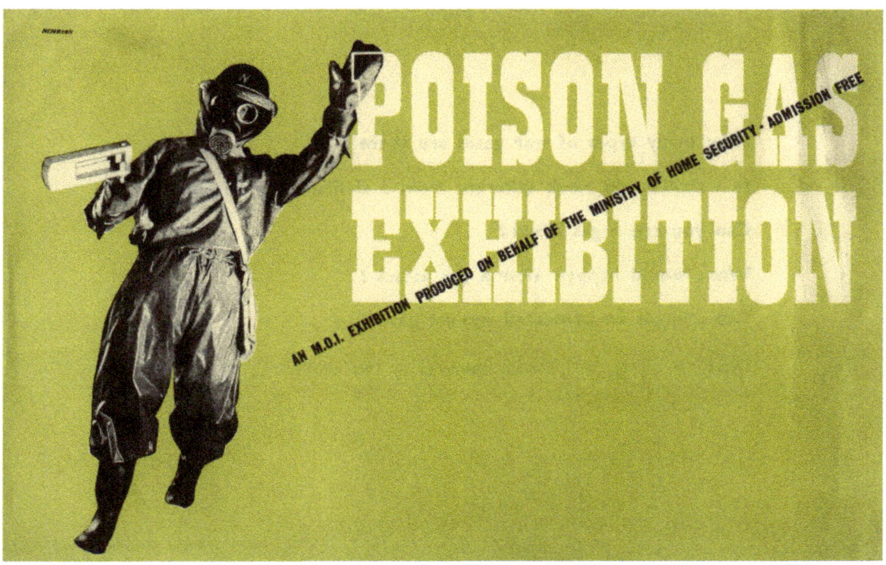

7.4 *Poison Gas* exhibition flyer by F. H. K. Henrion, 'a gas mask exhibition' produced by the Ministry of Information for the Ministry of Security and held at Charing Cross from August 1941. F. H. K. Henrion Archive, University of Brighton Design Archives. © Estate of F. H. K. Henrion. All rights reserved and permission to use the figure must be obtained from the copyright holder.

by features editor of the *Star* newspaper, Robert Sinclair.[82] Eyewitnesses reported that visitors were reading labels, but how much information remained lodged in their minds is impossible to ascertain.[83]

Poison Gas's installation at Charing Cross, although in a restricted area, was spatially innovative. It made use of a double-levelled circulation route through the small space, with display units viewed from a raised platform, a structure designed for reinstallation in other spaces and highlighted by *Display* as an unusual feature.[84] Each element was spotlit. The exhibition used photographs and drawings, plus typeset and silk-screened captions alongside physical exhibits to tell its story. A strangely uncanny model of a civil defence worker stood guard at the exhibition's entrance, clad in protective clothing and gas mask. The figure was echoed in a huge cutout photograph behind it, with three other 'horrific' figures, as *Shelf Appeal* described them, similarly dressed. The slogan 'Unlike high explosive bombs, war gas on the whole is not a killing weapon providing a gas mask is used' was followed by the challenge 'Are you ready?', using a narrative device to shift focus from fear to preparedness.[85] A box-frame inserted into the wall beyond the entrance altered the scale from life-size to miniature, showing tiny gas-carrying enemy aircraft models on strings representing the Blitzkrieg flying over a town, floating text warning of enemy threat.

Visitors' emotions were carefully controlled and directed. While anonymous, fully gas-suited figures gave an air of foreboding, across the way were friendly images that softened and personalised the struggle. Under the slogan 'THE PEOPLE TO TRUST' were three 'heroic-size' head and shoulder photographs of civil defence workers in tin hats, gazing into the distance, 'men and women with faces one trusts', *Shelf Appeal* described them. By showing these particular faces, displays were intended to instil confidence in the trustworthiness of all people set to defend them (Figure 7.5).[86] Photographic collages explained how to protect children of different ages, through images that viewers could identify with, while *Display* praised 'the big bold photographic treatment'.[87] The reproduction of photographs on such a notably large scale had been made possible through recent technological innovations.[88] *Shelf Appeal* also commended the use of photographs and commented on the merits of the exhibition as a 'form of public education', enabling 'a complex argument to be stated in a more detailed manner than would be permitted by the conventional methods of publicity' (original emphasis). *Poison Gas* succeeded in functioning, *Shelf Appeal* said, as 'a three-dimensional manual of defence against gas', emphasising its textual and instructional form.[89]

Learning civil defence: *Fire Guard* at Charing Cross Station

The extent to which these text-heavy exhibitions were breaking through to their audiences was in question. Various internal MOI reports noted

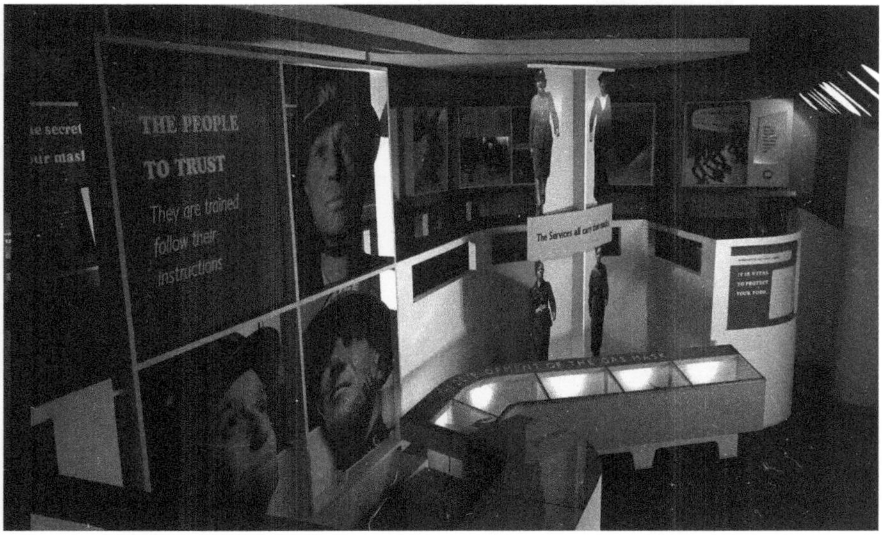

7.5 Interior view of *Poison Gas*. Ministry of Information official photographer © IWM D 4201.

that exhibitions were overly reliant on text and that, while visitors did read it, it was difficult to assess its impact.[90] One of the exhibitions mentioned was *Fire Guard*, which explained how to cope with incendiary bombs and promoted the role of firefighters on the home front, trying to drum up conscripts. *Shelf Appeal* explained that the exhibition set out to impress 'the average citizen' with the importance of firefighters' role in 'Europe's most modern army', referring to the Fire Guard organisation set up in summer 1941 to promote involvement in this unpopular activity.[91]

Peter A. Ray designed *Fire Guard*, adapting and reusing the two-level structure he had previously created for *Poison Gas*. Ray worked with F. H. K. Henrion, using collaged photographs and diagrams, posters, drawings and two massive abstracted wooden firefighter figures carrying buckets, designed by Bruce Angrave (Figure 7.6). *Shelf Appeal* described Charing Cross Station, site of *Fire Guard*, as the 'only satisfactory site at which to hold an exhibition with the purpose of quickly informing, instructing and reminding typical citizens'. A great merit of the station space was that it could carry on during blackout hours as its designers had devised a switch to turn out main lighting and leave only showcases, a 'Fire-bomb Fritz' model and stairway lights, as *Shelf Appeal* explained. The magazine also noted a recent Mass Observation report on home propaganda had revealed that exhibitions had been more successful in arousing people's interest and secured better results than many other better-known methods.

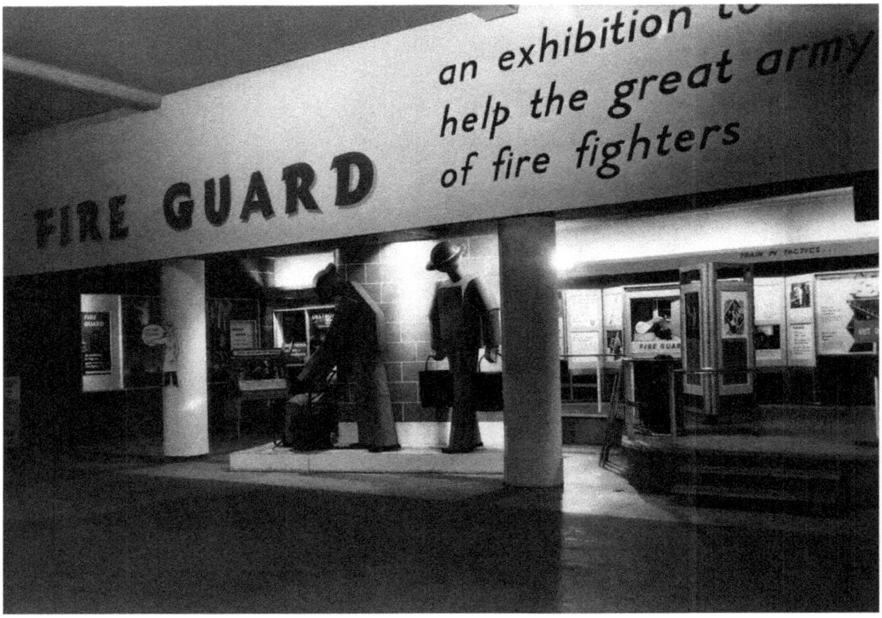

7.6 *Fire Guard* at Charing Cross Underground Station, summer 1941, designed by Peter A. Ray for the Ministry of Information. Ministry of Information official photographer © IWM D 11116. All rights reserved and permission to use the figure must be obtained from the copyright holder.

Being economical and aiming for self-sufficiency: *Battle for Fuel* exhibition at Dorland Hall

The need for fuel efficiency was crucial to the war effort but how should the government appeal to the public on such a seemingly unpopular theme? The major exhibition *Battle for Fuel* held at London's Dorland Hall in autumn 1942 attempted to address this, employing a range of devices to speak to different audiences. *Architectural Review* described the approach of *Battle for Fuel* as combining 'highbrow' and 'lowbrow'.[92] So-called 'low-brow' elements were seen on the ground floor, 'with jolly imitations of pit galleries and a real horse and miner'. Meanwhile, a 'highbrow' element, the *Review* said, included 'contraptions of metal and string' in the style of Moholy-Nagy or Gabo, designed by architect Peter Moro and hung in the stairwell, quizzing visitors 'Are you a fuel saver?' as they climbed the stairs. A huge collage of images of fuel use – domestic and industrial – set the scene.[93] Another room owed a clear debt to Herbert Bayer's concept of 'extended vision', to expand visitors' gaze, using photographic panels hung at eye level and overhead in a dark room, with dramatically lit panels controlled from an interactive panel.

A popular element across many government exhibitions was the introduction of cartoon characters, as friend or foe, to add narrative tension and humour to practical suggestions and to make displays more engaging and endearing.[94] In *Battle for Fuel* this device allowed designers to add jovial elements to serious and humdrum subjects. In a display entreating 'Always Fill the Oven', a cartoonish Hitler commented mendaciously on the sight of a single apple going into an oven – a conspicuous waste of energy – in verse:

> One apple per oven a
> German success is
> I'll give you an Iron Cross
> That used to be Hess's[95]

while visitors nearby were entreated 'Don't Wash under Running Taps', with cut-out hands by a sink and a cartoon Hitler clinging nervously to the taps, trying to coax the user to squander water: '… show me you're a girl whose clean hands help the Fuhrer'.[96] These suggested that wasted resources in the form of energy and water played into the hands of the enemy by depleting the stocks that could be used to fight.

Wartime exhibitions such as this used multiple strategies to engage, affect and instruct. At *Battle for Fuel* a room on the theme 'Warmth or Victory' showed photographs, expanded floor-to-ceiling, with slogans encouraging efficient fuel use, while another on 'Comfort or Guns' showed cartoons of domestic scenes dotted with practical suggestions for saving fuel such as 'Wash up in big batches', inviting viewers to consider the direct correlation between individual self-sacrifice and national triumph (Figure 7.7). Flow-charts elucidated the relationship between home fuel use and the grid; a hand-drawn cross-section of a house, with headshots of its inhabitants, indicated how each had wasted gas and electricity that day, cautioning visitors against replicating these behaviours.[97] For those needing statistics, Isotypes showed technical information, such as what the units of fuel – coke, paraffin and electricity – equated to in terms of hours of use, while quizzes and diagrams explained how to be a 'fuel saver'.[98] The most cinematic section of *Battle for Fuel* was a room carrying a photo-story-like display with expanded photographs accompanied by textual commentary showing people going about ordinary domestic tasks, while explaining their feelings about adjusting to new behaviours, such as shorter-than-normal baths or colder rooms, with a nod towards the kind of innovations connecting still and moving images championed by contemporaries.

Like many government campaigns on the home front, the emphasis here was on the impact of personal behaviours on the wider war effort, such as using unnecessary gas and electricity, and assorted means were used to appeal to the many viewers. The Queen visited *Battle for*

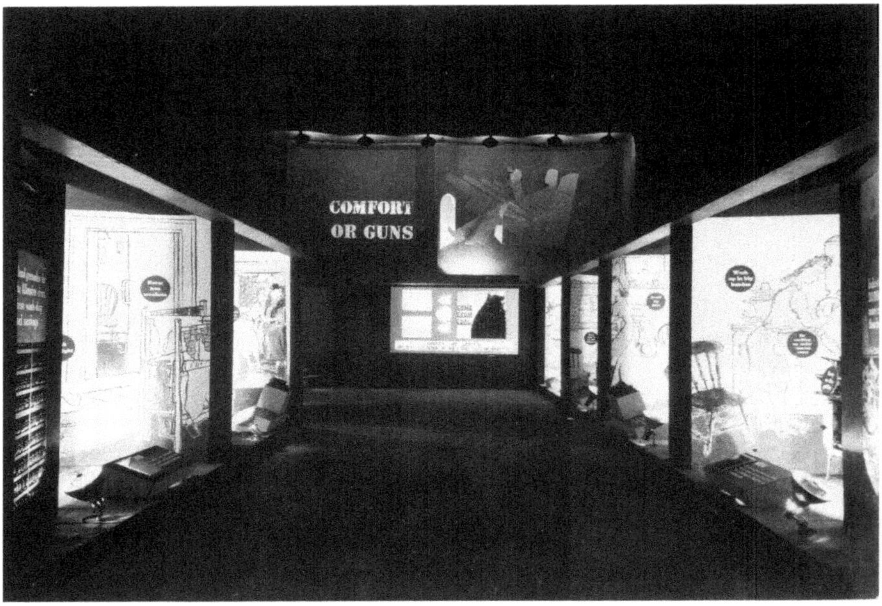

7.7 *Battle for Fuel* at Dorland Hall, autumn 1942. A display asking whether viewers were prioritising 'Comfort or Guns' showed cartoons of domestic scenes dotted with practical suggestions for saving fuel. Ministry of Information official photographer © IWM D10633. All rights reserved and permission to use the figure must be obtained from the copyright holder.

Fuel and the MOI recorded hundreds of daily visitors, reporting that as well as being important to the war effort it managed to be entertaining.[99] An MOI *Home Intelligence Division Weekly Report* confided that although some believed this exhibition 'the best of its kind', others were questioning the wisdom of spending so much money and using so much lighting on publicity of this kind.[100] Given tight budgets this lack of confidence was significant: exhibitions did not appear to be meeting the value for money criteria.

Appraising the impact of government exhibitions

The organisers of government exhibitions were attempting to appeal to a mass public in the mode of high-circulation magazines like *Picture Post* or *John Bull*, and other contemporary initiatives such as BBC information programme *The Brains Trust*, broadcast from 1941, where a panel of experts answered an audience's questions. At various points during the war, government officials and other bodies sought to evaluate the impact exhibitions were having. Despite the proliferation of government exhibitions of different shapes, sizes and scales, they were regularly omitted from discussions of the government's publicity campaigns and there continued to be lingering uncertainty as to the real impact they were actually having in shaping public opinion.[101]

A confidential internal MOI memo of September 1941 noted that the impact of their exhibitions was unclear and the case for public accountability for such high levels of government spending unproven. 'I have not been able to obtain any definite information as to what value we are getting for our money by way of public attendance', the official wrote. In some regions exhibitions were, they observed, 'definitely unpopular'; in a couple no exhibitions had been installed; and no work was being done to record public attendance, even at Charing Cross.[102]

Other bodies were also interested in the impact government exhibitions were having. Mass Observation reported regularly on the MOI's many free-to-visit propaganda exhibitions held across the country. In autumn 1941 they reported on visiting twenty-two at Worcester, Bolton, Portsmouth, Port Sunlight and Stockport.[103] They observed two main types: 'Photograph and Poster Exhibitions' and 'Exhibitions showing Practical Demonstrations'. The first type – 'Photograph and Poster Exhibitions' – they considered 'technically excellent', 'vivid and striking', but attendances were noted to be 'extremely poor', often with barely any visitors during an hour. Mass Observation attributed low numbers to lack of 'window dressing' and their being held in public libraries, museums or town halls that did not 'symbolise novelty and topicality in the ordinary way', with atmospheres 'unfavourable' and too 'studious'. The second type – 'Exhibitions showing Practical Demonstrations' – enjoyed comparatively good sites and shop windows, according to their observers. They benefited from 'concrete objects and processes', which drew bigger audiences and created lively, informal atmospheres.

The trade press were also watching to see how government exhibitions were playing to the public. In 1941, *Shelf Appeal* appraised all MOI 'exhibitions of *ideas*' presented in 'graphic form' to date. They observed their heavily textual, story-telling formats as developed through a process that 'parallels editorial production practice' but were not impressed with the exhibitions' visual qualities, bemoaning images as 'the cobwebbed files of the picture agencies and service photographs'.[104] The advantage of an exhibition as propaganda, the magazine thought, was in being able to 'dress up an argument in an attractive garb of colour and picture', and to 'slip their argument almost unobtrusively into his consciousness', exerting influence while the public were going about their daily lives. Although *Shelf Appeal* believed exhibitions to have the potential to appeal 'to all levels of intelligence simultaneously', several exhibitions missed the mark and failed to speak 'in the language of the people who read *John Bull*' (the popular magazine with a mass circulation of one million during the First World War).[105] Exposing again the patrician intention of the exhibitions – and the gaping disjuncture between those addressing and those being addressed – the magazine suggested exhibitions were 'too suave, too professional, too pleasing to be good propaganda', concluding that the MOI were deluded if it thought anyone would make a special journey to see such exhibitions.[106]

Early in 1943, the MOI's Exhibitions and Displays Division requested an MOI *Home Intelligence Special Report* to assess public reactions. This report focused on two exhibitions visited in February 1943, both of them forms of atrocity propaganda, relying on shock reactions for their impact: *RAF Exhibition of Bomb Damage on Germany* shown at Hastings and *The Evil We Fight* shown at Doncaster.[107] The Hastings RAF show was mounted on the first floor of Woolworth's, a shop chain that prided itself on being reasonably priced, fitting with the MOI's aspiration to appeal to a wide public in ordinary places. The exhibition was assessed on a weekday six days after opening, which reported it had attracted 72 visitors during half an hour in the morning and 110 in half an hour that afternoon, while on Saturdays, the report noted, there were about four times those visitors. Women were noted to be the majority, with men mainly in service. The most common response of visitors to the photographs of RAF bomb damage meted out on Germany was 'a general feeling of "grim satisfaction"' or a 'deep feeling of patriotism, and pride'. Interest in the subject matter was noted, as well as admiration for the photographs themselves.

The Evil We Fight, the second exhibition, referred to in the MOI report as 'Nazi Atrocity Exhibition', was held in Doncaster (Figure 7.8). It dispensed with the MOI's normally moderate tone, shocking visitors by showing the extent of Nazi atrocities and those responsible for them. The MOI's report noted that some visitors described being appalled and shocked by the images of 'half starved Greek children' and 'a pit full of half naked women obviously raped', while others thought it not extreme enough. The report recorded this as 'an outstanding event', praising 'the layout and clarity', but noting that some young people '"thoroughly enjoyed the gruesome details", without realizing their full significance'. A couple of months later, the MOI assessed the attitudes of 'housewives' towards official campaigns and instructions, concluding that working-class housewives were most receptive to instructions and that radio and cinema were favourite media for information and instruction. They thought neither posters nor leaflets 'cut much ice', and exhibitions were not even mentioned, showing how superficially they had penetrated the public imagination.[108]

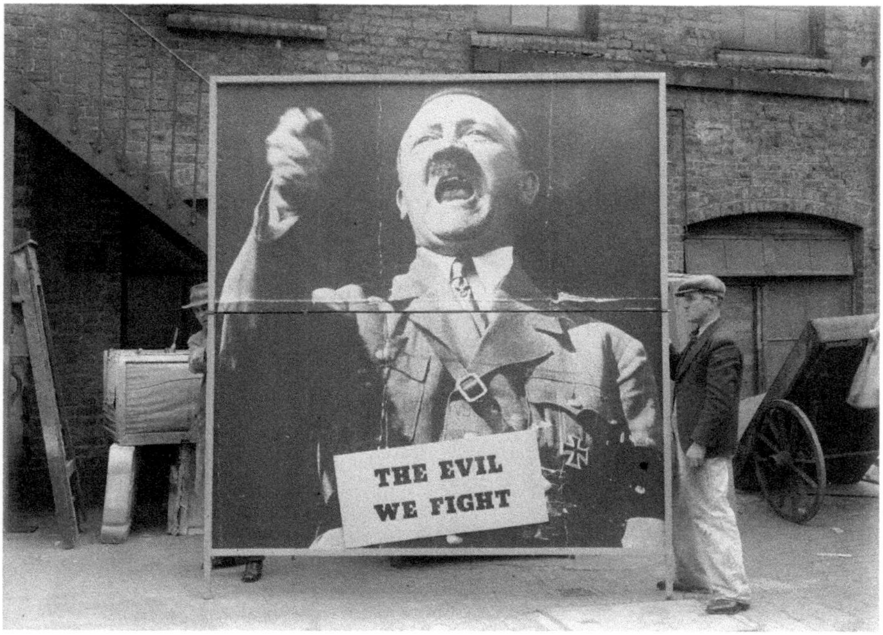

7.8 *The Evil We Fight* travelling exhibition showing Nazi atrocities. Ministry of Information official photographer © IWM D 20889.

Learning to grow: *Dig for Victory* exhibitions

The 'Dig for Victory' campaign (sometimes titled 'Grow More Food' or 'Grow Your Own Food') called for British people to grow in whatever spaces they had available, to provide access to fresh food and free up space on ships carrying essential wartime machinery. Spearheaded by the British Ministry of Agriculture, the campaign was fought across many media, from posters to booklets, films and exhibitions. Bronek Katz, F. H. K. Henrion and Charles Hasler collaborated on the MOI's travelling *Dig for Victory* display, showing how to make best use of outside space to grow food, demonstrating everything from sowing seeds to tackling garden pests.[109] Despite modest budget, material shortages and ordinary sites, the designers created a striking show. A mock grocer's stand carried onions and potatoes, each information display covered in a striped awning, as if part of a stall (Figure 7.9). Its introductory board was a wooden frame and poles doubling as enormous spades (Figure 7.11), with references to the subject of the poems hanging on the information panels they carried:

> *This is a spade to symbolise*
> *Our will to grow our own supplies ...*[110]

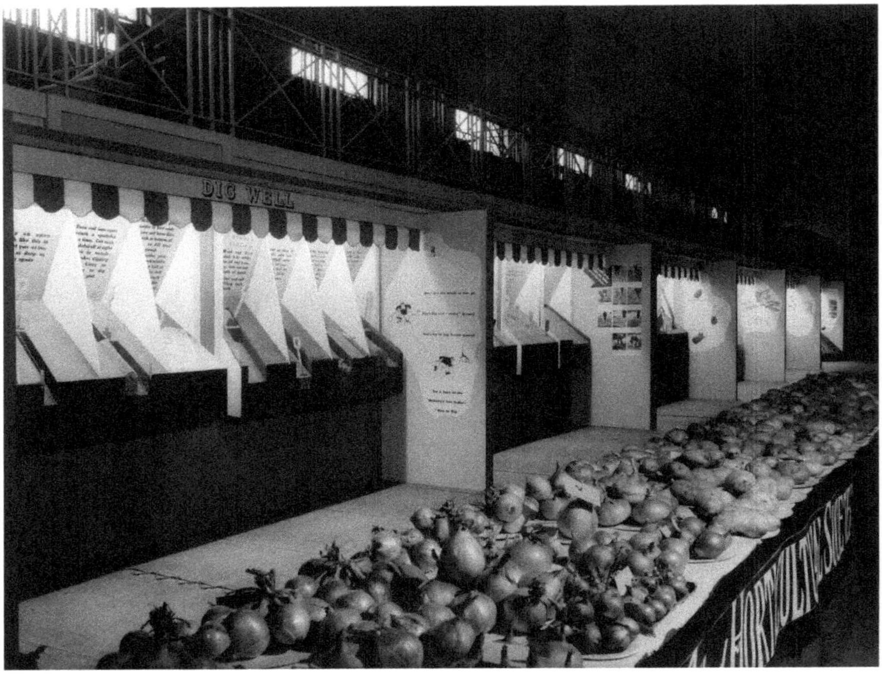

7.9 *Dig for Victory*, travelling display, designed by Bronek Katz, F. H. K. Henrion and Charles Hasler, showed how to make best use of outside space to grow food. F. H. K. Henrion Archive, University of Brighton Design Archives. (Uncatalogued). © Estate of F. H. K. Henrion. All rights reserved and permission to use the figure must be obtained from the copyright holder.

These set the scene, with a mix of quirky pictograms and whimsical rhymes.

Making a serious point about the connection between milk consumption and the capacity of ships, Panel 1 mused:

> *This is a cow who smacked her lips*
> *On food that came to her in ships…*

While Panel 4 continued:

> *These are the things that YOU can sow*
> *To supplement what farmers grow.*
> *With vegetables throughout our land*
> *We'll fell the scheme the Axis planned.*

The displays used a mix of drawings, photographs and models to provide information and instruction, some of them whimsical and quirky, many relying on anthropomorphic presentations of non-human characters. Cartoon slugs and other garden pests were denoted as 'the enemy' with swastikas suggesting that, by attacking the food supply, such pests were stealing crucial space on ships that had then to be given over to food (Figure 7.10).

Dig for Victory employed various means to 'force home' its message: 'surrealist shock-tactics, as well as cheap jokes', according to *Architectural Review*.[111] The Isotype system was once again adopted in the touring *Grow More Food*, the pictorial language a central means of presentation across MOI media (as it had been in interwar exhibitions, as noted in Chapter 3).[112] Imagery was shared between government campaigns in different forms as is apparent from a comparison of the exhibition's information board and posters associated with the campaign. Both shared spade structures (Figure 7.11). Abram Games's 1942 *Dig for Victory* poster entreated military personnel in barracks to supply their own 'cookhouse',

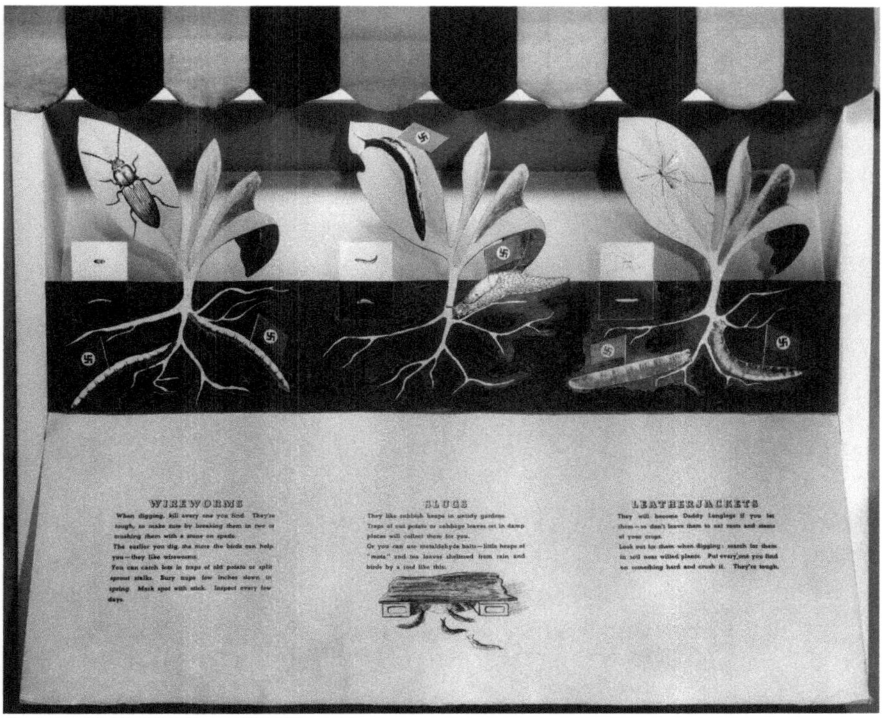

7.10 *Dig for Victory*: detail showing cartoon slugs and other 'enemy' garden pests denoted by swastikas. F. H. K. Henrion Archive, University of Brighton Design Archives. (Uncatalogued). © Estate of F. H. K. Henrion. All rights reserved and permission to use the figure must be obtained from the copyright holder.

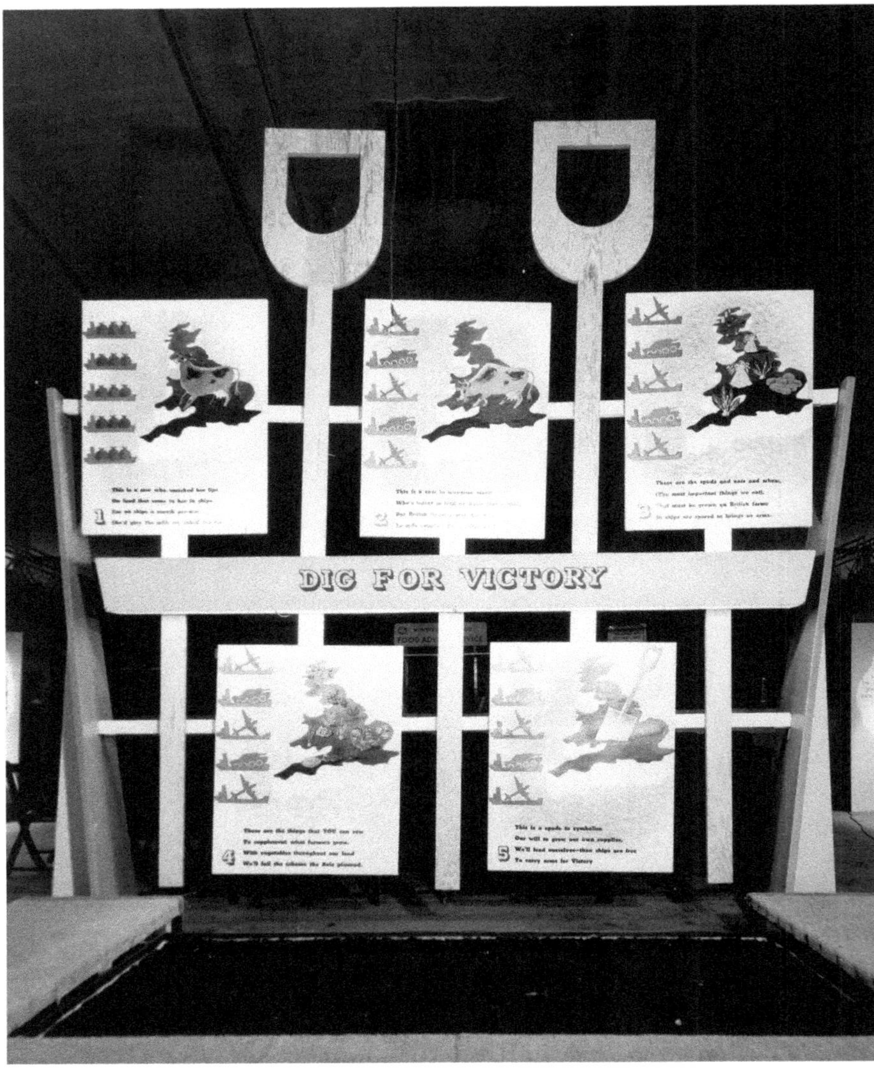

7.11 *Dig for Victory* pictograms explaining the impact self-sufficiency might have on the wider war effort. Crown Copyright Reserved. Image courtesy of the Museum of Domestic Design & Architecture, Middlesex University, www.moda.mdx.ac.uk. All rights reserved and permission to use the figure must be obtained from the copyright holder.

playing with imagery to show garden implements as the legs of a dining table and a vegetable patch as carpet, an eloquent and economical means of showing the potentially short distance between home-growing and eating (Figure 7.12).[113]

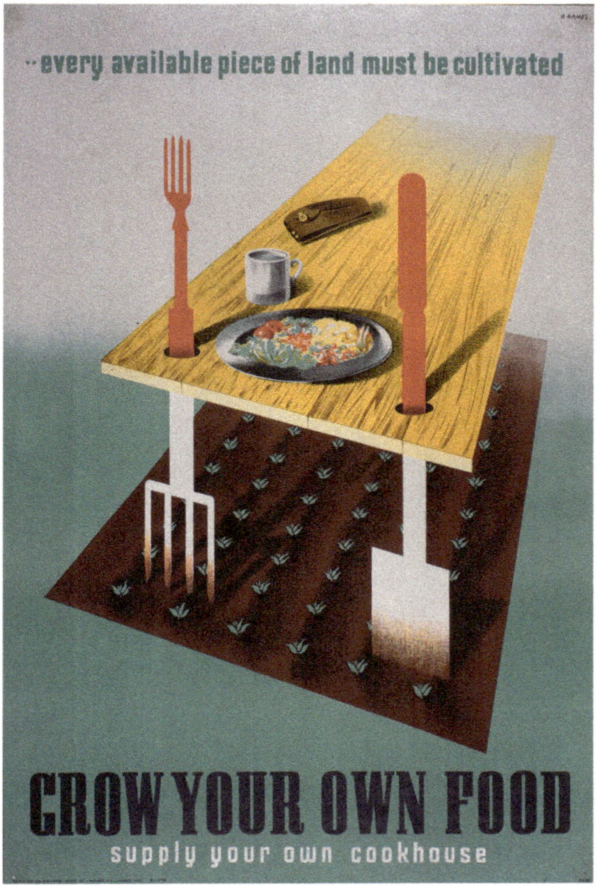

7.12 Abram Games's *Dig for Victory* poster, 1942, entreated military personnel in barracks to supply their own 'cookhouse'. Abram Games. © Estate of Abram Games. All rights reserved and permission to use the figure must be obtained from the copyright holder.

Learning self-sufficiency: *Off the Ration* exhibitions

Encouraging self-sufficiency in growing fruit and vegetables and rearing hens, rabbits and pigs for food was the focus of the Charing Cross 1942 MOI and Ministry of Agriculture exhibition *Off the Ration*. F. H. K. Henrion's display combined photographs, illustrations and demonstrations by land girls of real rabbits kept in cages with diagrams showing how one rabbit could produce twelve smaller rabbits (Figure 7.13).[114] Henrion recalled the publicity he created using black and white photographs 'with flat colouring and tones above and typography' (Figure 7.14). Of the displays themselves he recalled 'there were lots of

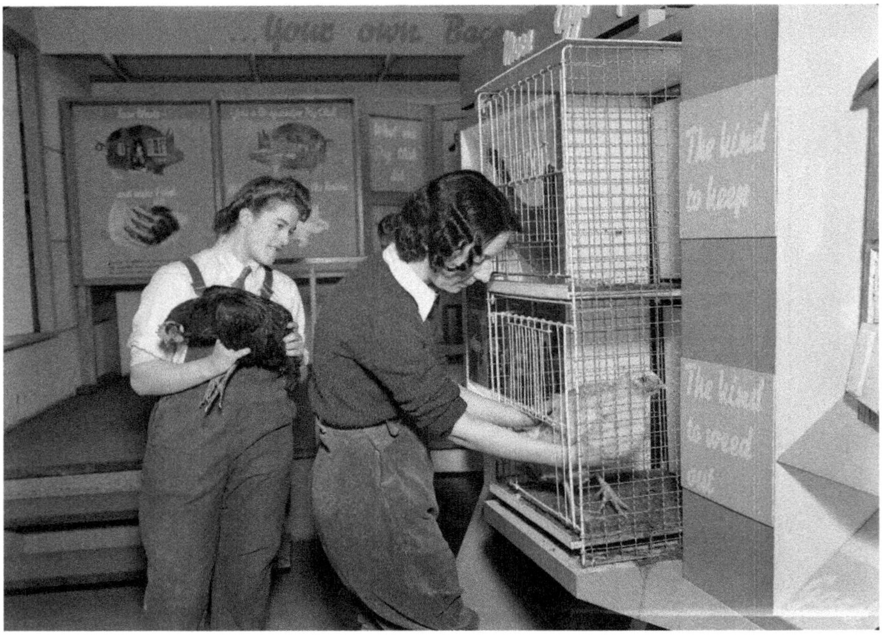

7.13 Land Girls M. Slingsby and A. Chesterton placing chickens into cages at *Off the Ration* exhibition, Charing Cross Underground Station, 1942. Ministry of Information official photographer © IWM D 7863.

people who went through the station and saw the exhibition' and that despite its small site 'we had live pigs, chickens, it was extremely successful, encouraging people to keep rabbits, pigs, chickens, and grow veg on plots'.[115] Despite these claims of success, it is almost impossible to judge the impact of such small displays; both because of very limited visitor data and because they were only one element in wider campaigns, expressed across multiple media. Travelling shop window displays for the same campaign also explained how to keep rabbits, giving statistics about rabbit reproduction and instructing how to feed them from waste, how to buy stock and to make hutches (Figure 7.15).[116] These displays succeeded in being visually appealing whilst informative, using a mix of photographs and illustrations to explain eating: photographs of rabbits with their stomachs cut away as illustrations to show what they ate, for example.

A larger open-air version of *Off the Ration* was held at London Zoo from August to October 1942, one of the few places in London that continued to offer entertainment to families during the Second World War. MOI photographs of the zoo installation show children visiting, suggesting displays were tailored to families, that children were part of the audience for

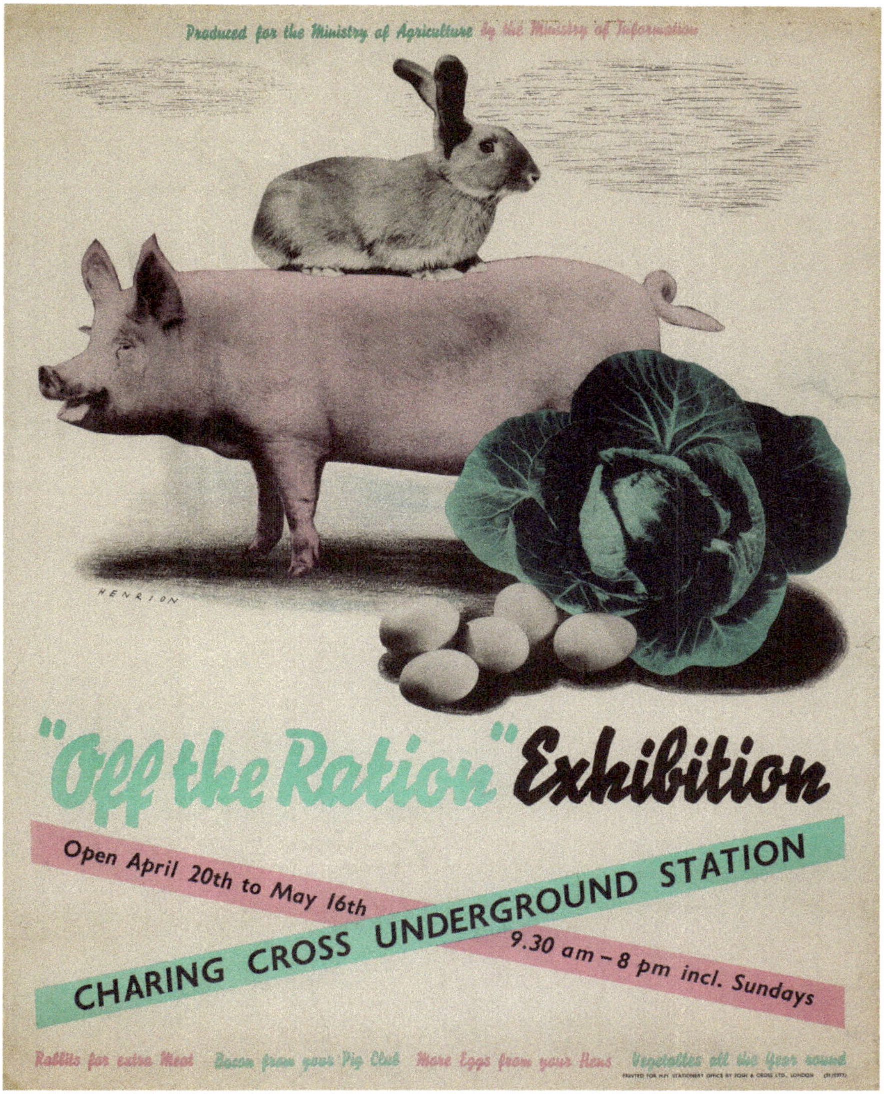

7.14 *Off the Ration* poster designed by F. H. K. Henrion for the exhibition at Charing Cross. F. H. K. Henrion Archive, University of Brighton Design Archives.

propaganda and recruited to play their part in the war effort. *Off the Ration*, again designed by Henrion and Bronek Katz for the Ministry of Agriculture, achieved an upbeat tone through colour, jaunty displays, cartoon-style presentation and live demonstrations of animal handling and bee-keeping,

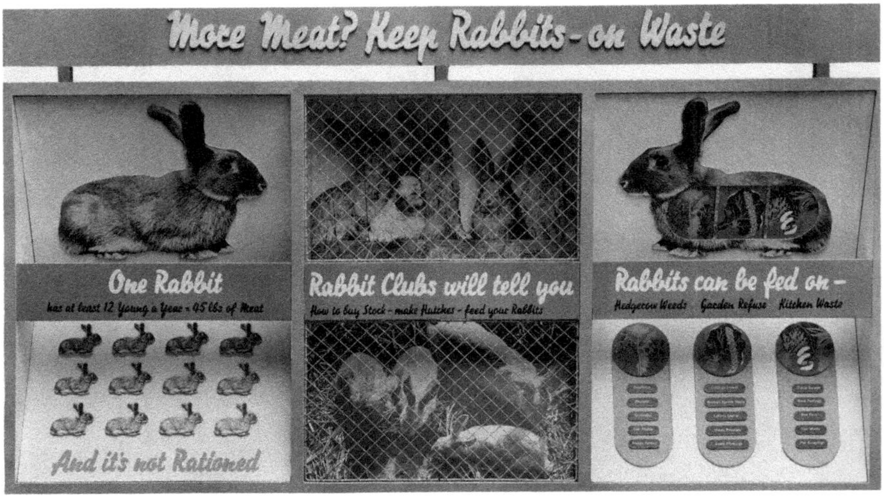

7.15 *Off the Ration* shop window scheme detail, designed by F. H. K. Henrion. F. H. K. Henrion Archive, University of Brighton Design Archives.

anticipating the festive atmosphere of the Festival of Britain a few years later (to which both Henrion and Katz would be major contributors).

Drawings of bears and chickens by children's illustrators Lewitt-Him (the collaboration between Polish-born Jan Le Witt and George Him) were accompanied by verses, keeping the mood light. A label attached to the lion's cage instructed:

> *Do not roar for meat like me,*
> *keep a modest cabbagery.*

While a cartoon hen, declared:

> *... I'm a modern British hen*
> *I've got a concrete house.*[117]

The humorous illustrations were interspersed with photographs, diagrams and instructive labels showing how to kill vermin and use tools. The serious intention to inform and instruct was for all ages.

Learning to care: Ministry of Information *Make Do and Mend* exhibitions

The MOI's 'Make Do and Mend' campaign for the Board of Trade encouraged people to get as much wear as possible from the clothes they owned. This was another of the government's most far-reaching campaigns, fought through multiple means – film, posters, pamphlets and also exhibitions. There is a sizeable literature on the campaign, analysing its successes and failings, but my focus here is on the practicalities of creating a three-dimensional contribution to the campaign. Typographer Charles Hasler designed touring exhibition *Make Do and Mend: Household & Clothes Economy* to explain how clothes were made, how their rationing operated and to give tips for extended use of clothing. An accompanying pamphlet for exhibition hosts explained how to mount the display, with photographs suggesting layouts, based on a small area typically around 35ft by 45ft (around 10m x 14m).[118] Wooden panels were made to be remounted by hosts at each new location (Figure 7.16). Exhibition panels arrived packed in fourteen cases with two-piece hinged screens, fascia boards, banners, tabletops; a central kiosk was the most complex display. Care of the material – photographic paper, mounting boards and plywood – was, the leaflet explained, considered of paramount importance given material restrictions.

Describing how work on exhibitions had shaped his career, typographer Charles Hasler, who joined the MOI's Exhibitions Branch in 1942, recalled: 'I more or less completed my typographical education in the MOI exhibitions division during the war after coming out of the army'.[119] Hasler believed MOI exhibitions were an interesting context for developing typography because of their visual and scripted format. He explained, 'Most creative typography then went into exhibitions as most of these [were] officially sponsored ... and for the first time were scripted'.[120] Hasler believed photography was also a key part of the exhibition design mix: 'the copy was typeset, proofed, and photographically enlarged, and mounted'. These images might be enlarged by as much as four times, line film used and printed on white document paper, staining or negative printing used where necessary.

Hasler drew on mid-nineteenth-century Playbill slab serif fonts from Egyptian fonts for *Make Do and Mend* and other MOI exhibitions, a font family he would return to a few years later as chair of the Festival of Britain's Typography Panel.[121] He believed that Egyptian letterforms worked well in exhibitions because of their constructional qualities and three-dimensional possibilities.[122] Aside from aesthetic considerations, choice of fonts for exhibitions depended on what was available in restricted wartime conditions. Typefaces soon got worn and damaged so that some, particularly hard-to-replace imported ones, ended up without enough characters.[123] To compensate for this, exhibition designers developed a practice of creating typeset captions photographically to enable lettering to be produced at any size and to preserve their characteristics.[124] The war restricted the colours available, as Hasler explained: yellow was out, for example, because it was derived from steel, meaning that technical innovation was necessary to allow for limited materials.[125]

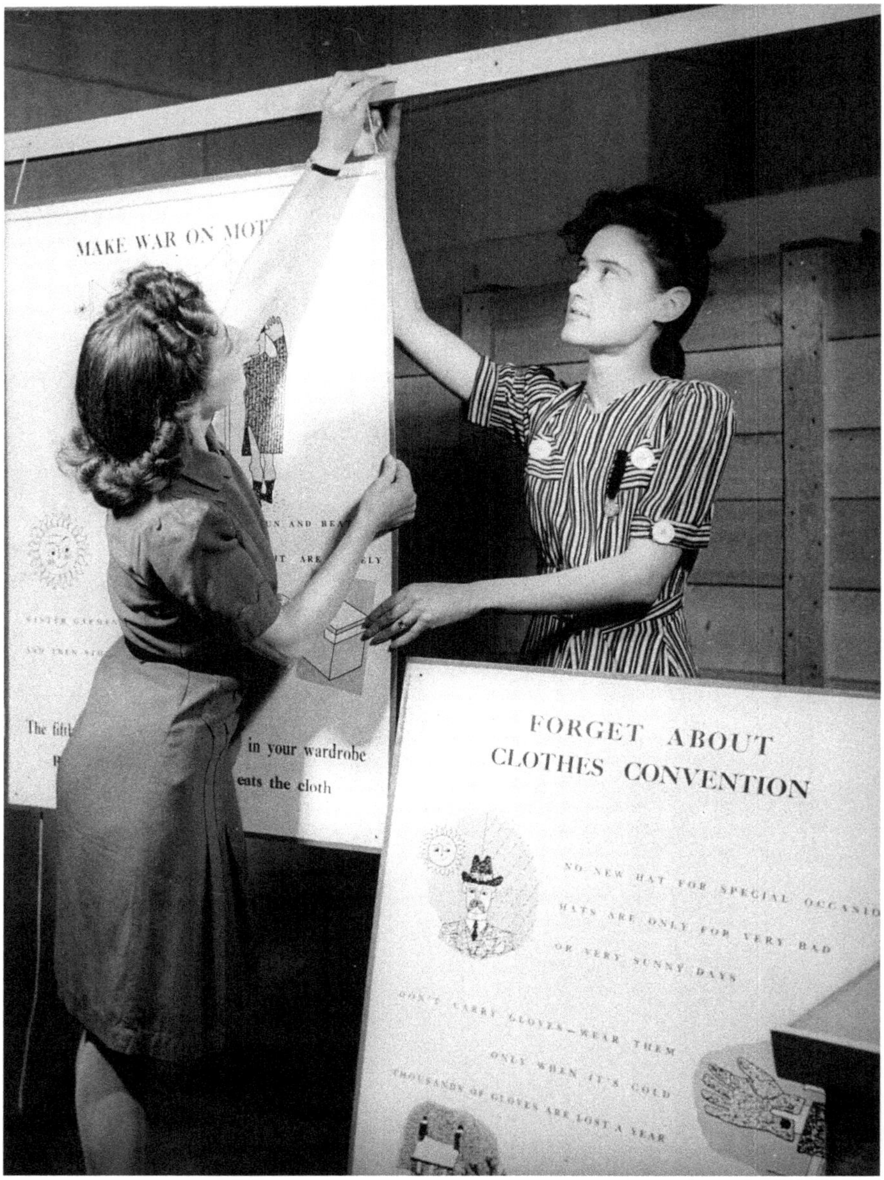

7.16 *Make Do and Mend: Household & Clothes Economy*, touring exhibition designed by Charles Hasler. Image courtesy of the Museum of Domestic Design & Architecture, Middlesex University, www.moda.mdx.ac.uk.

Poster panels incorporated a range of media including photographs, cartoons and line drawings, with a written commentary on subjects such as 'Make War on Moths' and 'Forget About Clothes Convention', 'no new hat for special occasions', 'hats are only for very bad or very sunny days'. Again, these were focused towards the urgency of now, towards overcoming immediate practical issues, and used familiar phrases, rhymes and comic devices to drive the message home. Moth grubs – growing into clothes moths – became 'the fifth column in your wardrobe', making the problem immediately relevant through a comic transfer that likened the pests to an enemy group working to undermine Britain's solidarity from within.[126] An MOI *Home Intelligence Report* of 1942 reported *Make Do and Mend* exhibitions as 'very popular', recommending more be organised by Women's Institutes, Co-operative Guilds and Mothers' Unions, to run alongside classes in mending.[127]

From Potato Pete to *Victory Over Japan*: exhibitions on the John Lewis bombsite

Aside from the popularity of Charing Cross Station for small exhibitions, the blitzed John Lewis site on Oxford Street in London's West End quickly became a recurring favourite for major exhibitions. A September 1940 air raid had flattened the department store, leaving it a decimated shell. Large and centrally situated, the site was adopted by the MOI for public exhibitions, attracting visitors going about their everyday lives, its extensive basement, originally used as the staff canteen, made it suited to housing major exhibitions including the MOI's *Britain's Aircraft*, *Victory Over Japan* and *The Army Exhibition*.

Potato Pete's Fair was one of the strangest of the series of government events held on the John Lewis bombsite. A free public information event dressed up as a celebration during its two-week run, the Fair aimed 'to teach 100,000 Londoners 198 ways to cook potatoes'.[128] It included 'Nursery Rhyme Land', in which popular rhymes were rewritten to have a potato angle, a 'Magic Potato', a 'U-boat Shy' with a ship-saving theme in support of the campaign for home-grown foods, a lucky dip, a 'cookery nook' and a cinema showed Walt Disney's popular Mickey Mouse cartoons.[129] *Display* described *Potato Pete's Fair* as 'the most unusual Christmas Bazaar'. 'Whoever thought … that a Christmas Fair could be built around the potato?' Despite grim wartime conditions and restricted materials, *Display* was convinced that everything in the Fair was planned as near as possible 'to traditional fairyland lines': 'Tom of spruce and fir cuttings from re-afforestation plantations were used to give a seasonal background and the scented air was most exhilarating'. *Display* declared, 'If ever the art of display has been put to the test it has surely been in that colourless everyday tuber, for, from pre-war ignominy, the potato has been so presented that it now occupies a high place in all our approaches to dietry [*sic*], and "Potato Pete" has become a national figure'.[130] Cartoonist David Langdon contributed work to this campaign, while Walt Disney had presented the Ministry of Food with a family of anthropomorphic root vegetables: Pop Carrot, Clara Carrot and the upper-class, monocle-wearing Carroty George.[131]

Taking up the potato theme, the Ministry of Food's Lord Woolton, working with the Potato Publicity Bureau, drove forward a campaign encouraging cultivation of potatoes for food. 'Edible farinaceous tubers of vegetable plant Solanum Tuberous', as a tongue-in-cheek piece in *Display* described them, became a focus for many exhibitions and window displays and the major element in 'Dig for Victory' campaigns, in all its forms. *Display* complained potatoes were not the most inspiring subject for moving and fabulous displays, 'prosaic, colourless, and uninspiring'. Nevertheless, the magazine noted shopkeepers up and down the country were embracing the challenge by mounting displays of potato salads and potatoes arranged in the shape of a 'V for victory sign'.[132] The patrician tone of these displays was conspicuous, explaining things that were likely already evident to the majority of viewers and even the most ordinary aspects of life being considered a valid focus for displays.

Demonstrations ran alongside exhibitions, with the Ministry of Food running a full publicity programme, where dieticians dressed in white overalls, to reinforce their professional standing, showed visitors how to cook healthily. A 'Mobilise Your Vitamins' display showed cartoon carrots and cabbages marching behind shopping baskets being transformed into finished dishes. Such displays were mobile, easy to install and dismantle, a means of 'establishing personal contact with the "man in the street" and the "woman with the shopping basket"'.[133] Demonstrations enabled friendly contact and sociable conversation, adding a lightness and human touch.

Inspiring citizens: *The Army Exhibition: The Equipment of a Division*

The most ambitious amongst the many exhibitions held on the John Lewis bombsite was *The Army Exhibition: The Equipment of a Division* (Figure 7.17). Designed for the War Office and touring onwards to three major venues, it explained how the army was organised; from communications to food, transport, clothes, camouflage, munitions, medical services and the rather laborious process of equipping a division.[134] Designers made the most of the varied vantage points across the site, with bombing exposing spaces above and below ground bridged in places, in the spirit of earlier exhibitions such as Gropius, Bayer and

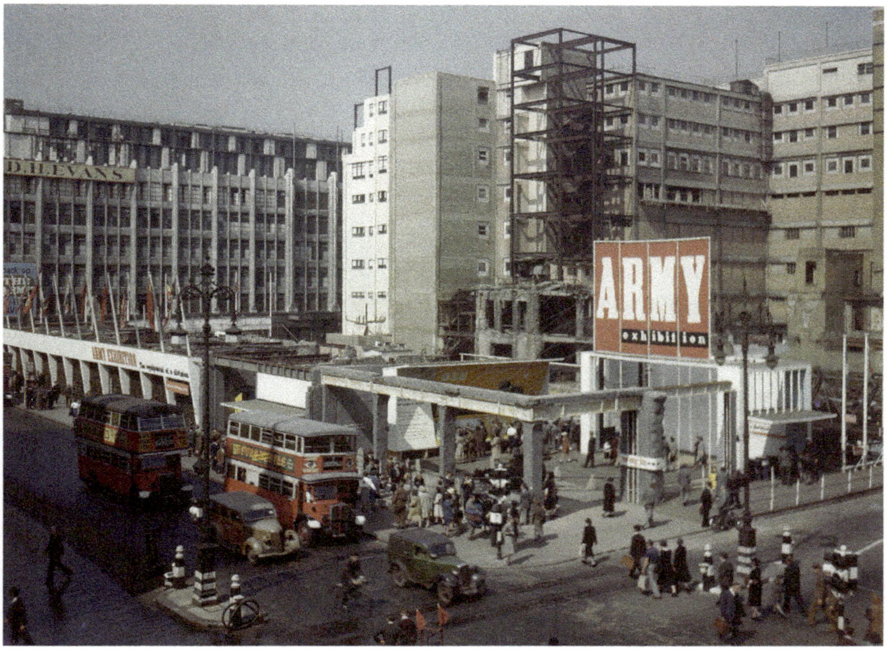

7.17 *The Army Exhibition: The Equipment of a Division*, mounted by the Ministry of Information and held at London's blitzed John Lewis department store site on Oxford Street in 1943. Ministry of Information official photographer © IWM TR 1147. All rights reserved and permission to use the figure must be obtained from the copyright holder.

Moholy-Nagy's *Ausstellungsstand der Baugewerkschaften* (*Exhibition of the Building Workers Unions*), Berlin 1931, which had used ramps to create views across and between displays (as discussed in Chapters 1 and 2). *Architectural Review* admired the layered aspect of *The Army* site: the magazine had been campaigning for the adoption of the English Picturesque in urban reconstruction projects, remarking on the exhibition's potential for 'unexpected vistas', with 'blasted walls and bared girders' achieving a 'picturesque unity'. 'The eighteenth century squire had to build them specially', the *Review* mused, 'to us the enemy's bombing has given them, and here is a way to make them a positive part of the urban scene'. The magazine's agenda for a new urban picturesque would be most conspicuously fulfilled at the Festival of Britain's South Bank Exhibition.[135]

Designed by a team including architects Bronek Katz (on the exterior), Frederick Gibberd (on the engineering section) and Peter Moro (on clothing and signals) and F. H. K. Henrion (on the concluding section), *The Army* boasted 23,000 exhibits ranging from full-sized 'Churchill tanks to optical lenses'.[136] An eye-catching yellow and grey introductory panel explained the exhibition's premise, with a huge map showing the African battle area viewed from the Mediterranean and the way men and machines were moving around a vast area (Figure 7.18). The colourful exhibition circuit of 56,000 square feet contained sections such as 'A gun's life – and yours', 'Every bit of fuel counts', 'To beat the dive bomber' and 'Every ounce of salvage counts', communicating its message about the might and modernity of the British army through graphs, full-sized military equipment, striking enlarged graphics and integrated text panels built out of the rubble of the open-air site under small, awninged display boxes and panels.

The exhibition's hands-on elements included guns children could hold and armoured vehicles to climb on. A striking 30-foot tower of jerricans, suspended in a metal frame built by designer Richard Levin (introduced in Chapter 1), illustrated the amount of fuel used by an armoured division in two and a half minutes with an eye-catching snapshot. Forty thousand visitors saw the exhibition in its first two days, and *Architectural Review* declared it 'the most ambitious and successful of all exhibitions so far staged' by the MOI and 'as up-to-date' as the 1938 MARS exhibition and 'yet in no way high-brow', referring to the particular criticisms levelled at MARS's esoteric presentation at the earlier New Burlington Galleries show, discussed in Chapter 3.[137] *The Army* toured to Birmingham, where it was again installed on a bombsite but lacked the visual impact of the multi-level London site, touring onwards to Glasgow and Cardiff.[138]

Justifying ongoing military campaigns: *Victory Over Japan*

Victory Over Japan, a large 'win-the-war' exhibition, as contemporaries described it, was two years in the making, one of a series of exhibitions justifying ongoing military campaigns.[139] Running alongside a broader campaign through news, films, photographs and articles, the exhibition was focused towards persuading the public of the importance of continuing to fight in Japan, once Germany was defeated.[140] Held on London's John Lewis Oxford Street bombsite, a 10,000 square-foot site that had hosted a string of government exhibitions focused towards materialising wartime abstractions, it opened in August 1945, days after the declaration of the end of war with the surrender of Japan following the devastating bombings of Hiroshima and Nagasaki.

Exhibitions allowed politicians to stage their military strategy. Prime Minister Clement Attlee was present at the opening of *Victory Over Japan*, with ambassadors and senior military officials in attendance, processions of Dominion troops, accompanied by the band of the Scots Guard. In his speech Attlee described the exhibition as a 'record' and a 'tribute',

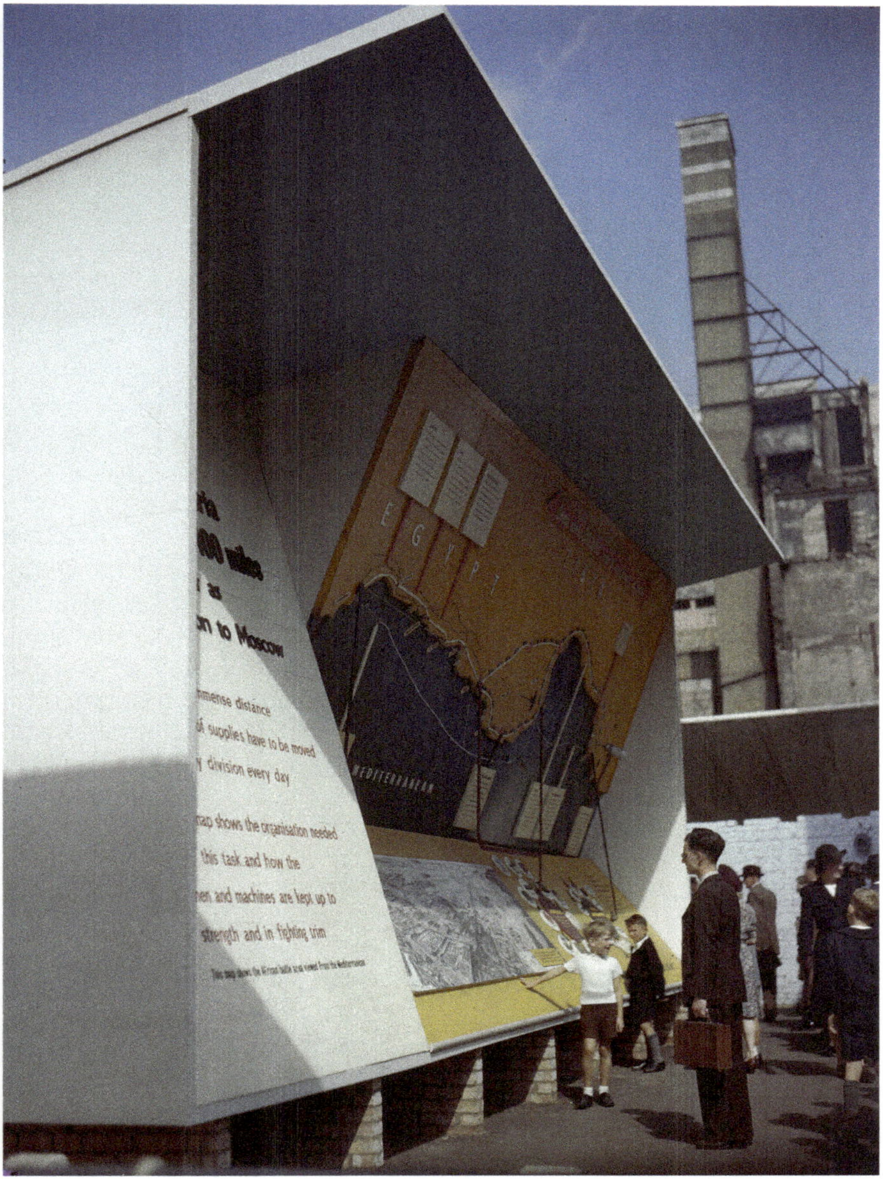

7.18 *The Army Exhibition*, by the Ministry of Information, introductory panel explaining the premise of the exhibition. Ministry of Information official photographer © IWM TR 1146. All rights reserved and permission to use the figure must be obtained from the copyright holder.

showing 'the nature of the enemy ... fanatically brave and barbarously cruel' and praising the Fourteenth Army fighting in Burma. The exhibition's purpose, Attlee explained, was so that those who had not had to suffer the hardships of fighting could see all that had been achieved and pay tribute to 'the spirit of selflessness and willing sacrifices by which our men were inspired'.[141]

Attempting to enable visitors to identify with soldiers fighting on the Eastern Front, *Victory* was highly experiential, using 'jungle realism', as *The Times* dubbed it. Visitors were exposed to simulated jungle conditions, an aspect not previously developed in government exhibitions. They entered through a dark and steaming jungle, so hot inside that one visitor recalled 'my collar soon became a damp rag round my neck'.[142] Giant mock cobwebs, created by enormous model spiders, brushed against their faces; jungle sound effects suggested running water, insects, birds, the 'chattering of monkeys' and 'wails of jackals and hyenas'; and the temperature was kept at 120 degrees.[143] This experiential approach to display was developing in other popular entertainments of the period, as I discuss later in this chapter. London pensioner Herbert Brush recorded visiting *Victory* in his diary for Mass Observation, describing queues of visitors stretching up Oxford Street. 'There is nothing like a "free show" to draw the people from all parts', Brush reflected. After the simulated jungle, Brush recalled photographs including one of the Emperor – 'an insignificant little man' – pictures of manufactures and a girl using a typewriter. 'I looked down on it from a platform above while a running commentary was made through a loudspeaker on the events of the war, and each point was marked as it was mentioned, by coloured lights'.[144] The exhibition's photographic central displays were ranged around a striking illuminated globe (Figure 7.19). The installation traced the history of the war in the Pacific, showing model submarines, suicide bombs, documents and photographic panoramas. Misha Black claimed 1.5 million people visited *Victory* during its four-month run, touring in its entirety to Birmingham, Leeds, Manchester, Glasgow and Cardiff, with four smaller replicas covering fifty other towns.[145] It was widely reported on in the media but its effectiveness as propaganda is, once again, tantalisingly hard to assess.

Exhibitionary exchanges with the US

A series of exhibitions highlighted Britain's close relationship with its new ally, the US, which had entered the war in late 1941, following the Japanese bombing of Pearl Harbor. These exhibitions included *John Olsson, the Story of an Average American* held at Charing Cross Station in 1943 (Figure 7.20). *John Olsson* celebrated the Anglo-American alliance, using expanded photographs and specially designed display furniture to

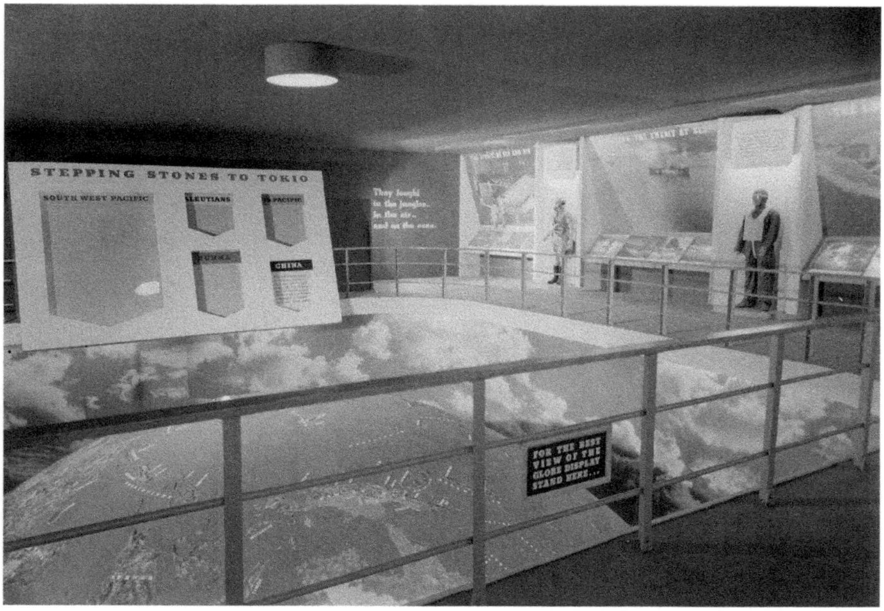

7.19 Interior view from the upper mezzanine of *Victory Over Japan*, mounted by the Ministry of Information at the John Lewis site on Oxford Street, in August 1945.

explain life in the US.[146] It was created jointly by the MOI and the Exhibit Unit of the US Office of War Information (OWI).

In tandem with exhibitions organised by the MOI at Senate House in London's Bloomsbury, as part of its wartime operations, the OWI Exhibit Unit developed wartime exhibitions for home and abroad, taking space at Senate House for its London-based operation. The Unit identified exhibitions as filling a particular niche: although slow compared to films, they remained relevant for longer and were particularly good for supplying background information and conveying 'emotional concepts'.[147] Henrion was, at the time, working across both the MOI and the OWI. As at the MOI, the central focus for OWI exhibits was 'to present American life in effective visual terms', with 'top flight pictures of people', showing 'their characterful determination', 'the scope and fruitfulness of the land', 'the impressive scale of our public works', plus transport, mineral resources, industry and armed forces, as OWI papers reveal.[148] *Young America* exhibition, designed by F. H. K. Henrion for the OWI, was shown in central London at Dean's Yard, Westminster in 1944. A series of photographic boards explained the structure of US society, with a particular focus on school, community, church and society.[149]

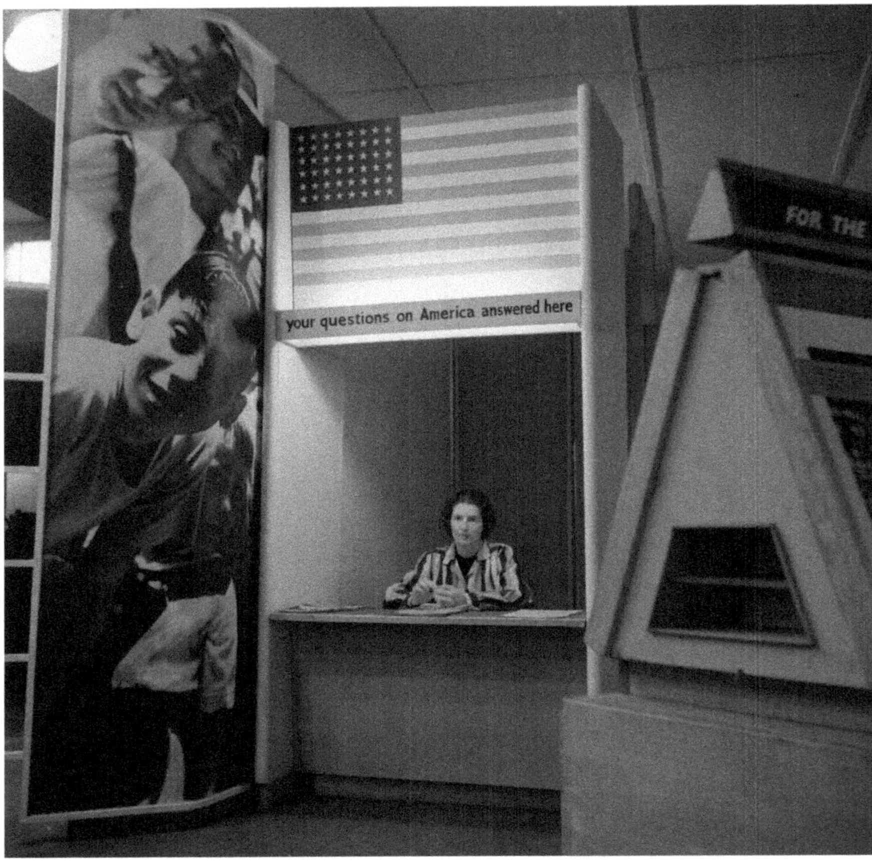

7.20 Kiosk where visitors could ask questions about life in America at *John Olsson – The Story of an Average American* at Charing Cross Station mounted jointly by the Ministry of Information and the Exhibit Unit of the US Office of War Information. Ministry of Information official photographer © IWM D 15217.

US exhibits that travelled to London included shows about the training of US Army Officers, the US Naval Academy at Annapolis, life on a Wisconsin dairy farm and a portrait of an anthracite miner. At the forefront of the OWI's planning around exhibitions, as at the MOI, was a consideration of what this form was capable of. Acknowledging that, compared to news or 'syndicate picture mediums', small targeted exhibits were slow, the OWI nevertheless considered that they worked 'longer once they reached the field' and that while other media had their place, exhibitions were 'best suited to supplying background information and conveying emotional concepts'. The OWI gathered files on subjects as diverse as 'American Types', showing airplane mechanics, dairy farmers and cowboys, community nursing, anthracite coal mining and irrigation,[150] and 'How America Lives', showing a family in a sleigh, ice fishing, baking bread and opening Christmas presents.[151] Many of the nearly four thousand photographs in 'How America Lives' were gathered from existing government sources.[152]

Roy Stryker and his team in the US Farm Security Administration's Historical Section prepared OWI exhibition photographs.[153] Stryker had become well known for spearheading this documentary photography programme from 1935 onwards, commissioning photographers including Dorothea Lange, Walker Evans, Ben Shahn and Louise Rosskam. These same photographs were recycled for wartime US exhibitions. Unlike their British counterparts, the OWI's exhibitions were purposely lighter on reading and more focused around pictures, with minimal textual accompaniment to be as accessible as possible. 'Pictures tell the story', an explanatory text explained. 'The picture – and especially the picture sequence – is a language understood by everyone: it knows no illiteracy. Words tell your story. Pictures make it real'. A pamphlet produced by the Historical Section gave practical hints for preparing 'exhibits-with-a-punch', to 'reduce exhibit design to its simplest "dos" and "don'ts" and to the simplest methods and materials'.[154]

The British Library of Information, an organisation representing British interests in the US, established a small section early in 1941 to organise and design US-based exhibitions, working with the British government's American Division and the MOI's Exhibitions Branch (Overseas Displays and Exhibitions Scheme) to oversee photographic displays touring the US on subjects such as *Women of Britain*, *Nutrition in Wartime*, *Bomb Damage* and *Social Services in Britain*, liaising over policy to ensure alignment.[155]

Britain at War, 1942, was one of the first photographic exhibitions organised by the MOI to be sponsored and held at MOMA New York, before touring round Canada. British magazine *Art and Industry* reported the exhibition aspired 'to cover the whole of the visual aspect of the war in terms of design' with sections including the Army, the Navy, the RAF and the Home Front.[156] It reported with admiration on Peter A. Ray's 'War-time Shapes' section, which used forty photographs portraying the shape of things brought about by war conditions. The images were clearly influenced by Moholy-Nagy's 'new vision', with functional objects of war such as 'static water tanks for AFS use', 'anti-blast window netting', 'bus light shades', the 'painted base of lamp-posts' and 'anti-incendiary bomb sandbags' seen through unusual angles and perspectives, making use of reflections, light and shadow.

Back the Attack and *Nature of the Enemy*: the US Office of War Information in London and Washington

The plaza outside New York City's Rockefeller Center became a regular site for OWI exhibitions at home. An exhibition of photographs introducing the United Nations was shown in March 1943, with a mock-up of the Atlantic charter and exploring the nature of the four freedoms in Roosevelt's speech: a vast sculpture of a serpent wrapped around a book labelled 'press' a metaphor for the threat to freedom of speech.[157] Soon after, the OWI installed its emotive *The Nature of the Enemy* at Rockefeller Plaza in summer 1943.[158] The exhibition was one element used by the US government to help justify entering the war. With strikingly vast photographs towering over visitors' heads, the exhibition dramatised wartime episodes, such as the bombing of Pearl Harbor. *The Nature of the Enemy* used waxwork models to create tableaux, mounted on raised platforms, to dramatise grotesque Nazi scenes including Hitler Youth and the concentration camps (Figure 7.21). Visitors were offered the chance to imagine they could interact directly with the war at a 'Buy a Bond and Bomb Berlin' stand. The exhibition was considered so successful that it was reproduced for travelling.[159]

Looking at an OWI installation on home soil shows the different framing of the two exhibition-making bodies. The presentation of *Nature of the Enemy*, in being focused mainly through vast photographs and waxwork tableaux, was quite different from the compound display trickery of MOI exhibitions, which drew on lots of photographic techniques, objects and playful texts in tandem. It is evident, however, that there was much cross-fertilisation between exhibitions across all of these contexts and that those developing OWI exhibitions looked not only to the examples of their allies, including Britain, but also to recent German exhibitions mounted under the Nazis. The holdings of the OWI archive at the US Library of Congress (LOC) contain material about the *Ausstellungsstand der Baugewerkschaften* (*Exhibition of the Building Workers Unions*) in Berlin of 1931 by Bayer, Gropius and Moholy-Nagy (as discussed in Chapters 1 and 2) and Bayer's *Die Neue Linie* exhibition, mounted to introduce the journal of the same name, launched in 1929. It contains documentation of Nazi exhibitions, such as an exhibition of leisure-time and vacation activities, designed by Herbert Bayer and held in Hamburg in 1934.[160]

More significantly, the archive contains a large illustrated album of designer Willi Hackenberger and Herbert Bayer's Nazi exhibition *Das Wunder des Lebens* (*The Wonder of Life*) sponsored by the Reich Committee for Public Health Service and held in Berlin in 1935.[161] This exhibition was obsessively focused on eugenics and bloodlines and included Hitler's family tree, images of healthy 'Aryan' types and famous men who were

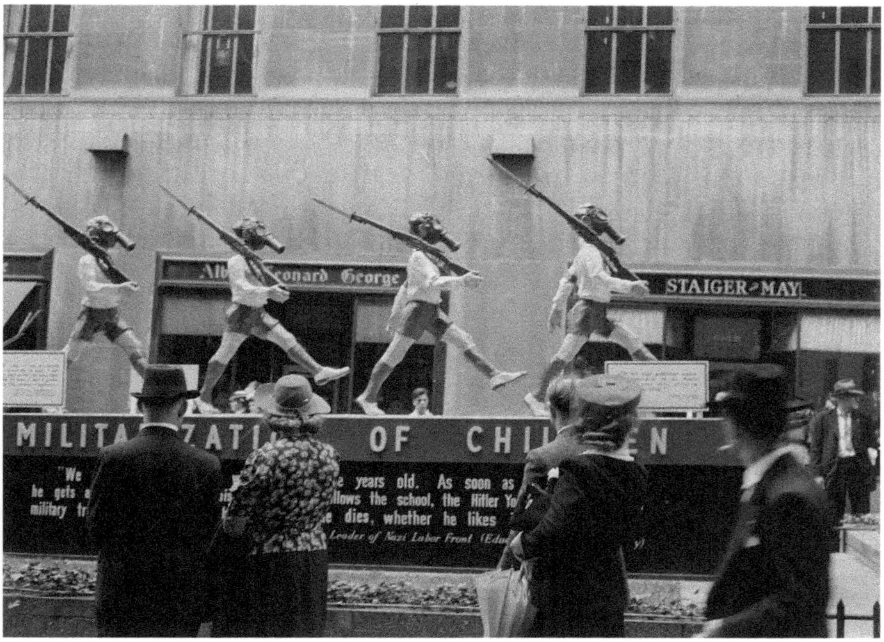

7.21 Spectators looking at exhibits at the *Nature of the Enemy* exhibition, Office of War Information at Rockefeller Plaza, New York, summer 1943. Arthur Siegel. Image courtesy of the Library of Congress Prints and Photographs Division, Washington, DC. All rights reserved and permission to use the figure must be obtained from the copyright holder.

members of large families. Pictorial material presented Jewish and other 'undesirable' people, who, according to Nazi ideology, biologically threatened German public health and graphics pointed to the relative fertility of 'inferior' versus 'superior' beings. Clearly Bayer's earlier experience of developing exhibitions, including those in the Nazi context, had provided the techniques and skills of creating three-dimensional argumentation that would be operative once he was called upon to produce US wartime propaganda. Certainly the core formal elements of such exhibitions – expanded illustrative photographs, persuasive text, enacted in and through space – were common across the ideological divide.

The OWI's exhibition *Back the Attack* was directly influenced by the London *Army Exhibition*, as is evident from the many images of the London show held in the archival files at Washington's Library of Congress.[162] It was mounted in the shadow of the Washington Monument in September 1943, a collaboration with the Adjutant General's Office, and set out to unveil the US army's war effort, with a reconstructed field post office and adjutant general's department tent that visitors could tour around. However, *Back the Attack* was lacking in display acumen, putting documents and objects on show without apparent regard for creating visual excitement. While the Anglo-Soviet relationship was being celebrated in UK exhibitions (as discussed in Chapter 6), the OWI designed the American–Soviet war exhibition in June 1943, installed at New York City's Museum of Science and Industry. This was structured through floor-to-ceiling collages of expanded photographs showing Russian life and people and some reconstructed war elements including reproduction Russian dugouts.[163]

OWI–MOI exhibitions in Paris: *La Guerre et La Paix and Pacific '45*

The MOI and the OWI collaborated on displays to be shown in France, mounting a joint propaganda exhibition at the Grand Palais, Paris from April to June 1945 named *La Guerre et la Paix* (*War and Peace*). The ambitiously large exhibition's purpose was to show French people – a few months after liberation – and others a 'total picture' of the Allied war effort and to explain what they had learnt from participation in war. The exhibition was jointly funded by Britain and the US and consisted of around five thousand images including photographs, maps, charts, diagrams, models and objects. It served as a focus for distributing literature and showing documentary films in the Palais's 1000-seat movie theatre, with sections focusing on 'the price the people paid' during the occupation and 'culture is carried on'.[164]

In this technologically advanced exhibition, the opening statement was projected onto a screen:

> IN THESE HALLS IS A RECORD
>
> – A record of a free people's struggle against the attack on their liberty
> – A record of their labors and their unity in the face of destruction
> – A record of their triumph in the worst of all wars

The words were accompanied by an animated montage showing continuous moving images of marching soldiers, workers and ships being launched. Contemporaries reported the 'definite feeling of exhilaration' felt by visitors to the exhibition, with 'countless' people visiting.[165]

Pacific '45, an OWI exhibition about the war against Japan, was held nearby, in a former Ford showroom on Paris's Champs-Élysées. Former *New York Magazine* Art Director Francis Brennan (OWI's wartime chief of visual presentation for strategic planning and graphics chief for military publications and propaganda) transformed the space into the open doors of a landing ship, as if beached on a Pacific Island. Visitors entered up a ramp to see illustrated panels, a huge mural of an island invasion, illuminated dioramas about the Pacific war and a huge plastic globe. The exhibition was again strikingly advanced technologically: a synchronised floormap with soundtrack explained the American strategy in the Pacific since Pearl Harbor. A United States Information Agency press release boasted that 'nothing like it has been seen in exhibit-conscious Paris'.[166]

MOMA's *Road to Victory* in Britain

The Road to Victory: A Procession of Photographs of the Nation at War, the major exhibition mounted at MOMA New York in 1942, reinforced public support for America's entry into the Second World War. It was also the US wartime propaganda exhibition most admired and reported on in Britain. Media theorist Fred Turner cites *The Road to Victory* as highly significant as a central focus for the development of a US 'democratic personality' during the Second World War, when psychologists started to engineer a public personality for the US citizen to counter concerns over the impact of fascism.[167] Originating at MOMA New York, with the support of the OWI, *Road to Victory* went on tour to Britain the year after. One British critic described the MOMA exhibition as the consummate example of 'synchronisation' between forms, for its use of a textual language of 'almost pictorial descriptiveness', alongside striking expanded photographs.[168]

As with other US government exhibitions discussed earlier in this chapter, the images in *Road to Victory* were chosen by photographer Edward Steichen from existing photographs held in US government collections. Many had been repurposed from the 1930s Farm Security Administration programme's chronicles of the Great Depression.[169] To make them into large photographic murals, negatives were enlarged in sections, on strips of photographic paper forty inches wide, pasted on to wallpaper covering gallery walls, seams airbrushed, retouched by hand and varnished.[170] Steichen worked with exhibition designer Herbert Bayer and poet Carl Sandburg, who wrote a special text. Entering the exhibition at MOMA, visitors were faced with panels of buffalo and Native Americans and Sandburg's opening panel that read, in biblical evocation,

> In the beginning was virgin land and America was promises – and the buffalo by thousands pawed the Great Plains ...[171]

These words inducted visitors into an environment where they could move among images and sound at their own pace.

The MOMA exhibition went on to tour in the US to San Francisco, Cleveland, Detroit and Chicago. A modified version of *Road to Victory* travelled to London's Dorland Hall, opening in March 1943 with the revised title *America Marches with the United Nations*. In the London context, the exhibition became a vehicle for signalling US solidarity with Britain and for persuading British audiences of the merits of the embryonic United Nations, with Dorland Hall's high-ceilinged space given over to explaining the mechanics of the new international organisation. The exhibition's entrance, designed by Milner Gray, welcomed visitors with a message as if spoken by the people of American: 'To the people of Britain ... you kept open the road to victory. America is proud to march beside the nation which gave our world hope and time'. Small versions of the same photographs of Native Americans were displayed beside very similar opening words:

> In the beginning was virgin land – and the buffalo by thousands pawed the Great Plains.

The strikingly enlarged photographic display of landscape and people that had dominated MOMA's rooms, creating an internal 'road' through which visitors wound, was shown in London but the photographs were smaller and wall-mounted, making it far less striking and impactful.[172] As in the MOMA show, the primary impact of *America Marches* was visual, its photographs illustrating scenes of everyday life in the US and elements of the diverse landscape, interspersed with excerpts from Sandburg's original script used at MOMA, but edited down for its London showing. An example of Sandburg's text accompanied an extensive collage at Dorland Hall, showing photos of people in a range of interiors accompanied by the words:

> Many people, many faces,
> in their homes,
> their home towns,
> their churches, shops,
> schools

Sandburg's words were comforting, poetic and light on information, unlike the MOI's exhibition texts, which set out to impart complex information with extensive descriptions and labelling. After its London showing, *America Marches* was revised yet further for presentation to travel around Britain.[173]

Alongside creating exhibitions for home audiences, the MOI's Overseas Displays and Exhibitions were circulated around the world, particularly from 1942 to 1943. To avoid heavy wartime shipping costs, the MOI developed a system of sending designs as small prints on thin tracing paper and photographs as negatives, to be transported by air then constructed into full-sized three-dimensional colour displays by press offices in each location. As well as sending photographic exhibitions to the US, displays were created for Latin America, the USSR, China, Portugal, Sweden, the Middle East, 'French North Africa' and 'Empire countries', as *Architectural Review* described them.[174] Shop windows were used to display material in Latin America and Portugal, while exhibitions of enlarged photographs of bomb damage in enemy territory were sent to the US and USSR.

A large-scale exhibition about the campaigns leading up to the liquidation of the Italian Empire was sent by fast transit to Cairo for distribution to India, Abyssinia, Palestine, Iraq, Iran and New York (for re-transmission to Latin America and China), to Canada, South Africa and Russia. Shop spaces were used to host exhibitions originating in Britain. The Brussels Bon Marché department store played host to the touring 1943 exhibition *RAF in Action*, developed by the MOI with the Air Ministry Exhibition section. This was sent onwards to the US, Latin America, Cairo, Canada, New Zealand and India, telling a story about RAF activities through photographs and props including flying equipment, rubber dinghies, a model bomber suspended from the roof set against a 40-foot canopy mural of cumulus clouds and a 12-foot high plaster figure of the winged 'Victory' with a quote from Churchill commemorating 'the few'. Display panels illustrated key RAF victories such as the bombing of the German city of Duisburg.[175]

Displaying the 'four feathered freedoms': *For Liberty*

Aside from the MOI's regular use of the John Lewis bombsite for its exhibitions programme, the site was taken over in 1943 by the AIA for *For Liberty: Paintings on War, Peace and Freedom*, a major exhibitionary 'demonstration'. Although the AIA's pre-war politics had positioned the group as adversarial and oppositional to official policy, particularly on the government's non-intervention in the Spanish Civil War, once the Second World War started several of these same artists and designers were employed in government. The foremost of these was AIA founder Misha Black, who found himself at the epicentre of the government's exhibitions programme, whilst continuing to lead the AIA. F. H. K. Henrion, who worked on *For Liberty*, was also employed at the MOI. He talked about the anomaly of the MOI working closely with the AIA in an interview, explaining that 'suddenly during the war to be anti-Fascist became fashionable … so while they were very unpopular before the war, the AIA became even government backed in some of their exhibitions'.[176] This was not the only example of a pragmatic alliance between Leftist activists and official propaganda bodies during wartime. (As discussed in Chapter 6, the same had happened through *Allies inside Germany*.)

In *For Liberty*, a free-to-enter exhibition held from March to April 1943, the AIA explored the 'four freedoms', the universalist values set out by US President Roosevelt to Congress in his State of the Union Address of 1941. These were freedom of speech and expression; freedom to worship; freedom from want; and freedom from fear. This 'is no vision of a distant millennium. It is a definite basis for a kind of world attainable in our own time and generation', Roosevelt declared.[177] These values and their materialisation in three-dimensional form as an exhibition provided a platform from which the AIA could position themselves, by articulating what they believed was at stake, whether the war was lost or won.

Artists were being criticised by some contemporaries for not using their work to 'fight' the war. *Art and Industry*'s editorial of April 1942 asked 'Is British Art Fighting?', asserting that 'British art today should be in the forefront of the battle, encouraging, stimulating, inspiring, driving; encouraging to the faint-hearted, stimulating to the weary, inspiring to the whole nation, driving on to ever increasing effort, a scourge to the sluggard and the selfish, evidence of a flaming will to win. Put it to work'.[178] Art critic Herbert Read responded, 'The function of art is to provide us with values worth fighting for. Art is a persistent search for truth'.[179] *For Liberty* took up Read's proposition that art's function should be about representing values that spurred people to fight. Its catalogue, likely written by Misha Black, explained that the art on display showed 'that the function of art in wartime is not only to record what is happening and to give enjoyment and recreation but to stimulate and encourage by vividly representing what we are fighting for'.[180] It attempted to reinforce the commonalities of a like-minded community, rather than simply explaining the progress of war, which was the focus of government exhibitions such as *The Army* held on the same site that year (discussed earlier in this chapter).

For Liberty was sponsored by Leftist broadsheet *News Chronicle*; their addition to the catalogue amplified its focus on safeguarding values: 'the future Britain against intellectual poverty'.[181] 'Time will cement the alliance between newspaper and artist now being brought into being. It is an alliance with wide horizons and of incalculable power'. Asserting the idea of exhibitions as cementing alliances, creating strength and resilience, the foreword stated: 'today, in this exhibition, the power is used to forge the weapons of mind and spirit essential for victory over Fascism', boldly suggesting that those who had the tools to build an exhibition today had the toolkit to create the new world of tomorrow. In this rhetorical vision, an exhibition put on show a common set of ideas and values and had the agency to forge a bond between newspaper and artist, between members of the AIA network and between artists and the public.

From their founding in 1933, the AIA had focused on mounting exhibitions away from established cultural institutions, 'in the thick and bustle of every-day life: in tube stations and public shelters, in the busy shopping streets of the East End and West End of London', as they described it.[182] Being part of the bustle of everyday life was akin to a missionary instinct and continued to be talismanic for the AIA, even in the midst of war. This aspiration (discussed in more detail in Chapter 4) was taken to its limits as practice in *For Liberty*, with the choice of a bombsite, by its very definition a place in flux, temporarily suspended between past use and future reconstruction. Here, on Oxford Street, was a department store – a space of capital – blown open to the elements; its normal value as commercial space suspended. The bombing had revealed, at ground level, a structure, 'like the ruins of a Greek temple', as journalist Kingsley Martin described the exposed concrete columns of John Lewis.[183]

Contemporaries believed that wartime West End bombing, while causing devastating damage, had opened up novel opportunities for replanning this quarter. It also allowed for novel creative interventions. The use of a space recently bombed by a highly visible act of Nazi aggression was a bold act of creative defiance, which the AIA turned to positive advantage with publicity announcing 'German Bombs Provide Exhibition Site'.[184] Henrion, who designed the exhibition's installation, recalled Oxford Street 'teeming with service men on leave'. The entrance was striking: the Fire Brigade painted the site in bright, saturated colours, which Henrion recalled as becoming 'like a Graham Sutherland painting, orange and blue, etc' (Figure 7.22).[185] This made a strong impact on visitors and the press, 'this trumpet call of colour made a direct emotional appeal of such poignancy that it will not be blurred easily by any exhibition work done since', *Architectural Review* reported.[186] *Display* announced that the department store's windows had been reborn, 'girders and stanchions are painted red; wrecked walls a vivid yellow', imagining Henrion's painted interventions as creating a phoenix from the ashes.[187] Aside from brightly painted colours, Henrion's entrance represented the four freedoms as doves – 'the four feathered freedoms', as *Display* described them. *Architectural Review* praised the stylised doves as 'surrealism stripped of all that so often appeared to be bogus'.[188] Henrion also designed publicity for the exhibition: posters, letterheads and the catalogue (Figure 7.23).[189]

Downstairs, in what had been the John Lewis staff canteen, 150 paintings and sculptures by AIA members were exhibited. Taking up the idea of the exhibition as not only documenting but making manifest invisible values and truths, the catalogue explained that artists were formulating and expressing ideas, as well as 'illustrating and interpreting fact'.[190] It was crucial to this presentation that painters, sculptors, designers and writers collaborated together across medium. This, the AIA stated, was 'a new feature … which carries AIA policy a step further', with members '<u>working to a theme and arranging the works in such a way that they became part of a whole scheme and not separate units. We hope to develop this new technique in future</u>' (original emphasis).[191] In this statement was an assertion of the power of exhibitions to create bonds between people across types of creative work – art, design and literature – and, indeed, across backgrounds. In line with this, the exhibition's epicentre was the 'Four Freedoms Room', showing twelve specially created paintings by different artists, including John Tunnard and Carel Weight, linked by a poem that responded to the exhibition's theme, commissioned from Leftist poet and commentator Cecil Day-Lewis. The poem concluded

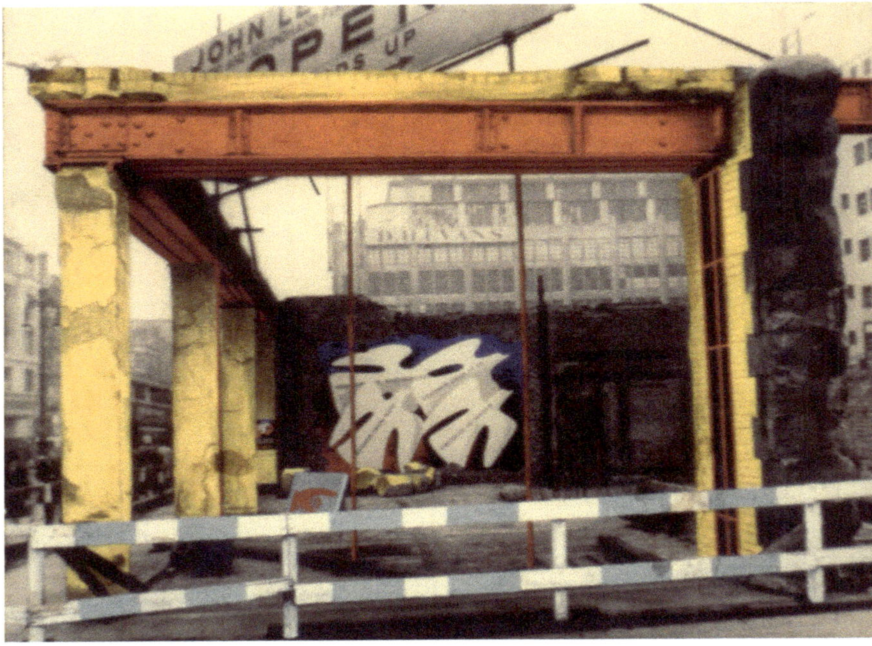

7.22 *For Liberty* exhibition's entrance, designed by F. H. K. Henrion and organised by the AIA, admired by *Architectural Review* as 'this trumpet call of colour'. Photographer unknown. F. H. K. Henrion Archive FHK/3/23, University of Brighton Design Archives. © Estate of F. H. K. Henrion. All rights reserved and permission to use the figure must be obtained from the copyright holder.

with the optimistic words, 'And our heirs shall unfold, like a cluster of apple-blossom, in a fine tomorrow'.[192] The line was inscribed under Betty Rea's sculpture *New World*, showing sculpted heads of four children looking joyfully upwards and placing children's futures as a central concern of the exhibition.

The displacement of an art exhibition from established galleries and museums to a West End bombsite, untethered from constraints of institution and unbounded by walls, was mirrored in the trajectories of *For Liberty*'s makers, many themselves displaced as refugees from Central and Eastern Europe and living outside the shelter of established relationships in Britain. Aside from Henrion, these included Polish painter Feliks Topolski, who arrived in 1935 and contributed a painted panel shown on *For Liberty*'s staircase. Hungarian-born sculptor Peter Lambda, who had arrived in Britain in 1938, contributed an imposing portrait bust of Soviet premier Joseph Stalin celebrating the Anglo-Soviet relationship as allies and Hungarian-born sculptor Peter László Peri contributed the sculptures *Fascist Rule* and *This Is How We Are Fighting*.

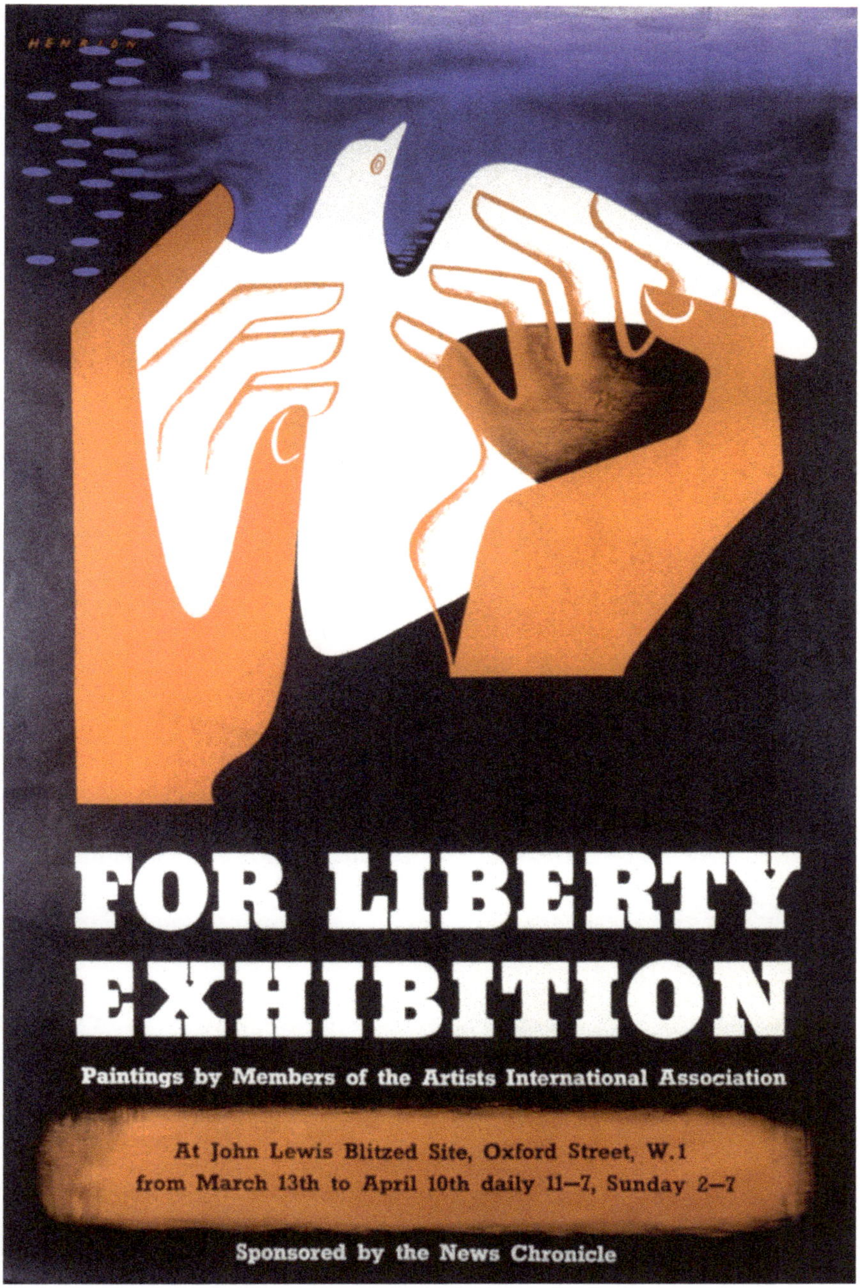

7.23 *For Liberty* poster, designed by F. H. K. Henrion. F. H. K. Henrion Archive, University of Brighton Design Archives, Ref: FHKH/6.

The position that exhibitions afforded their makers – in providing a mouthpiece and a sense of local connection – parallels that characterised by photography theorist Ariella Azoulay in her description of stateless people becoming members of the citizenry of photography through connecting with the public through photographs.[193] Austrian Expressionist painter Oskar Kokoschka (whose work in Britain I discuss at length in Chapter 6) made a painting for *For Liberty*'s 'Four Freedoms room' provocatively entitled *What We Are Fighting For?* This was the final and most bitterly critical of the paintings Kokoschka made in England, attacking the behaviour of both Allied and Axis powers by depicting an emaciated corpse lying in the centre surrounded by a scene of bloody devastation.[194]

Minister of Information Brendan Bracken opened *For Liberty*. He had likely identified as politically expedient the exhibition's message about future peace, which chimed with the MOI's efforts and allowed them to ally themselves with this influential group of artists.[195] In the AIA's own estimation, *For Liberty* was their most successful exhibition, attracting 36,000 visitors in one month, and panels toured onwards to the Peter Jones Art Gallery at Sloane Square.[196] The press, particularly right-wing newspapers, criticised the exhibition as producing propaganda for political or social aims while critic Jan Gordon, in *The Studio*'s 'London Commentary' column, praised the exhibition as a constructive, 'practical' exhibition that could 'suggest to artists ways in which they may develop their own possibilities', bringing together paintings with a 'deliberately propagandist tendency' following specific themes, although Gordon considered it mainly unsuccessful as propaganda.[197]

For Liberty offered artists, many marginalised by dint of displacement, a mouthpiece from which to express their views on war and to make a direct address to the British population. The organisation of *For Liberty* was led by women including painter Beryl Sinclair and sculptor Betty Rea, with dozens of women showing work as part of *For Liberty*, finding in the AIA a freedom and egalitarianism that was lacking in other exhibiting institutions.[198] The exhibition offered this group an opportunity to explore and represent the values that drove the war effort from a site in central London: a bold act of defiance.

Ursula and Ernö Goldfinger's exhibitions for the Army Bureau of Current Affairs

Aside from the solidarity exhibitions discussed in Chapter 6, architect Ernö Goldfinger and his wife Ursula Goldfinger continued to collaborate on making exhibitions. From 1943 to 1947, the Goldfingers designed nine exhibitions for the Army Bureau of Current Affairs (ABCA), the army education scheme set up in 1941 and directed by William Emrys

Williams. Williams, co-founder of Penguin Books, was passionate about life-long learning. While running the ABCA, he was Secretary of the British Institute of Adult Education (BIAE). He also organised the *Art for the People* exhibitions scheme, founded in 1934, to tour important works of art to small towns, intending them to be seen by working-class audiences.[199]

Through the ABCA, Williams hoped to ensure army men and women were fully briefed about developments affecting them. He deployed exhibitions as a key medium for this process of democratisation, considering them useful for wartime promotional campaigns and for sharing current affairs to inspire soldiers.[200] The ABCA produced dozens of easily transportable photographic exhibitions (up to 120 per year), to be toured to army commands for a fortnight at a time on topics such as *Battle of Egypt*, *Convoy to Russia* and *Soviet Armed Forces*, employing many different designers.[201] Williams appointed the Goldfingers to start developing a series of exhibitions covering a wide range of subjects: from traffic to planning your home.[202]

In their ABCA work the Goldfingers could address issues of national importance but with relative freedom to bring their Communist-aligned political perspectives. They designed nine ABCA exhibitions on subjects as diverse as *LCC Plan for London*, *Health Centres* (1944),[203] *Planning Your Kitchen* (1944) and *Traffic* (1944), highlighting the inefficiency of traffic crossings.[204] Their *Cinema Exhibition* of 1943 explained the workings and intention of the mass medium of cinema. Across several strikingly designed panels, each combining an arresting combination of primary colours as structural elements, with clever photomontages and a visually appealing mix of typefaces, many of them hand-drawn by Goldfinger, they explained the range of uses of film, from entertainment to applications in scientific research, propaganda and publicity.[205] Reviewing wartime exhibitions in *Architectural Review*, architect G. S. Kallmann described the Goldfingers' *Cinema Exhibition* as 'exquisitely balanced, but perhaps unknowingly the onlooker may be more attracted by the pattern than by the story',[206] and going on to suggest that Goldfinger's exhibitions were less successful than other contemporary information exhibitions because he was a 'less compromising' designer, who prioritised ideological focus over other considerations.[207] As I discuss in Chapter 8, the ABCA continued to commission the Goldfingers after the war including exhibitions on *Planning Your Home*, *Planning Your Neighbourhood* (about re-planning Shoreditch) and *A National Health Service*.[208]

Cashing in on atrocities: *The War in Wax*

Avoiding sensationalism while evoking reaction and emotion had been a guiding principle of MOI exhibitions, as was articulated in government planning documents. The level to which atrocities committed by Britain's enemies in prisoner-of-war camps and other acts of violence should be exposed in MOI exhibitions was discussed during the planning of *Victory Over Japan*. Their conclusion was that atrocities should 'neither be pushed nor developed'.[209] Commercial exhibitions were not subject to such constraints, unashamedly attracting and confronting visitors with the horrifying spectacle of atrocities in the process of being meted out. The *Daily Express* held a 1945 exhibition of photographs from German concentration camps. Mass Observation diarist Herbert Brush visited, writing that there were 'awful pictures, enough to make one feel ill, but everyone wanted to see them. Heaps of human skeletons' but that a five-minute visit had been long enough.[210]

Up the road from the MOI's *Victory Over Japan*, the sensationalist, privately mounted exhibition *The War in Wax* showed from 1945 in a disused shop at 60 Oxford Street, attracting large crowds. Keeping its potential to shock to the fore, the exhibition's hoarding announced 'The World's Most Modern WaxWorks', including 'the Horrors of the German Concentration Camps All in Life-Like and Life Size Figures', with 'Over 100 figures'. Entry cost 6d for the Main Hall and Children's section, while visiting 'the Concentration Camp' cost visitors an additional 6d (Figure 7.24). A leaflet further explained that the exhibition included 'Life size and life-like figures of Churchill, Roosevelt, Stalin, The Allied Generals, Hitler, Mussolini, Quislings, Etc' and showcased important historic episodes including the 'Casablanca Conference', 'Desert Victory', 'Burma Road', 'Woman at War', 'The Fall of Kharkov' and 'The Horrors of the German Concentration Camp'.

Londoners had seen exhibitions of horrifying waxworks. By 1943, Madame Tussaud's wax museum had been trading on London's Baker Street for over a century, containing the popular 'Chamber of Horrors' exhibit, showing wax tableaux of notorious murderers and infamous historical figures.[211] I have not found images of *The War in Wax*'s interior installation, but it likely simulated the sensationalist exhibition on Baker Street, by acting as a wartime showcase of horrific acts and atrocities represented with waxworks of Nazi leaders and suffering in the concentration camps. In a confusing swerve away from such horrors, and clearly intended to attract family audiences, *The World in Wax* also boasted of 'A Fascinating and Delightful Children's Section of Mechanical moving figures including Cinderella, Snow White, Etc'.[212]

The production from midway through the war of waxwork simulacra of key leaders in the global conflict had allowed for wartime displays intended to bring the horror of war close to the British public, addressing head-on public fascination with horror. Manufacturers reproduced infamous figures in wax. *Display* featured a Hitler waxwork figure created by Gems, 'swinging to and fro' from a gibbet in front of the slogan 'Don't

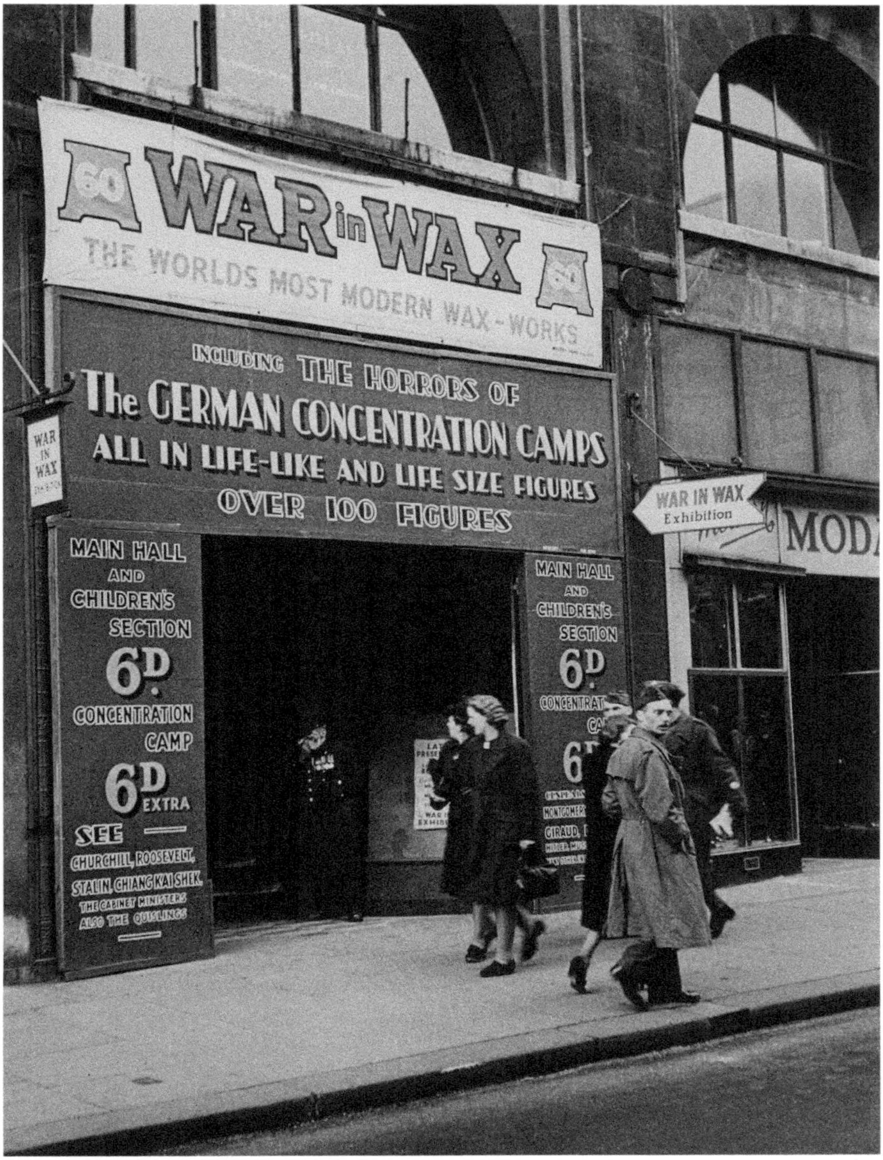

7.24 *The War in Wax* exhibition, at 60 Oxford Street, 1945, photograph taken by Austrian emigré photographer Wolfgang Suschitzky, which he described as 'the only obscene photograph I have ever taken', image courtesy FOTOHOF archive.

let him have the factory – Give him the works'; a tableau created as 'factory propaganda', which the magazine reported as 'the first tableau of its kind in the country'.[213] At *The War in Wax* visitors were confronted with the entangled orders of simulacra: real politicians, currently meting out appalling horrors on the world, shown alongside fairytale characters, offering the possibility that all were mere fictions.[214] These displays, with mechanical fairytale figures, were influenced by the robots and animatronics pioneered for entertainment in the United States at the same time. A humanoid robot nicknamed 'Elektro' had featured at the 1939–40 New York World's Fair: walking by voice command, speaking a few words, smoking cigarettes – with its robot dog, 'Sparko' that could bark, sit and beg to humans.

Display reported that models such as those at *The War in Wax* 'put the accent on reality', by allowing the exhibition to be more lifelike, but that there was 'no showing this for the churlish and delicate stomach' given that 'reconstructed in detail are scenes of brutality practised by Germanic sadists on helpless and long-suffering victims of Nazi persecution'. While some might have recoiled from this, for *Display* this was a draw: 'Each and everyone who experiences even a moment's complacency, who harbours selfish thoughts at the expense of those less fortunate on the Continent, should see this exhibition'. 'Here is an instance of private enterprise leaving officialdom at the starting post; no form of propaganda is there better than visual presentation'.[215] Not all were as enthusiastic as *Display* about the benign intentions of the *War in Wax*. Austrian émigré photographer Wolfgang Suschitzky, chancing upon the exhibition's hoarding, photographed it, later explaining how appalled he was by the sight of such an exhibition: 'Walking one day along Oxford Street toward Charing Cross Road, I came upon this shop front. I was too revolted to go in and investigate'. This, Suschitzky said, was 'the only obscene photograph I have ever taken', referring to the shockingly salacious treatment of a devastating subject.[216]

Exhibitions had proliferated during wartime across multiple contexts of diverse scales, to aid in fighting the war, with the involvement of many designers. Their impact on people's behaviours and emotions remained difficult to assess but the characteristics and quirks of the form itself were becoming a distinctive visual and spatial language. Architect Gerhard Kallmann, in a major feature for *Architectural Review*, described all these exhibitions as starting to draw from the same tedious pattern book, with each exhibition using every element, developing a 'distressing slickness, a stylism which petrifies all liveliness until a sickly *sauce exhibitionnaise* with all the well-known ingredients is poured over the lot', likening these elements to an unappealing British salad cream that smothered the taste of the exhibition's individual and distinctive elements. Despite this, the form was still considered useful by the government and endured as an element of postwar official communication, as the next chapter shows.

Notes

1 Black pointed to a focus on exhibition design in a special 'England' number of *Graphis*, under the heading 'The British Exhibition', *Graphis*, March–April 1946, pp. 212–13; Black, 'Propaganda in Three Dimensions', p. 129.
2 Black wrote that 'a nucleus now exists in this country able to compete on equal terms with their foreign counterparts'. Black, 'Design for Ceremony and Exhibitions', in Herbert Read (ed.), *The Practice of Design* (London: Lund Humphries, 1946), p. 196; Black, *Exhibition Design*, p. 11.
3 Black, 'Propaganda in Three Dimensions', p. 129.
4 Ellul, *Propaganda*.
5 TNA INF1/132 internal memos from December 1940. The fact that Black was Russian and had not yet become naturalised meant he could only be employed on a fee basis.
6 Designer Peter Ray was initially employed as a consultant to design exhibitions but ended up designing graphics as head of the MOI's Story Presentation Unit.
7 It is not clear from the records when Katz arrived but after the war he stayed in Britain, working in architectural practice with Reginald Vaughan, including making a major contribution to the Festival of Britain in 1951.
8 Eventually Black was granted citizenship in 1950: TNA HO 334/344/14923, 'Naturalisation Certificate: Moisei Tcherny. From Russia. Resident in London. Certificate BNA14923 issued 7 September 1950. Note(s): Alias: Misha Black'.
9 Gill Levin, interview with the author over Zoom, 28 May 2023.
10 By the Department of Overseas Trade (through British Industries Fairs and representation at international trade fairs and exhibitions) and the Lord Privy Seal's Department and the Post Office, as noted in an article in *Display* magazine, vol. 27, no. 1, April 1945, p. 2. Highet describes the Post Office and Privy Seal exhibitions in his 'Note on Exhibition Technique' TNA INF1/132.
11 See Henry Irving, 'The Ministry of Information on the British Home Front', in Simon Eliot and Marc Wiggam (eds), *Allied Communication to the Public During the Second World War* (London: Bloomsbury Academic, 2019), p. 24; Ian McLaine, *Ministry of Morale: Home Front Morale and the Ministry of Information in World War II* (Abingdon: Routledge, 2021).
12 *Display* cover of June 1940 (arranged by Whiteley's 'display manager' Mr H. Storey); Selfridges news, *Display*, June 1940, p. 74.
13 *Display*, July 1940, cover, pp. 98, 112.
14 Richard Harman, editor of *Display*, in a lecture to DIA, reported in *Display*, April 1940, pp. 28, 30; a point repeated in *Display*, June 1940, p. 74.
15 *Display*, August 1940. The second theme was the Royal Navy, *Display*, July 1940, pp. 99, 115, 130.
16 Michael H. C. Baker, *London Transport in the 1930s* (Hersham: Ian Allan, 2007), p. 10.
17 Written for MOI by A. G. Highet, 'A Note on Exhibition Technique', p. 3. TNA INF1/132, undated. Henry Irving (of MOI Digital project https://moidigital.ac.uk) confirms that Highet had contributed to MOI's planning phase, coming from a background in PR for the GPO.
18 Calculated costing up to £2 per square foot, basing his estimate on the 150 displays mounted soon before the war by the Post Office and the Lord Privy Seal's Department.
19 In December 1940, as indicated in a note written on TNA INF1/132 'Home Planning Committee: Programme for Exhibitions', p. 5.
20 Brian Foss, 'British Artists and the Second World War. With particular reference to the War Artists' Advisory Committee of the Ministry of information' (PhD thesis, University College London, 1991), p. 3; Brian Foss, *War*

Paint: Art, War, State and Identity in Britain 1939–1945 (New Haven, CT and London: Yale University Press, 2007). During the Second World War Clark also led the Arts Panel of the Council for the Encouragement of Music and the Arts (CEMA); see Sue Breakell, 'Exhibiting "The Taste of Everyday Things": Kenneth Clark and CEMA's Wartime Exhibitions of Design', in Liz Farrelly and Joanna Weddell (eds), *Design Objects and the Museum* (London: Bloomsbury, 2016). For Clark's relationship with design advocacy see Sue Breakell, '"The Exercise of a Peculiar Art-Skill": Kenneth Clark's Design Advocacy and the Council of Industrial Design', *Visual Culture in Britain*, vol. 16, no. 1, 2015, pp. 42–66.

21 Kenneth Clark, 'War Artists at the National Gallery', *The Studio*, 1941, pp. 2–12; also *War Pictures at the National Gallery, London* (National Gallery, 1944). Dame Myra Hess also held a series of concerts at the National Gallery in 1939–44 in aid of the Musicians' Benevolent Fund.

22 Kallmann, 'The Wartime Exhibition', p. 103.

23 TNA INF1/132, part of document entitled 'Home Planning Committee: Programme for Exhibitions', written November 1940, p. 4 under heading 'suitable sites'.

24 TNA INF1/132 'Home Planning Committee: Programme for Exhibitions', part 1 (addendum).

25 As Stuart Hall described it in his analysis of *Picture Post*, 'The Social Eye of Picture Post' [1973], *Typography Papers*, 8, 2009, p. 82.

26 TNA INF1/132, Displays and Exhibitions Division, 28 Nov 1940, Home Planning Committee, Programme for Exhibitions.

27 Black, 'Propaganda in Three Dimensions', p. 129.

28 Turner, *The Democratic Surround*, p. 6. See also Fred Turner, '*The Family of Man* and the Politics of Attention in Cold War America', *Public Culture*, vol. 24, no. 1, 2012, pp. 55–84.

29 See, for example, Simon Eliot and Marc Wiggam (eds), *Allied Communication to the Public During the Second World War* (London: Bloomsbury, 2019) or Robert Mackay, *Half the Battle: Civilian Morale in Britain during the Second World War* (Manchester: Manchester University Press, 2002).

30 TNA INF1/132, Memo from Gray to Aynsley, 28 July 1941. Also, Charles Hasler and F. H. K. Henrion both discussed the techniques they had used to make MOI exhibitions.

31 Richard Hollis interview with F. H. K. Henrion in 1986 Imperial War Museum Catalogue 9592/ 1986-11-12: www.iwm.org.uk/collections/item/object/80009378.

32 TNA INF1/132, Displays and Exhibitions Division, 28 Nov 1940, Home Planning Committee, Programme for Exhibitions.

33 Report of the MOI's Home Planning Committee. 'Home Planning Committee: Programme for Exhibitions', TNA INF1/132, written November 1940.

34 'Home Planning Committee: Programme for Exhibitions', TNA INF1/132, pp. 2–3.

35 Categories and definitions: 'Non-structural exhibits', TNA INF1/132.

36 Lucy Noakes, *Dying for the Nation: Death, Grief and Bereavement in Second World War Britain* (Manchester: Manchester University Press, 2020).

37 Categories and definitions (using original emphases): 'Non-structural exhibits', TNA INF1/132.

38 TNA INF1/132, 'Home Planning Committee: Programme for Exhibitions', p. 3.

39 A few months later Noel Coward would write a popular song about it, which opened 'London Pride has been handed down to us; London Pride is a flower that's free'.

40 IWM, D1753.

41 'Courage Of London in War-Time', *The Times*, 18 December 1940, p. 6. The Blitz began in earnest in London in September 1940, lasting eight months, and in November 1940 the Blitzkrieg had started to attack other large, provincial cities in Britain, continuing until May 1941.
42 *Display*, March 1941, p. 273. The Topical Press Agency was founded in 1903, employing almost 1,500 representatives worldwide by its peak in 1929, selling the work of a team of photographers based in London. As well as contemporary press photography, the company also focused on sales of stock images.
43 TNA INF1/132, National Archives memo to Aynsley from Milner Gray of 15 October 1941 saying that with Treasury approval of the ATS Large Touring exhibitions, the USSR Large Touring Exhibition and the use of the Charing Cross site for the period of a year, the volume of sanctioned work for which the Branch was responsible had so increased that further additions to staff have become necessary.
44 *Display*, March 1941, p. 273.
45 *The Times*, 18 December 1940, p. 6.
46 Typography historian Robin Kinross describes the use of photographic elements to communicate urgency as 'unEnglish', given that this was characteristic of modern European graphic design beyond Britain before the 1940s, discussing Henrion's poster 'Aid the Wounded' of 1942; 'Design in Central-European London', *Typography Papers*, 8, 2009, p. 107.
47 Walter Benjamin, in his essay 'On the Concept of History' wrote, 'For every image of the past that is not recognised by the present as one of its own concerns threatens to disappear irretrievably', with Benjamin's belief in the importance of the quality of contemporaneity (*Aktualität*) central to his writings.
48 Stuart Hall has written about *Picture Post* as showing the tendency to 'the democratization of the subject in photography'; Hall, 'The Social Eye', p. 82.
49 *Display*, March 1941, p. 273.
50 *The Times*, Issue 48802, 18 December 1940, p. 6.
51 'Synecdochic', in the sense that a part of the population was suggested to stand in for the whole, to use a phrase drawn from Buchloh, 'From Faktura to Factography', p. 111.
52 *Display*, February 1941, pp. 266–7.
53 Hall describes the 'quality of usualness' in 'The Social Eye', p. 82.
54 John Berger in *Understanding a Photograph* (London: Penguin, 2013), pp. 20–1 asserts that reading a photograph is only possible through knowing what is beyond the frame. Walter Benjamin criticised the new style of photographic documentary flourishing in Germany in the 1920s: 'It has even succeeded in making misery itself an object of pleasure, by treating it stylishly and with technical perfection'; Hall, 'The Social Eye', p. 83 quoting Benjamin, 'The Author as Producer', *New Left Review*, vol. 62, July–August 1970.
55 *The Times*, Issue 48802, 18 December 1940, p. 6.
56 *Display*, March 1941, pp. 272–3.
57 *Shelf Appeal*, July/August 1941.
58 *Display*, March 1941, pp. 272–3.
59 See Erden Ertem's PhD on the picturesque campaign in *Architectural Review*, 'Shaping "The Second Half Century": The *Architectural Review*, 1947–1971' (MIT, 2004).
60 *Display*, February 1940, pp. 454–8 and 465.
61 *London Can Take It*, directed by Humphrey Jennings for the GPO Film Unit in 1940, was distributed throughout the US by Warner Bros.
62 Allan Sekula, in 'On the Invention of Photographic Meaning', in Victor Burgin (ed.), *Thinking Photography Communications and Culture* (London: Palgrave, 1982) discusses how people need to learn photographic literacy – that photographs are not intrinsically legible or significant.

63 MOI's Photograph Library, which collected and organised photographs, was based at Senate House. For further information about MOI's Photograph Library see https://moidigital.ac.uk/blog/photographs-motorbikes-and-ministry-information-interview-anne-olivier-bell/.
64 An important precursor to these magazines was Parisian illustrated magazine *VU* (founded in 1928), which, as Tim Satterthwaite shows in *Modernist Magazines and the Social Ideal* (London: Bloomsbury, 2020), assembled photographs in essay form. *Picture Post* was launched by Hungarian refugee Stefan Lorant in 1938. Other, less successful picture magazines included *Weekly Illustrated* and the socialist *Clarion*.
65 As discussed in Hall, 'The Social Eye', p. 70.
66 Raymond Williams uses this term in *Marxism and Literature* (Oxford: Oxford University Press, 1977), p. 128.
67 Kallmann, 'The Wartime Exhibition', p. 106.
68 *Display*, August 1941, p. 81; also mentioned in 'These Exhibitions of Ideas Are Planned', *Shelf Appeal*, July/August 1941.
69 Kallmann, 'The Wartime Exhibition', pp. 105–6.
70 Francis Bird, *Press, Parliament and People* (London: William Heinemann, 1946).
71 A parallel process was followed in the United States Office of War Information where photographs gathered by the FSA were reused to show 'ordinary' families.
72 IWM D Series of photographs, D1750–1766.
73 Mass Observation, File report 531, 'Exhibition at Charing Cross Underground Station, December 1940'.
74 *Display*, February 1941, pp. 266–7.
75 *Display*, March 1941, p. 273.
76 *Architectural Review*, October 1943.
77 F. A. Mercer and Grace Lovat Fraser (eds), *Modern Publicity in War* (London and New York: Studio Publications, 1941), p. 50.
78 Images from *Life Line*, UoBDA Henrion collection.
79 IWM D series photographs D17703 and D17705.
80 The front cover of *Display*, October 1941, called it a 'gas mask exhibition'. With another version made to tour to thirteen sites around the country, according to *Shelf Appeal*, July/August 1941. The travelling version was discussed in 'The Ministry of Information's Exhibition "Poison Gas"', *Art and Industry*, November 1941, pp. 142–3, as moving to each site for six weeks. MOI's *Poison Gas* exhibition was hosted at the Laing in 1941, reported on in local newspapers.
81 Back of leaflet about the exhibition held at UoBDA Henrion collection, 'Exhibitions'.
82 *Poison Gas* general contractors, according to *Art and Industry*, November 1941, were Beck and Pollitzer (Contracts, Ltd) and the displays constructed by Leon Goodman Displays Ltd, according to 'These Exhibitions of Ideas Are Planned', *Shelf Appeal*, July/August 1941, and also noted in *Shelf Appeal*, September 1941, p. 17.
83 *Shelf Appeal*, September 1941, pp. 16–17 stated that MOI had evidence (from staff watching) that the text matter was read by visitors both in Charing Cross and in the travelling version.
84 *Display*, October 1941.
85 Supplied by Gems Ltd, as featured in *Display* of January 1941, p. 249. *Shelf Appeal*, September 1941, p. 16.
86 *Shelf Appeal*, September 1941, p. 17.
87 Front cover of *Display*, October 1941.
88 Richard Hollis interview with F. H. K. Henrion in 1986 Imperial War Museum Catalogue 9592/ 1986-11-12: www.iwm.org.uk/collections/item/object/80009378.

89 *Shelf Appeal*, September 1941, pp. 16–17. *Poison Gas* compared favourably to the recent MOI exhibition *Life Line*, *Shelf Appeal* believed, as the former was focusing on making the subject interesting whereas in *Life Line* the designers only seemed interested in its form.
90 TNA INF1/132.
91 'Britain Shall Not Burn', *Shelf Appeal*, November 1941. Photograph of *Fire Guard* in UoBDA, Henrion collection, D4201 has a label on the back explaining the exhibition.
92 Kallmann, 'The Wartime Exhibition', pp. 95–106.
93 IWM D10625 and 10635.
94 Richard Farmer, *The Food Companions: Cinema and Consumption in Wartime Britain, 1939–45* (Manchester: Manchester University Press, 2011). Jo Fox discusses the mode of engagement of home front propaganda in 'Careless Talk: Tensions within British Domestic Propaganda during the Second World War', *Journal of British Studies*, vol. 51, no. 4, October 2012, pp. 936–66.
95 IWM D10623.
96 IWM D10629.
97 IWM D10273.
98 IWM D10235.
99 Caption to IWM D10826.
100 *MOI Home Intelligence Division Weekly Report*, no. 108, Part II, 11. SECRET 29 October 1942: https://moidigital.ac.uk/reports/home-intelligence-reports (accessed 24 May 2022).
101 For example, *Shelf Appeal*, July/August 1941, p. 2 where details of departmental spend on publicity omits exhibitions.
102 TNA INF1/132 'Exhibitions and Photograph Displays', Note from Mr Woodburn to Mr Welch of 24/9/41.
103 Mass Observation File Report 869, September 1941 was based on exhibitions observed in August 1941. See XIII Exhibitions and Demonstrations [n.p.].
104 *Shelf Appeal*, July/Aug 1941.
105 Howard Cox and Simon Mowatt, 'Horatio Bottomley and the Rise of *John Bull* Magazine', *Media History*, vol. 25, no. 1, 2019, pp. 100–25.
106 *Shelf Appeal*, July/August 1941. *Life Line* appeared on the cover of this issue of *Shelf Appeal* with an illustration by exhibition designer Peter A. Ray to illustrate a general survey article about the MOI exhibitions scheme entitled 'These Exhibitions of Ideas Are Planned to the Last Drawing'. Poster for *Life Line* is in MODA collection Hasler collection CH/5/3/6/27.
107 MOI digital, Home Intelligence Reports, www.moidigital.ac.uk/reports/home-intelligence-reports/home-intelligence-special-report-inf-1-293/idm140465678899040/.
108 MOI digital, Home Intelligence Reports, www.moidigital.ac.uk/reports/home-intelligence-reports/home-intelligence-special-report-inf-1-293/idm140465679558080/.
109 Museum of Domestic Design & Architecture (MODA) Charles Hasler collection CH/1/2. Hasler had joined MOI in 1942.
110 Image in MODA Charles Hasler collection CH/1/2.
111 Kallmann, 'The Wartime Exhibition', p. 104.
112 *Young America* of 1944, which gave Britons insight into the lives of their US allies, used Isotype to explain the percentage of US population across each job category. During the Second World War Otto Neurath got involved with the British MOI in contexts including isotypes for films such as Paul Rotha (dir.), *World of Plenty*, 1943: https://media.dlib.indiana.edu/media_objects/w95050635.
113 Abram Games, *Over My Shoulder* (London: Studio, 1960), pp. 8, 12.

114 UoBDA Henrion collection, Exhibitions Box. Ministry of Information No. IWM D7868 – 'MOI "Grow More Food" Exhibition at Charing Cross'.
115 Richard Hollis interview with F. H. K. Henrion in 1986 Imperial War Museum Catalogue 9592/ 1986–11–12: www.iwm.org.uk/collections/item/object/8000 9378, REEL 2.
116 IWM D8534.
117 IWM D10038 and IWM D 17481.
118 MODA, CH/1/2/1 collection.
119 MODA CH, letter to Neil Morgan, 22 June 1984, p. 2.
120 MODA CH, letter to J Groves, 4 February 1985, p. 4.
121 According to MODA archive, he joined the MOI Exhibitions Branch in 1942.
122 MODA CH, letter to Neil Morgan, 22 June 1984.
123 MODA CH, letter to Neil Morgan, 22 June 1984, p. 2.
124 As Hasler explained in 'Type-Face Letter Forms', *Architects' Journal*, 6 September 1951, MODA CH/1/2/1.
125 MODA CH, letter to Neil Morgan, 22 June 1984, p. 2.
126 MODA CH/1/2/1 collection, 'Make War on Moths' panel.
127 www.moidigital.ac.uk/reports/home-intelligence-reports/home-intelligence-special-report-inf-1-293/idm140465682632112/.
128 Ashley Havinden, *Advertising and the Artist* (London: Studio Publications, 1956).
129 *Display*, April 1945, p. 4. Entry was free but admission to some special features was by 2d ticket redeemable at any London greengrocer's for 2d worth of potatoes. There had been widespread fascination amongst avant-garde artists from the 1920s with the mass cultural form of cartoons, as Esther Leslie discusses in *Hollywood Flatlands: Animation, Critical Theory and the Avant-Garde* (London: Verso, 2002).
130 *Display*, vol. 27, no. 1, April 1945, pp. 3–4. See also BFI, Potato Pete Fair – 08275.
131 Farmer, *The Food Companions*, p. 47. David Langdon obituary, *Guardian*, 22 November 2011.
132 *Display*, March 1943, p. 182.
133 *Display*, vol. 27, no. 1, April 1945, pp. 2–3.
134 As Peter Moro explained to Louise Brodie in interview, BL Sound Archive [Part 14 of 15].
135 Kallmann, 'The Wartime Exhibition', p. 100. See also Atkinson, '"The First Modern Townscape"?'.
136 Michael Havinden et al. (eds), *Ashley Havinden: Advertising and the Artist* (Edinburgh: National Galleries of Scotland, 2003), p. 12.
137 'The Equipment of a Division', *Architectural Review*, August 1943, p. x.
138 Birmingham iteration was in March 1944, MODA Hasler collection CH/1/2 and TNA INF2/44/256 – Army Exhibition at Cardiff – general view of the exhibition; TNA INF2/44/57 – Army Exhibition at Cardiff; TNA INF 2/44/258 – Army Exhibition at Cardiff. IWM sought to acquire items from the exhibition for their collections, as a letter of 17 October 1943 from L. R. Bradley at IWM to C. Bloxham at MOI confirms (IWM Ministry of Information folder, CN2/1/GOV/65/1–18).
139 As *The Times* described it in '"Victory Over Japan" Exhibition', *The Times*, 21 August 1945, p. 6. Jenna Lundin Aral, 'Information as Spectacle: Second World War Exhibitions by the Ministry of Information', *Journal of Design History*, vol. 31, issue 1, February 2018, pp. 46–65.
140 Three small exhibitions at Charing Cross in 1944 had developed this theme: *Jungle Front*, *Ocean Front* and *Our Eastern Job*. TNA INF1/966, Memo 14 May 1943.
141 '14th Army's Work in Burma', *The Times*, 22 August 1945, p. 7.

142 Herbert Brush, quoted in Simon Garfield, *Our Hidden Lives: The Everyday Diaries of a Forgotten Britain 1945–1948* (London: Random House, 2004).
143 '"Victory Over Japan" Exhibition', *The Times*, 21 August 1945, p. 6; 22 August 1945; 24 December 1945.
144 Quoted in Garfield, *Our Hidden Lives*, p. 82.
145 IWM MOI folder, CN2/1/GOV/65/1–18, letter from Director IWM to H. Cecil Taylor, 24 August 1945.
146 *Tools for British Civilian Use*, an exhibit consisting of 24 printed cards produced by the British MOI in collaboration with the US OWI, illustrated the 'enormous production capacity of the United States during World War Two' 1942–5. US LOC LOT 4750G Tools for Civilian Use.
147 US National Archives RG208 'Description of Exhibit Unit Program', 3 February 1943.
148 USNA, Maryland: OWI 208-EX (records of OWI/ the historian relating to the Domestic Branch 1942–5), Box 3, Entry 6A.
149 UOBDA F. H. K. Henrion collection.
150 LOC LOT 9445 G, American Types.
151 LOC LOT 9763 G, How America Lives.
152 Including Office of Emergency Management Wm. Nelson (now OWI Gilmore); Bureau of Reclamation; Interior Department; FSA's Roy Stryker (now OWI); National Park Service; Interior Dept; War Department; Department of Agriculture. Images also came from the Feature Picture Section and Mr Flynn's News Picture Section of the Bureau of Overseas Publications, OWI.
153 USNA, Maryland: OWI 208, entry no. 7, box no. 29, Office of Facts and Figures 1941–2.
154 LOC Prints and Photographs collection, LOT 8898 F Exhibit panels, posters, etc., produced under the auspices of the US Farm Security Administration. Included in this file were photographs of Herbert Bayer's 1930 *The New Line*.
155 TNA INF1/132 Memo: 'Proposals for the organisation of MOI Exhibitions in the USA and for the operation of an exhibitions branch from New York', 1941. American Division had requested (Milner Gray memo 17 April 1941 NA INF 1/132) that the MOI had a staff member solely responsible for keeping contact with American exhibitions and sending small exhibits to America. These duties were carried out by the Senior Assistant responsible for distribution. See also David A. Lincove, 'The British Library of Information in New York: A Tool of British Foreign Policy, 1919–1942', *Libraries & the Cultural Record*, vol. 46, no. 2, 2011, pp. 156–84.
156 'War-Time Shapes. Ministry of Information's New York Exhibition', *Art and Industry*, April 1942, pp. 94–7.
157 LOC LC-USW 3–19418-E.
158 LOC LC-USW 3–31093-C and USNA RG208.
159 USNA RG208 785063, letter from Robert Carson to Robert R. Ferry, 1 June 1943.
160 LOC LOT 5490 F.
161 LOC LOT 8426.
162 LOC Prints and Photographs collection, LOT 7312, no. 1.
163 LOC Prints and Photographs collection, LC – USW 3–31090_C.
164 USNA RG208 Box 3 Entry 6A 735006.
165 USNA RG208 785063, 'Exhibits – Pacific Show'.
166 USNA RG208 785063, 'Unclassified to Bourne Washington from Stoffel Paris'.
167 Turner, '*The Family of Man* and the Politics of Attention', p. 62.
168 Kallmann, 'The Wartime Exhibition', p. 106.
169 Turner, *The Democratic Surround*, pp. 102–9.
170 Monroe Wheeler, 'A Note on the Exhibition', *The Bulletin of the Museum of Modern Art*, vol. 9, nos. 5–6, June 1942, pp. 18–20.

171 MOMA 1942 0040 1942 05 18 42518–34.
172 Turner, *The Democratic Surround*, p. 104.
173 *America Marches* was illustrated in *Art and Industry*, 1952, p. 112. There is a sequence of photos in IWM's D series D13797 etc. America Marches travelled as far as Australia, as shown in USNA RG208 Progress Report, May 1944.
174 Kallmann, 'The Wartime Exhibition', p. 97.
175 Front cover of *Display and Signs*, February 1945, from the Brussels Bon Marché store.
176 Richard Hollis interview with F. H. K. Henrion in 1986 Imperial War Museum Catalogue 9592/ 1986–11–12: www.iwm.org.uk/collections/item/object/80009378, REEL 2.
177 Franklin D. Roosevelt Annual Message to Congress, January 6, 1941; Records of the United States Senate; SEN 77A-H1; Record Group 46; National Archives.
178 Editorial, 'Is British Art Fighting?', *Art and Industry*, April 1942, p. 89.
179 Herbert Read, 'Is British Art Fighting?', *Art and Industry*, July 1942, pp. 2–4.
180 *For Liberty* catalogue, p. 2.
181 Catalogue, p. 5.
182 BL, *AIA Bulletin* no. 81, January 1944.
183 Martin, *Editor*.
184 Reverse side of MOI distributed 'For Liberty' photograph in Henrion collection, UOBDA.
185 Richard Hollis interview with F. H. K. Henrion in 1986 Imperial War Museum Catalogue 9592/ 1986–11–12: www.iwm.org.uk/collections/item/object/8000 9378.
186 Kallmann, 'The Wartime Exhibition', p. 103.
187 *Display*, April 1943.
188 Kallmann, 'The Wartime Exhibition', p. 100.
189 In the same year as *For Liberty*, Henrion was commissioned to paint classroom murals for the Thorncliffe works college, Chapeltown, Sheffield, where Churchill tanks were built: RIBA 72699–72704. Adrian Shaughnessy, *FHK Henrion: The Complete Designer* (London: Unit Editions, 2013).
190 Catalogue, p. 12.
191 BL, *AIA Bulletin*, no. 81, January 1944.
192 Cecil Day-Lewis poem, 12 March 1943, was reproduced in *For Liberty*'s exhibition catalogue, p. 12. Day-Lewis was also employed by MOI during the war.
193 Ariella Azoulay, *The Civil Contract of Photography* (New York: Zone Books, 2008).
194 Wolff, *The Aesthetics of Uncertainty*, pp. 108, 111.
195 *Display*, April 1943, p. 5.
196 Peter Jones exhibition held from 14 December 1943 to 14 January 1944 was reported in *AIA Bulletin*, no. 81 and visitor numbers reported in 'The First Ten Years', *AIA Bulletin*, no. 78, July 1943.
197 Jan Gordon, *The Studio*, November 1943, p. 163. For a further exploration of *For Liberty*, see Harriet Atkinson and Jane Dibblin's documentary film *Art on the Streets* (2023).
198 Katy Deepwell, *Women Artists between the Wars: 'A fair field and no favour'* (Manchester: Manchester University Press, 2010).
199 Kallmann, 'The Wartime Exhibition', p. 98.
200 Dunnett and Stamp, *Ernö Goldfinger*, p. 61.
201 'Rebuilding Britain', *Display*, April 1943. S. P. Mackenzie, 'The Coming of ABCA, 1941–1942', in *Politics and Military Morale: Current Affairs and Citizenship Education in the British Army 1914–1950* (Oxford: Oxford University Press, 2011).
202 Erin McKellar, 'Living, Working, Playing: Ernö Goldfinger's Planning Exhibitions, 1943–46', in Gaia Caramellino and Stephanie Dadour (eds), *The Housing Project:*

Discourse, Ideals, Models and Politics in 20th Century Exhibitions (Leuven: Leuven University Press, 2020).
203 *Builder*, 29 December 1944 (see Dunnett and Stamp, *Ernö Goldfinger*, p. 61 and an image on p. 72).
204 Elwall, *Ernö Goldfinger*, p. 65.
205 Sketches for Cinema are held at RIBA GolEr/ 401/4 'ABCA – Cinema'.
206 Kallmann, 'The Wartime Exhibition', p. 104.
207 Kallmann, 'The Wartime Exhibition', p. 106; Elwall, *Ernö Goldfinger*, p. 61.
208 Referenced by F. R. Yerbury in *Modern Homes*, 1947, p. 73.
209 TNA INF1/966, 17 April 1944.
210 Herbert Brush visited on Tuesday 22 May 1945; Garfield, *Our Hidden Lives*, p. 29.
211 Kate Berridge, *Madame Tussaud: A Life in Wax* (New York: W. Morrow, 2006).
212 Janina Struk, *Photographing the Holocaust: Interpretations of the Evidence* (London: I. B. Tauris, 2004), p. 56.
213 *Display*, August 1942, p. 76 and *Display*, August 1943, p. 50. Manufacturers included Gems Ltd and Stagg Display, Birmingham.
214 Baudrillard famously gave the example of Disneyland to demonstrate the function of the third order of simulacra and the production of a hyperreality that lets us believe that we can tell reality from representation, the real from the imaginary and the copy from its original; *Simulacra and Simulation* (London: Semiotext(e), 1981).
215 *Display*, August 1943, p. 50.
216 Quoted in Charmian Brinson and Richard Dove (eds), *Working for the War Effort: German-Speaking Exiles in the Performing Arts in Britain after 1933* (Elstree: Valentine Mitchell, 2021).

8

Exhibitions as welfare

After the Second World War, designer Beverley Pick noted with satisfaction that exhibitions had become 'a powerful new propaganda medium capable of reaching a very broad section of the population', with greater impact than many more established forms of advertising. If exhibitions had succeeded in fighting the war, Pick was excited by their potential for heralding peace.[1] Contemporaries echoed Pick's sentiments, seeing exhibitions' power as a newly potent form of propaganda.[2] Civil servants and journalists had weighed up exhibitions' potential usefulness within the domestic wartime propaganda campaign, as discussed in Chapter 7, concluding that they still had value for communicating official messages to the public once the war was over. Elected in 1945, Clement Attlee's Labour government used exhibitions for a range of causes, from justifying postwar military strategy to showing the workings of the embryonic National Health Service, all as part of an effective multi-form propaganda machinery created to channel the state's power and to reach home and foreign audiences alike.

This final chapter proposes that in the immediate postwar period, exhibitions funded through the public purse operated as welfare, supporting and instructing the British population, initially through the Ministry of Information (MOI) and subsequently the newly constituted Central Office of Information (COI). It analyses exhibitions' role in the period immediately following the war as a communications medium supporting government processes of domestic postwar recovery and reconstruction and the development of the British welfare state. Government exhibitions also explained investment in the national nuclear programme and excused processes and transitions of decolonisation. In the immediate postwar period, the British government also continued to use exhibitions to connect with foreign audiences: as vehicles of 'soft power', to project an identity as model social democracy, in preparation for a new type of global conflict in the form of the Cold War, and to recalibrate Britain's standing on the world stage, despite waning international influence. Through exhibitions, the government created a presentation of Britain's unity in diversity, compelling home audiences to engage with 'British' values in an increasingly uncertain world.

The majority of exhibitions discussed in this chapter were government-led and focused towards domestic audiences in Britain. They adapted the visual and spatial tropes developed during the war to create a tempered Modernism, suited to speaking to the British public. These include *Germany Under Control* (the final major exhibition held on London's John Lewis bombsite, showing British government control of postwar Germany), *The Health of the People* (marking the centenary of the first Public Health Act and celebrating the new National Health Service), *The Nation and the Child* (illustrating milestones in education and welfare reform), *The Miner Comes to Town* (recruiting for miners in central London) and a series of small government exhibitions describing economic recovery including *How Goes Britain*, *The Atom Train* (showing potential applications of atomic energy) and *Colonial Empire* (justifying Britain's continuing Empire). Many of those who designed these exhibitions would later come together as part of the mass team creating the eight nationwide exhibitions of the 1951 Festival of Britain, the major postwar government promotion of Britain as social democracy.[3]

This chapter ends by considering how far exhibitions continued to have currency for political activists such as the AIA in the postwar period. It discusses the AIA's last exhibition as political organisation: *The Mirror and the Square* held at New Burlington Galleries in 1952, their final major exhibition before the requirement for artist members to subscribe to a common political clause was removed, making the AIA a more conventional exhibiting organisation thereafter. This exhibition and the debates enacted around it were emblematic of exhibitions disintegrating as arguments with the fragmentation of political positions in Britain, as the international storm-clouds of the Cold War gathered. Given this chapter's focus is exhibitions' political and official entanglements in Britain, I choose not to concentrate on the work of other artists' groups whose work was focused through exhibitions immediately after the war, the most prominent of which, the Independent Group, has received significant attention elsewhere.[4]

Showcasing postwar reconstruction: *Germany Under Control*

The British government's motivation for the continued use of exhibitions in the early postwar period matched cultural theorist Stuart Hall's description of the BBC, which acted as an instrument of cultural education, ultimately intended to control and subordinate the population.[5] Hall draws on political theorist Antonio Gramsci's concept of a key function of state as raising the great mass of the population to a particular cultural and moral level, with popular culture at the centre of the state's sphere of activity, understanding the BBC's role as adapting citizens to 'the necessities of the continuous development of the economic apparatus of production'.[6] In much the same way, major exhibitions continued to be used by the British government, alongside broadcasts and other media, to persuade the population of the efficacy of their efforts, in this case towards recovery and reconstruction.

Germany Under Control, held on the John Lewis bombsite, was mounted to show how the government was exerting military power over Germany in the aftermath of war.[7] It sought to show, 'in miniature … graphically and effectively', the birth of Nazism and its aftermath and the current 'New Era' in Germany, stressing the reconstruction efforts and the 're-education' of the German people happening in the British military zone. During the period of planning for *Germany Under Control*, the MOI closed and was replaced by the government's new Central Office of Information (COI), which continued to prioritise the making of exhibitions as one form of 'material', as they described them. *COI Review* explained: 'our fundamental process is the making of material – of magazines and books, of films and exhibitions, of the whole range of visual, written, and spoken material'. At the COI, the creation of exhibitions as communications material sat once more alongside the production of magazines, books and films.[8]

Germany Under Control was only partially complete by the time it opened in June 1946, with a military band and music played on a cinema organ at the ceremony attended by military and religious dignitaries. Herbert Morrison, President of the Council and Minister in charge of the COI, spoke, sharing his enthusiasm for exhibitions, which had continuing significance to the COI as a means for 'enlightening the British public'.[9] The pamphlet *Germany – Our Way* was distributed free of charge, while a cinema showed short films reinforcing the exhibition's messages and the BBC scheduled television programmes to send these ideas about a changing Germany even further.[10] Those who shaped the exhibition's initial development were largely from military backgrounds, with limited expertise in visual presentation due to the mass resignation of the MOI design team on its closure earlier the same year, as COI exhibitions lead Cecil Cooke explained. The exhibition used many of the same elements as previous MOI exhibitions, with charts, maps, graphs, drawings and photographic collages central, but lacked something of the visual panache of earlier shows. One room spectacularised Germany's collapse in 1945, using murals and photographs set in spotlit niches to dramatise different aspects of the collapse affecting transport and production and offering a rogues' gallery of the Nazis who had brought the country down.

As an afterthought, designer James Holland (who had been central to MOI wartime exhibitions) created a 'popular' side, to bring more interesting objects including a tank and a Volkswagen car shipped from Hamburg.[11] 221,000 people paid to visit the exhibition on Oxford Street before it toured to fifteen cities around the United Kingdom.[12] Demonstrators were positioned around the site to answer visitors' questions, which, they reported, ranged from 'Why are we sending new potatoes to Germany from the Channel Islands instead of having them in England?' to 'Why are the Jews still kept at Belsen concentration camp?'[13] Internal government correspondence showed misgivings about the exhibition's 'bad presentation' and overuse of specialist language.[14] But the major public criticism was that although it was interesting to see how Germany was being managed, the exhibition did not explain what the country was gaining from the £88 million spent on it. Mass Observation diarist B. Charles visited *Germany Under Control* in April 1947 while it was touring, recording that he found it interesting but 'I can't, however, imagine WHY it is being held, unless it is to try to prove to the British people the enormous amount of work being done by the British in their Zone'.[15] The government's messages were evidently not reaching every visitor, throwing up questions as to the effectiveness of this form.

Exhibitions and the early welfare state: *The Health of the People*

The experience of war had transformed social attitudes in Britain. A major strand of the COI's postwar exhibitions work was focused on developing public understanding of the transformations towards the 'welfare state', a society based on social consensus. A series of government exhibitions picked up themes from the Beveridge Report. *Social Insurance and Allied Services*, as the report was titled, had been published in 1942, catching the public imagination with its focus on attacking the 'evils' of 'Want, Disease, Ignorance, Squalor and Idleness', describing steps to universalise good healthcare, education and housing and to create employment.[16] Attlee's government set about transforming Beveridge's vision into a reality. With Minister of Health Aneurin Bevan leading, the central plank of this vision was to create a universal health service, with exhibitions allied to the cause. *Printed Advertising* magazine explained that exhibitions had a key role in the mixed ecology of government communication for social education, declaring that 'Films, exhibitions, and press and poster advertising, used with skill and vision, can form an integral part of the citizen's social education'. They pointed to the government's use of health exhibitions to give guidance on guarding health, with Bevan using exhibition openings as occasions at which to showcase the government's programme of modernisation.[17]

Marking the centenary of the first Public Health Act, the COI mounted *The Health of the People* exhibition at their large Marble Arch site from May to June 1948, opened by Princess Elizabeth.[18] Held in the immediate aftermath of the National Health Service Act 1946, which brought medical care under public administration, the exhibition set out to explain the positive difference that new health services would make.[19] Designed by C. F. Garney, working with a scriptwriter and taking guidance from technical experts, it relied on the popular trope of using historical comparisons to highlight the path towards progress over the century from 1848 to 1948. The BBC used the exhibition as the occasion on which to broadcast a discussion of the NHS in 'What's Your Worry?', a slot on radio's *Woman's Hour*.[20]

COI Review explained how the research team of designer and scriptwriter had worked together to assess potential for 'visual presentation', helping visitors conceive of industrial 'squalor' and the 'inhumanity' of early nineteenth-century Britain.[21] On arrival at *The Health of the People*, visitors were confronted with the woodblock of a disorderly Victorian crowd with the slogan 'Industry drew the workless into the cities', a full-scale model of an early nineteenth-century cellar slum with people dossing down on a sooty floor and several scenes from '100 years ago', such as children as chimney sweeps in dioramas, paper sculptures representing Victorian health reformers and enlarged cartoons from Victorian magazines. All of this represented the 'before'. After passing through this section, all subsequent displays showed the 'after'.

To spell out the changes to healthcare, *Health of the People* used another popular trope for explaining wider change: locating a fictional family within it, whom visitors could follow through the exhibition's whole narrative.[22] In this case, the fictional 'Average family' of 'Home Town' – 'John Average (employed)', 'Mary Average (housewife)', 'Katie Average (schoolchild)' and 'Billy Average (infant)' – were shown living under the new health system.[23] Statistical panels, illuminated to reveal facts in a sequence, compared the population's health in 1848 and 1948. A range of visual means was used to try to make the information playful and engaging, including typography, text panels and display structures made from strung wires. A polemical narrative was set out on panels, one of which read 'Chadwick believed that sanitation would put doctors out of business. He was wrong …' A diorama dramatised 'the health team, consisting of the family doctor, dentist nurse, etc', another the protection of towns and the water supply. These were text-heavy displays with a serious intention.

An extensive painted mural showed women using maternity services, doctors operating in pristine theatres and scientists working in state-of-the-art laboratories, dramatising the new service as orderly and efficient. Another display explained infrastructure as key to good health, including water supply, sewerage and refuse collection. Sculptor Richard Huws designed a mechanical man, 'Godfrey', which animated the body's mechanisms for digestion, breathing and blood circulation, while a photographic display showed images of children being well looked after at nursery, school and home, to inspire trust in the structures of the state to envelope every aspect of family life, including in the home.[24] Following the story of the Average family, with a narrative thread running from the exhibition's start to finish, softened and personalised the exhibition's didactic conclusion: a 'word to John & Mary Average' explaining what visitors should do to 'master the simple rules of health', with seven rules spelt out and an interactive, press-button display showing the sixteen aids newly available as part of the NHS.[25] Near the end, a display celebrated '100 years of progress' from the Public Health Act of 1848 to the present, with expanded photographs of happy, healthy families alongside the acknowledgement that 'much remains to conquer still' (Figure 8.1). All of this was intended to promote the multiple benefits of the NHS to ordinary, working people.

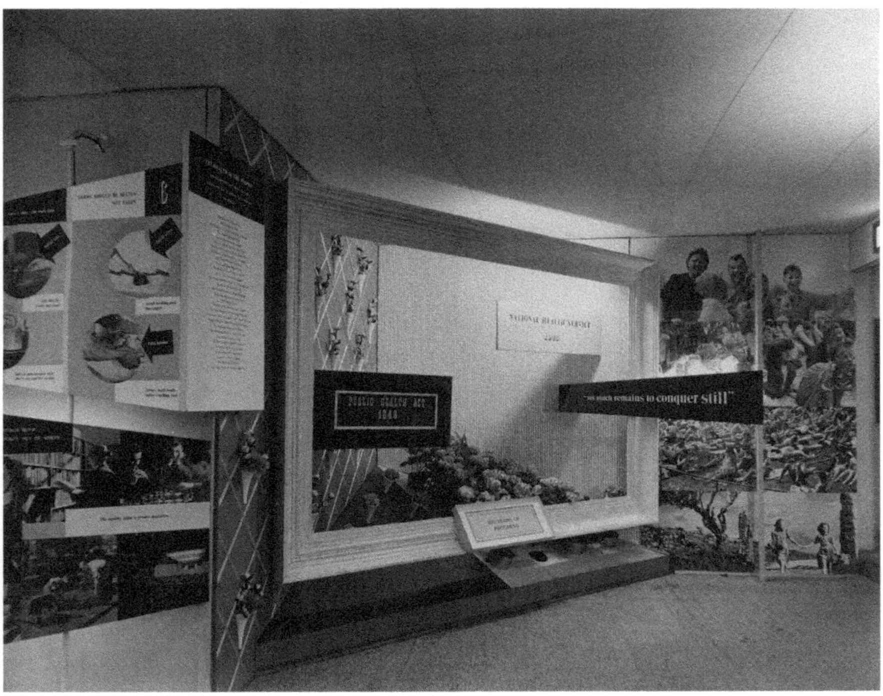

8.1 *Health of the People* exhibition, mounted by the Central Office of Information at Marble Arch: display about 100 years of progress from the Public Health Act of 1848 to the present. Image courtesy of the Wellcome Collection.

While the government was mounting exhibitions on this subject, in 1946 Ernö Goldfinger and his architectural partner Colin Penn developed an exhibition on the National Health Service for the ABCA, which they licensed out.[26] Twenty panels introduced the new Service and contrasted 'bad environment', which caused ill-health, with 'good environment' as 'the basis of health'. They showed 'medical services to-day', the current distribution of doctors and medical services and explained the NHS Bill 1946, including its financing and how services might work in future. Several panels also showed what health centres might look like, developing a long-standing interest of Goldfinger's in the structure of spaces for social use. The Education Department of the Admiralty bought 150 sets of Goldfinger's exhibition for distribution around the country. The Ministry of Health refused to take it, however, admitting in correspondence that 'frankly, this is not an exhibition which this Department would want to use, or with which to have its name associated with in any way' and explaining they objected to the exhibition's mixing of 'humorous drawings' with 'the serious treatment ... adopted for the subject as a whole', anticipating that nurses, health visitors and midwives would resent it. The Department offered factual corrections and said they thought the layout of the proposed health centre 'unsuitable'.[27]

Picturing milestones in social reform: *The Nation and the Child*

The COI's exhibitions promoting social change were shown across several public contexts, including commercial ones. *The Nation and the Child*, a display created by the COI with four government ministries, illustrated milestones in welfare and education reform. It was mounted at the *Daily Mail* Ideal Home Exhibition in 1948.[28] The exhibition mapped the period 'from the days when children were sent down coalmines and up chimneys to the present day when everything ... is done to educate and prepare them for work in which they will be happy'.[29] Typographer Charles Hasler oversaw the design, using Blado Italic, Legend and Poliphilus typefaces. He was soon after to become the chair of the Festival of Britain's Typography Panel. Dorothy Rogers created paper sculptures to animate historic scenes and architect Gordon Bowyer playfully simulated children's toys to create display vehicles and visual interest: a pile of children's bricks with playful fonts and paper drum. The figure of an emaciated child, sculpted by George Fullard, formed part of a display on nineteenth-century child slavery (Figure 8.2). This acted as a foil to the exhibition's narrative of progress a century on.

The exhibition's playful exterior was not enough to mask its strongly didactic intention, with design bureaucrat Paul Reilly criticising the display's 'long-winded' text, smothering its 'worthy message with too

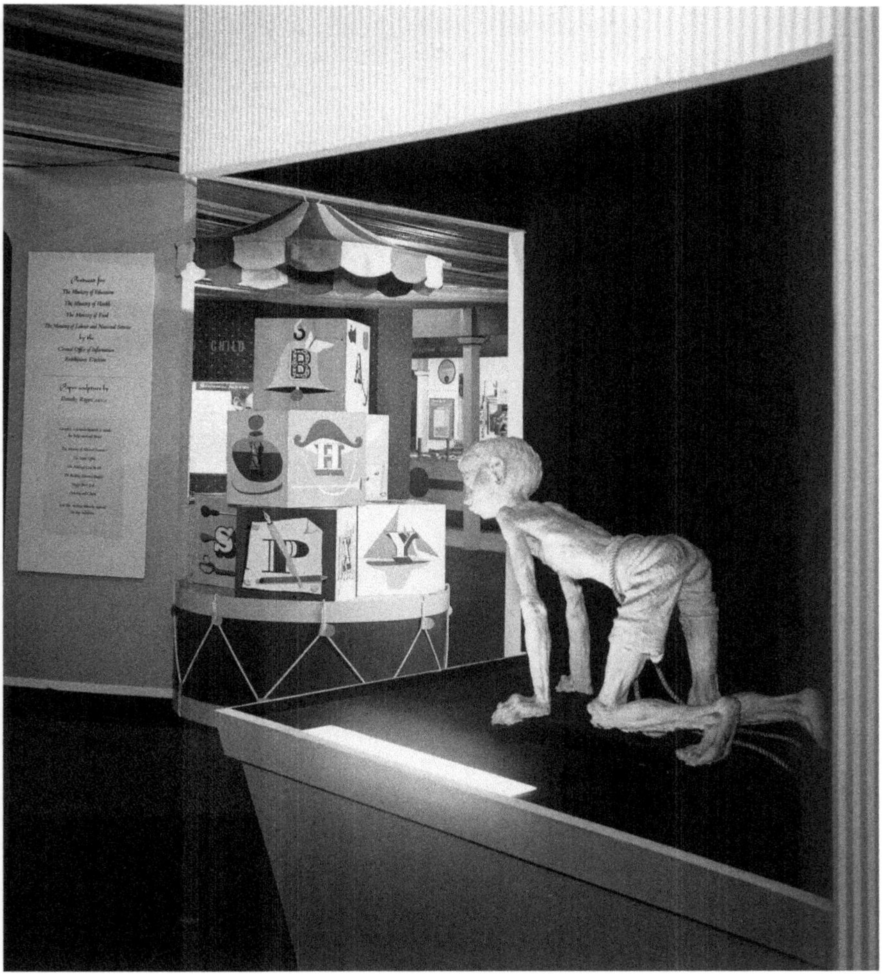

8.2 *The Nation and the Child*, with typographer design by Charles Hasler, using Blado Italic, Legend and Poliphilus typefaces, and an emaciated child sculpted by George Fullard. Crown Copyright Reserved. Image courtesy of the Museum of Domestic Design & Architecture, Middlesex University, www.moda.mdx.ac.uk. All rights reserved and permission to use the figure must be obtained from the copyright holder.

many words'.[30] Text was integral to the communication of messages of this form of exhibition, but the verbosity of overly text-reliant exhibitions was becoming a regular problem, limiting their usefulness. Given the designers' intention to have a wide public reach and engagement, to represent subjects in visual and verbal form, and in three dimensions, the exhibitions, in being overly wordy, and with the visual elements appearing more like window dressing, were failing to distinguish themselves sufficiently from more effective written forms such as magazines, books or pamphlets.

Bolstering declining industries: *The Miner Comes to Town*

The government had identified attracting people back into declining industries such as coalmining as potentially a key driver of economic recovery, starting a recruitment campaign to encourage this. The campaign's coincidence with the middle of a fuel crisis led the *Guardian* to comment that this was 'not the happiest moment for this kind of propaganda'.[31] As part of this campaign, the Coal Board and Ministry of Fuel organised *The Miner Comes to Town*, held at the COI's exhibitions centre at London's Marble Arch in autumn 1947. As well as finding new recruits for mining jobs, the exhibition introduced mines and showed the nation's well-being as dependent on coal. Signalling its significance to the government's agenda, the exhibition was opened by no fewer than five senior ministers: Prime Minister Clement Attlee, Minister of Fuel and Power Emanuel Shinwell, Chancellor of the Exchequer Stafford Cripps, Minister of Health Aneurin Bevan and Minister of Labour George Isaacs, with Lord Hyndley, chairman of the Coal Board, and Will Lawther, president of the National Union of Mineworks, in attendance.

The Miner Comes to Town's large COI design team included many who would go on to become significant architects and designers: architect Peter Moro and Gordon Cullen and designers Robin Day, Ronald Avery, Ian Chapman, Ronald Dickens, Pauline Behr and W. F. Manthorpe. Misha Black was Supervising Designer, with James Holland as Chief Designer, and R. J. Harrison as Chief Architect.[32] Robin Day had started his professional career as a furniture-maker's apprentice, but his practice diversified following studies at London's Royal College of Art, which included mural painting, publicity design, display and interior design, skills that he drew on heavily in later practice.

Exhibitions were a crucial early training for Robin Day, as Milner Gray explained in a feature about Day for *Art and Industry* magazine: 'It is in the field of exhibition design that perhaps the widest variety of design skills are brought together – if the effort is to be successful – into one complete and inseparable whole', Gray wrote. 'Planning, structural design, lighting, two-dimensional display and typography, generally the work severally of experts working as a team, are handled personally by Day in his exhibition work, resulting in a particularly well co-ordinated and personal effect'.[33] 'It is as an exhibition designer that Robin Day first made his mark', Gray went on, emphasising the role of exhibition design in Day's development as professional designer.[34]

Day and architect Peter Moro, who had met whilst working at the Regent Street Polytechnic, would go on to design a series of exhibitions together. Both considered exhibitions to have provided important, interdisciplinary training for their later design careers, something Moro alluded to on various occasions.[35] Moro reflected on his experience designing exhibitions, which forced him to experiment with how interest and atmosphere in architecture were generated by means of space, materials and lighting.[36] 'It was really wonderful to be able to design at a time when no buildings went up'; working on 'propaganda', 'information' and 'topical' exhibitions 'gave you an opportunity to use your design skills and imagination'. Exhibitions were, for Moro, 'a very good laboratory for trying out architectural devices'.[37] They were 'great fun because it's immediate, it's not permanent, and you can experiment ... with form and colour, but you can also experiment with leading people without arrows or ropes or barriers, or whatever ... a sort of laboratory for architecture' and 'psychological aspects' of architecture.[38]

The visual highlight of *The Miner* was its striking entrance – a 'mine with no coal' – a pit-head bandstand erected in steel scaffolding in Hyde Park, managed by a miner from a colliery near Manchester, with forty-five miner guides, accompanied by six pit ponies from County Durham that lodged overnight at the Buckingham Palace stables (Figure 8.3).[39] Inside, displays were inventive and visually engaging, walls lined with small lozenge-shaped portraits of miners and aerial photographs of coalfields with the exhortation: 'Our present problem is how to get more coal ... Our immediate future depends on more coal ... Our whole future depends on coal'.[40] A relief map showed mining areas, while Day and Moro's 'Coal By-products Tree', with stylised foliage forming a canopy and a multi-coloured trunk, showed by-products of coal in chemical flasks.[41]

Trade journal *Printed Advertising* commented on the instructional quality of MOI and COI work, 'exhortatory campaigns are the most difficult of all to assess for impact. They have certainly been the least popular with the public, even in the height of the war when exhortation found a more tolerant audience. As living conditions improve they are likely to become more unpopular still'.[42] This was another acknowledgement that the effectiveness of these labour-intensive and expensive events was difficult to quantify, with didactic or 'exhortatory' displays looking likely to have limited impact after the return to affluence post war. Meanwhile the *Guardian* newspaper, unconvinced by the impact of *The Miner*, commented, 'Persons not schooled in the arts of propaganda may wonder whether the West End is really the best place at this moment for a brand-new 10-ton Meco-Moore cutter-loader, a Joy power-loader, and all these modern mining machines and craftsmen'. The clear incongruity of this spectacle did not deter the government from continuing to use this form of impetus to postwar reconstruction.

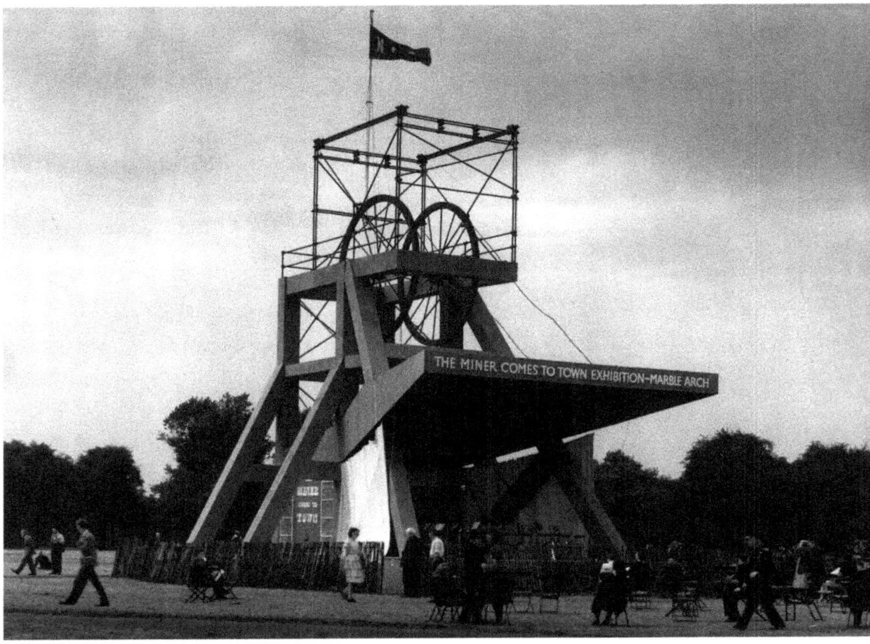

8.3 *The Miner Comes to Town*, organised by the Central Office of Information, designed by Peter Moro and Robin Day. Alfred Cracknell. Image courtesy of Architectural Press Archive / RIBA Collections.

Driving economic recovery: *How Goes Britain, Raising the Standard, Britain Goes Ahead* and *Britain in the Balance*

A series of government exhibitions were focused on materialising the abstractions of British economic recovery. *How Goes Britain*, organised by the COI at Charing Cross Station in January 1948, was intended to showcase effective recovery from privations, including shortages of food and materials. A painted mural on the outer wall showed ships set in a decorative seascape, with the striking title in bold letters (Figure 8.4). Visually playful devices mapped the situation, incorporating explanations through Isotype (Figures 8.5 and 8.6). *Raising the Standard*, 'an economic exhibition', opened at Charing Cross later that same year, in September 1948. Organised by the COI for the Economic Information Unit, it showed Britain's buoyant trade through photographic collages and charts. The exhibition repeated the device of tracking one family to show the progress of industry in postwar recovery, charting how this was impacting on the standard of living of different families and showing the role hard-working individuals could play in aiding recovery.

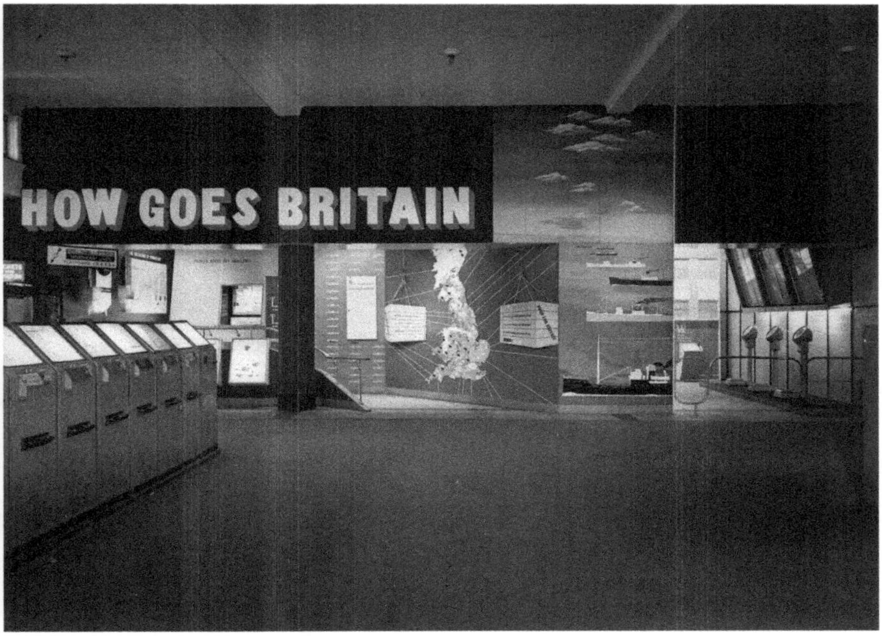

8.4 *How Goes Britain*, held at Charing Cross Underground Station in January 1948, showcased Britain's effective recovery from privations. Topical Press. U42045 © TfL from the London Transport Museum collection. All rights reserved and permission to use the figure must be obtained from the copyright holder.

A collage at the exhibition's entrance was labelled 'this was one family's share of the nation's imports of food and raw materials in the first four months of 1948'. At each stage the impact on individuals was represented; three photographs of the same family living in different economic circumstances, shown side-by-side, presented the direct impact on each of them with and without Marshall Aid loan and increased exports.[43] Also held at Charing Cross, *Britain Goes Ahead* opened in January 1948 to show evidence of postwar recovery. Details included how industry was innovating to deal with material shortages by producing synthetic materials and laminates, with mounted samples, using striking bold type and signage.[44] *Britain in the Balance*, later that year at Charing Cross, highlighted financial recovery, attracting 129,361 visitors over thirteen weeks.[45] Exhibitions such as these were offered as a focus for visitors to feel reassured by government expertise in overseeing a recovering economy and to witness housing planning and reconstruction, with the huge boom in planning exhibitions after the war aimed at a wide audience including children.[46]

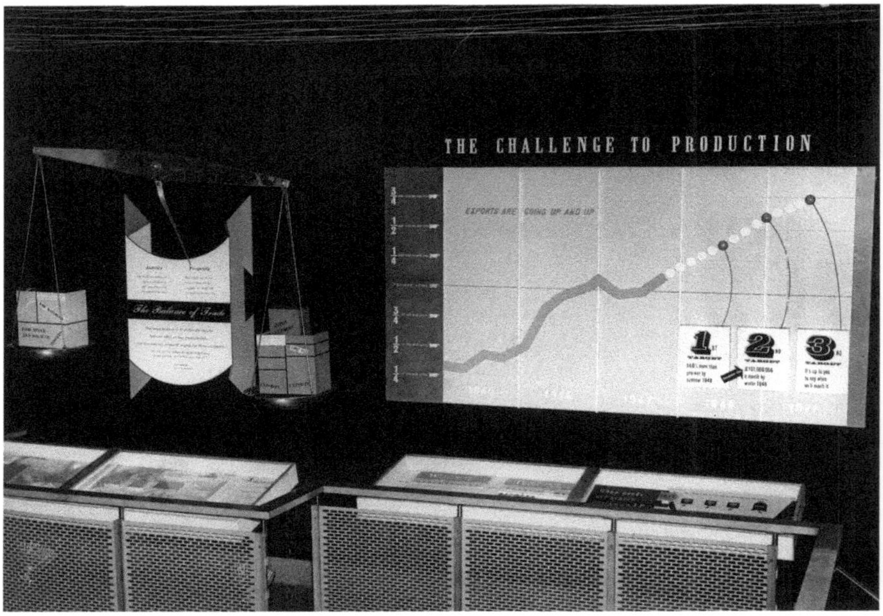

8.5 *How Goes Britain*, Charing Cross Underground Station in January 1948. Topical Press. U42044 © TfL from the London Transport Museum collection.

Exhibitions for the early Cold War: the *Atom Train*

With the US's detonation of atomic bombs over the Japanese cities of Nagasaki and Hiroshima in 1945, a new threat had been unleashed on the world: that of a war to end all wars. While the vast majority of Britain's postwar atomic defence project was developing in secret, Britain's scientific establishment believed increasingly that the public needed to be educated about the potential of nuclear technology, for both good and ill.[47] Exhibitions were one element of a widespread nuclear education programme developed by government and others. London's Science Museum hosted the *Atomic Energy and Uranium* exhibition in 1946 to explain atomic energy science, while *The Daily Express Atomic Age Exhibition* held at Dorland Hall in 1947 was focused towards warning about the cultural and social implications of the technologies, signalling the potential for mass destruction of the bomb. It opened with 'The Science Behind the Atomic Bomb' and continued to 'Atomic Bombs in Action', 'If Britain Were Atom-Bombed' and 'Atomic Energy in Peace'.[48]

Taking up this duty of public education, the government Ministry of Supply, working with the new Atomic Scientists Association (ASA), set

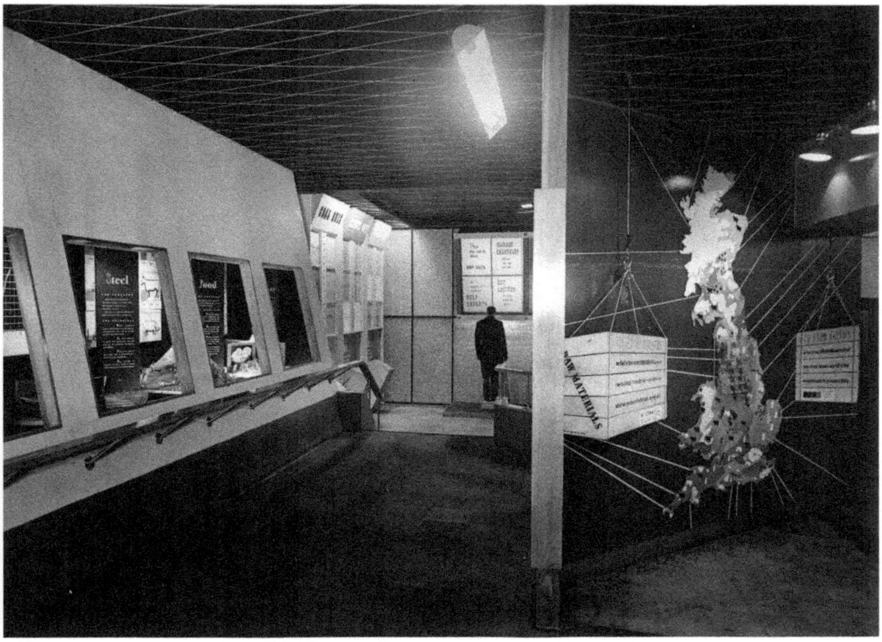

8.6 *How Goes Britain*, Charing Cross Underground Station in January 1948. Topical Press. U42048 © TfL from the London Transport Museum collection.

up an Exhibition Committee chaired by Prime Minister Clement Attlee. Together, they commissioned an ambitious, touring exhibition to share key facts and practical applications of atomic energy across medicine, industry and agriculture with the public.[49] Exhibitions that were both reproducible and mobile had been shown to work well in the context of Britain's war.[50] This was the chosen format for *Atom Train: The Travelling Exhibition on Atomic Energy*, which toured Britain for six months aboard two converted train carriages (Figure 8.7).[51]

Physicist Dr Joseph Rotblat developed *Atom Train*'s concept. Born in 1908 in Warsaw, part of the Russian Empire, Rotblat had arrived in Britain on the eve of the Second World War, taking up a research position at Liverpool University and becoming a stalwart of the ASA. Rotblat's work on splitting the atom had led him to the conclusion that it was possible to produce an atomic bomb and in 1943 he was given permission to withdraw from the Manhattan Project, fearing its consequences.[52] When invited by Rotblat to open the *Atom Train* exhibition, in the interests of maintaining secrecy, Attlee pleaded other pressing engagements, his advisers saying it was better 'in the present international situation' for the Prime Minister

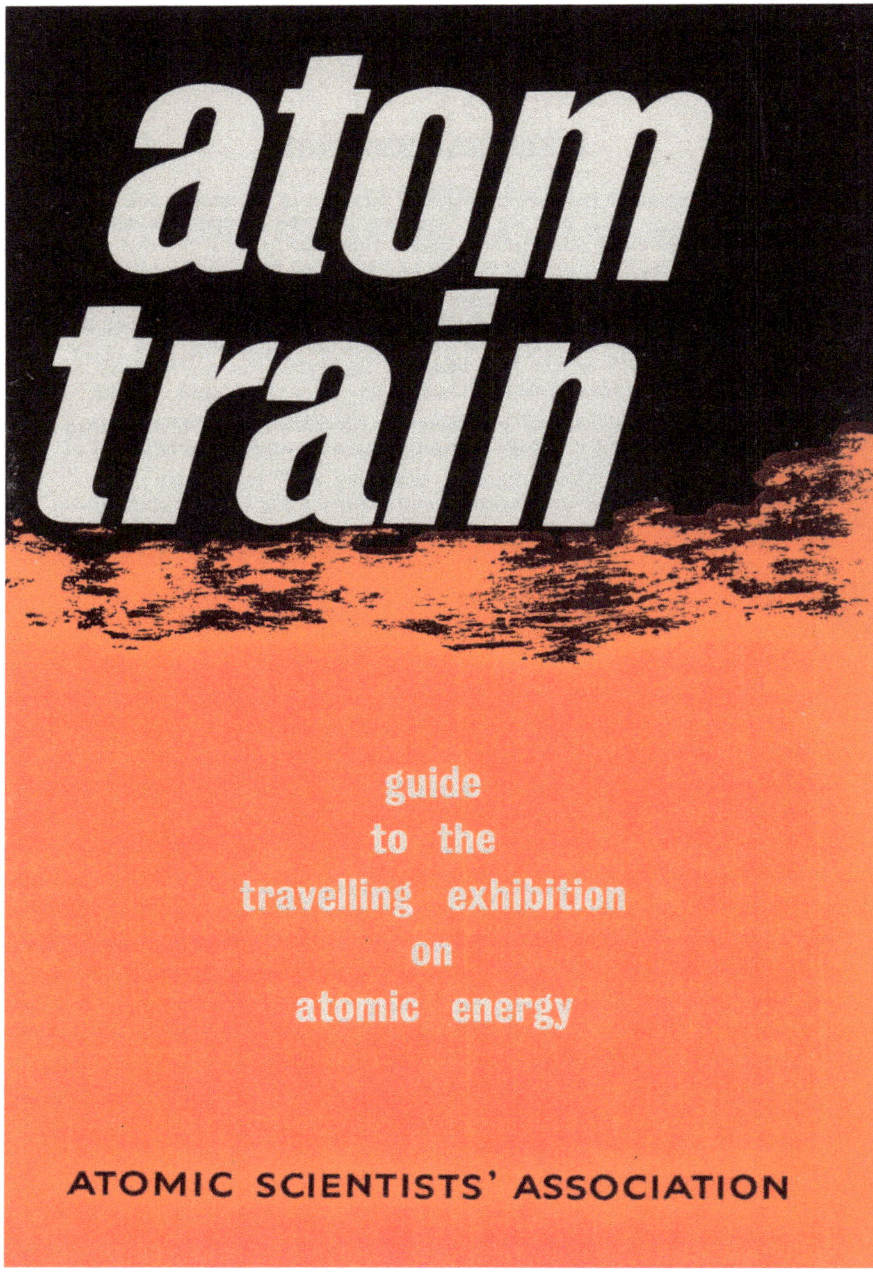

8.7 Catalogue of *Atom Train*. The Papers of Professor Sir Joseph Rotblat, Churchill Archives Centre, Cambridge. RTBT 10/17/4/13. Image courtesy of Churchill Archives Centre.

not to visit it publicly and considering the exhibition 'not so very remarkable'.[53] Instead it was opened by physicist Sir James Chadwick, who gave a sobering speech about people needing to understand atomic energy as 'on the side of human survival', lest it lead to the disappearance of 'the civilised world'.[54]

Atom Train was designed by architect Peter Moro and interior and furniture designer Robin Day, showing 'delicate instruments', with a script by Adrian Thomas and typography by Charles Hasler who applied 'gold lettering' to the 'smart black coaches' of the train on which it travelled.[55] An advertisement for the exhibition, showing an eager-looking mother and child, hinted at the target audience and explained that *Atom Train* 'is to show you the vast possibilities of atomic energy for good and evil' (Figure 8.8).

The *Daily Mirror* explained the attraction of the exhibition: 'In two railway carriages brightly decorated in what might be called the modern teashop-cum-cocktail-bar style, you can, if you pay a shilling, see The Atom Train Exhibition'.[56] The ambiguous nature of nuclear technologies was overtly acknowledged in the exhibition: designer Peter Moro explained that the train showed both 'the good uses' of the atom and 'the destructive uses', 'in photographs and pictures and diagrams ... in a graphic way', including 'skeletons'.[57] Its catalogue explained, 'Everyone knows that this new power can be used for destruction; much less is known as yet of its possibilities for good'.[58] The catalogue concluded that atomic energy has 'vast possibilities' for 'good and evil': 'there is no secret about it'.[59]

A balance sheet approach, showing nuclear science's potential for 'good' or 'evil', created the exhibition's structuring tension. Climbing aboard the *Atom Train*, visitors were immediately confronted by an image entitled 'Atomic Energy for Good or Evil', with human and skeleton hands pointing towards each other, a ringed atom in between. The potential for 'good' was taken up in Gordon Bowyer's colourful mural 'The Brighter Side', indicating 'the benefits mankind might, if he chose, gain from atomic energy', showing a picnic rug sitting comfortably in a landscape (Figure 8.9).[60] To add to the potential 'evil', the exhibition cautioned by including blackened roof tiles from Hiroshima and photographs of the decimation of Hiroshima and Nagasaki; a piece of molten, radioactive sand from New Mexico; and a chart showing what would happen if an atomic bomb dropped on central Manchester or London, based on evidence from Japan.[61] In keeping with this, ceilings and walls in this section were black. The *Guardian* described the apparently harmless specimens of radioactive minerals shown in the exhibition 'winking in lovely iridescent hues in their case', suggesting they looked beguilingly 'a great deal less lethal' than 'the pebble which slew Goliath'.[62] To counter this benign appearance, visitors were provided with Geiger counters allowing them to test for themselves that fragments were still radioactive.

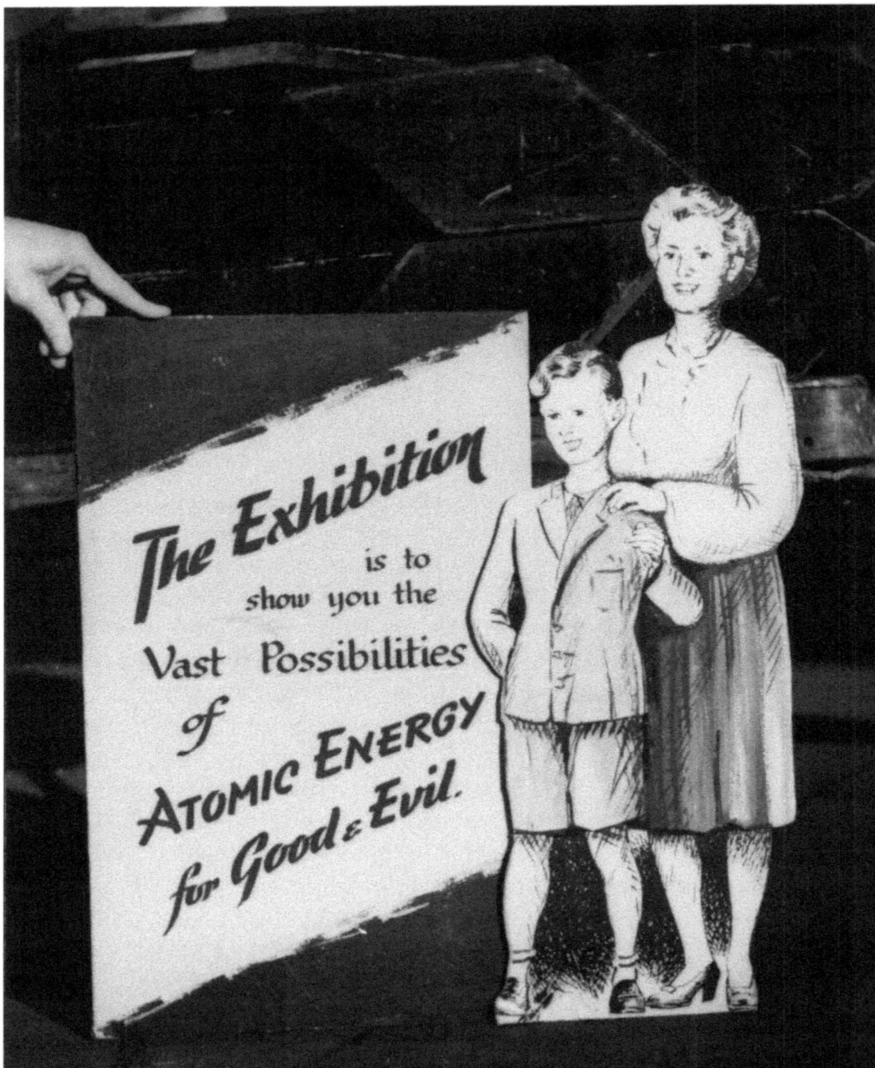

8.8 *Atom Train* advertisement. The Papers of Professor Sir Joseph Rotblat, Churchill Archives Centre, Cambridge. RTBT 10/17/4/10. Image courtesy of Churchill Archives Centre. All rights reserved and permission to use the figure must be obtained from the copyright holder.

Atom Train's argument was developed through scriptwriter Adrian Thomas's narrative, which visitors followed as they progressed round, anticipating the highly directional route set out at the South Bank site of the Festival of Britain a few months later, through which visitors

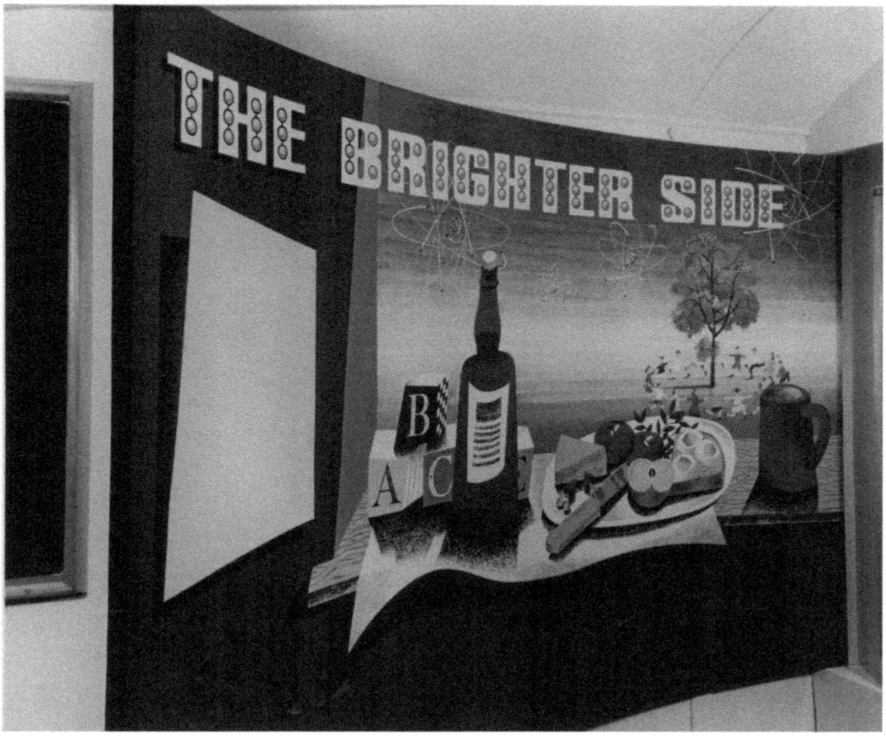

8.9 *Atom Train*: Gordon Bowyer's colourful mural 'The Brighter Side' indicated the benefits that atomic power might bring. The Papers of Professor Sir Joseph Rotblat, Churchill Archives Centre, Cambridge. RTBT 10/17/4/10. Image courtesy of Churchill Archives Centre. All rights reserved and permission to use the figure must be obtained from the copyright holder.

were invited to enact the 'story' of Britain.[63] Describing his approach to scripting exhibitions, Thomas explained the interplay between scriptwriter and designer in a specimen script to show how visual and textual elements worked together, as interdependent.[64] Explaining his approach as close to writing for other commercial forms, 'the exhibition script writer must', he said, 'have the feature writer's capacity for research, the scenarist's ability to visualise the dramatic possibilities of the story he is telling and the advertising copywriter's ruthlessness when it comes to the condensation and editing of his text'.[65] Words took space in exhibitions as explanations, exhortations, narratives and labels.

Visitors about to disembark from *Atom Train* were faced with a photographic mural entitled 'The Choice', which used a mechanised display (described as a 'ghost change') to show the choice between construction – a bright atomic future (a photograph of happy children playing) – and nuclear obliteration (a photograph of a crying baby amongst the ruins of Hiroshima) with the text 'the choice: destruction – construction'.[66]

Atom Train visited many sites around Britain including Whitehaven in Cumbria, a town situated close to Sellafield. This was a large-scale nuclear site, which had opened a few months earlier for the production of radioactive plutonium, to be used in Britain's nuclear bomb. By visiting Sellafield, the exhibition was intended to educate the local population about this new industry. The Sellafield site was in transition: in the early 1950s it would be redesigned to become the world's first commercial-sized nuclear power station. After its tour of Britain, *Atom Train* travelled further afield; from 1948 to 1949 it toured the Middle East, visiting Beirut and Cairo, with parts going to Scandinavia.[67] The *Daily Mirror*'s 'Cassandra' columnist quipped wryly that this was 'a cheerful collection of working models and pedagogic diagrams' showing how 'the world can almost certainly blow itself to bits unless we all go mad and use the atom for sensible purposes', which had been created in order to rid atomic scientists of their moral responsibilities.[68]

Atom Train's focus on Britain's nuclear programme was echoed in the themes and designs of the later Festival of Britain 1951. The Festival's major Exhibition of Science at South Kensington was conceptualised and organised by Polish British mathematician and philosopher Jacob Bronowski. Previously Scientific Deputy to the British Chiefs of Staff Mission to Japan in 1945, Bronowski had written the influential report *The Effects of the Atomic Bombs at Hiroshima and Nagasaki*. His 1951 book *The Common Sense of Science*, a philosophical discussion of the potential of science to benefit nations and, at the same time, to be used for malign purposes by its politicians, was produced in the same year as his work on the Festival of Britain opened to the public.[69] The connection between Britain's atomic programme and the national culture of official exhibitions was underlined by the transformation of the ship that carried the Festival of Britain's Sea Travelling Exhibition – HMS *Campania* – for use, after the Festival closed, to carry scientists and members of the navy to Montebello in Australia for the first British nuclear test, codenamed Operation Hurricane, in October 1952.

Justifying empire: the *Colonial Exhibition*

The British government continued to use exhibitions and displays as one of many devices through which to legitimise its mastery over its rapidly fragmenting empire. Cultural historian Raphael Samuel has conceptualised why they might have relied on such devices, describing how 'economically, and politically, Britain's dire circumstances at the end of the war put a premium on the success of … schemes of colonial development'.[70]

Exhibitions enabled the government to show people in Britain an image of life and work in British colonies, at the very moment when Britain was being forced to retreat from Burma, Ceylon and India.[71] Through the late 1940s, the COI's Photographs Division continued creating visual stories about the British colonies for display at home and abroad. These photographic 'Picture Sets', as they called them, attempted to describe in twenty photographs and six hundred words 'the whole shape and appearance and life of a Colony – its people, its land, its products, its climate' or else to tell 'in twelve modestly captioned pictures the story of those battles against superstition, poverty and disease which constitute the finest pages of British Colonial history'. Examples included displays about Britain's colonial education policy under the title 'Colonial Empire: Battle against Ignorance' or images of a civil servant and his family in their modern house at Accra within a display on 'Introducing West Africa'.[72]

In 1948, the Colonial Office started a campaign to arouse greater interest in British Territories by means of lectures and films, resulting in Britain's inaugural 'Colonial Month' in London of 1949, a series of exhibitions and displays designed to 'convey ... knowledge of colonial life and problems and progress' and to counter the 'gross ignorance about the great estate' for which Britain was responsible. One resulting event was the *Colonial Exhibition*, held at Marble Arch to 'expose the visitor to the temperature and oppressive greenery of a West African forest' and intended 'to jerk him into an appreciation of some of the difficulties of life in Africa'. The *Guardian* opined that it was important not to give a misleadingly 'melodramatic' or 'picturesque' impression of the 'millions of peaceful farmers and herdsmen who constitute the great majority of colonial peoples', concluding there was 'solid information' on the 'manifold civilising tasks which Britain is carrying out in her dependencies'.[73]

At the London opening of the *Colonial Exhibition*, King George VI spoke to a hall filled with representatives from thirty colonies, a speech broadcast to the world by Pathé News.[74] He thanked these 'fellow citizens' for their loyalty during the war and contribution to victory, saying he 'wished them to feel at home here', 'to profit by their stays amongst us' and praised 'societies and private persons' who had shown 'kindness' and hospitality 'to our colonial guests', learning more about these fellow members of the Commonwealth and giving visitors knowledge of British 'manners and customs'.[75] Other broadcasters promoted 'Colonial Month' events. The BBC's 'Art of the Colonies', a television programme showing a selection of exhibits from an *Exhibition of Colonial Art* held at the Royal Anthropological Institute, reinforced the exhibition's message about Britain's close relationship with its colonies. Many newspapers celebrated the links forged through the exhibition.[76] Not everyone was so celebratory: it was a visit to the *Colonial Exhibition* that provoked Milton Brown, a Nigerian living and working in London, to protest, by writing the article 'An African at the Colonial Exhibition' in the *Daily Worker*, pointing out the inconsistencies between the Africa presented in the exhibition and the realities of life on the continent.[77] Empire was not a subject tackled at the major nationwide exhibitions of the Festival of Britain held a few years later in 1951. As Jo Littler argues, this was not because Empire was not thought about but more because Empire was a subject of tense consideration and negotiation within the Festival.[78]

Exhibitions as political statements in the early Cold War: *The Mirror and the Square*

The last exhibition I discuss in this book was not mounted to present new government policy or to justify strategy. Instead, my focus moves to the final major exhibition of the AIA before its dissolution as a political organisation. This was *The Mirror and Square* held at London's New Burlington Galleries in 1952. The arguments around the exhibition allow insights into the increasingly divided political loyalties and artistic ferment of early Cold War Britain. The AIA's founding ideals of 1933, to fight fascism and imperialism and to support anti-war causes by activating viewers to a shared political position, were increasingly hard to translate into a cohesive agenda in the postwar world. With the stepping back of prominent AIA founding members, the previous political bonds between these erstwhile artist-activists were splintering. By the late 1940s, AIA exhibitions became more loosely used as the justification of an emerging social contract.[79] Whilst some members remained committed to campaigning for peace, others were more preoccupied with building professional networks and selling work. The AIA continued to hold exhibitions with a broad range of styles and approaches, including traditionally social realist subjects. Members publicly aligned themselves with political causes at home and abroad, visiting new socialist states such as Yugoslavia in 1947, to see at close quarters how the new country was developing, but their previously clear commonalities were becoming less tangible.[80]

In the early 1950s, the AIA focused on international co-operation, connecting with the emerging United Nations Educational, Scientific and Cultural Organisation (UNESCO).[81] As the peace movement gained momentum in Britain, following the outbreak of the Korean War, AIA artists engaged closely with it. A letter to the *Manchester Guardian* in May 1951 signed by Victor Pasmore, John Berger, Leslie Hurry and Patrick Carpenter announced that the AIA was to hold a 'peace exhibition'.[82] Artists for Peace, a group made up almost entirely from the membership of the AIA, was formed in 1951, with independent Marxist art critic John Berger prominent.[83] The group mounted three art exhibitions in London – in 1951, 1952 and 1953 – to assert and represent their anti-war position.[84]

The particular splintering that would become evident in 1952 through the AIA's *Mirror and the Square* exhibition was already apparent at the group's regular meetings several months earlier, in discussions over how to respond to the gathering storm-clouds of the early Cold War. At a meeting in early 1950, painter Patrick Carpenter gave an impassioned speech about the armaments race, the danger of a Third World War and the need to abolish the atom bomb, proposing a motion to all members present that the AIA should actively support the activities of the British Peace Committee. Painter Victor Pasmore seconded the motion.[85] Painter Beryl Sinclair, the AIA's Chair, strongly opposed it, however. Sinclair spoke out, criticising the Peace Committee's political dishonesty, her position supported by prominent members of the AIA's advisory committee. Carpenter's proposal went to a vote, resulting in an almost even split in the room for and against his proposition – to support the Peace Committee – a split exposing the ideological gulf at the heart of the AIA.

One thing members did agree upon, however, was that they should mount an exhibition in the cause of peace, an idea that eventually took form as *The Mirror and the Square*. Artist-member Stephen Bone, part of the conservative wing of the AIA keen to downplay the organisation's Leftist past, led on the exhibition's concept as an 'Abstract versus Representational Art' exhibition, as he described it, with all works 'graded … according to the degree of abstraction or realism'.[86] This idea, with its intention of teaching the public to appreciate modern art, was unanimously agreed.[87]

The Mirror and the Square was held at London's New Burlington Galleries in December 1952, with 290 artworks, attracting contributions from a wide range of artists, from very established to newly emerging artists, working across a range of visual forms and styles in painting and sculpture. The venue had long been a hub of London's avant-garde, hosting a series of landmark exhibitions including the work of International Surrealists in 1936, Picasso's *Guernica* during its tour of Britain in 1938 and the Modern Architectural Research Group or MARS's manifesto exhibition of new architecture in 1938 (as discussed in Chapters 3 and 5).

The works in *The Mirror and the Square* were organised across two large galleries and a lobby at New Burlington Galleries. The overwhelming majority were paintings, with thirty-five sculptures and some mobiles. The huge show drew works from established painters such as Augustus John, Matthew Smith and Stanley Spencer, as well as the up-and-coming generation: Patrick Heron, Roger Hilton and Terry Frost. In terms of style, there was a great plethora of forms. Many works of social realism focused on industrial Britain, a number with mining themes (such as Josef Herman's *Miners* and Derek Chittock's *The Price of Coal*). There were abstract reliefs (by Mary Martin), representational works (by Michael

Ayrton), semi-abstract works (by Victor Pasmore), Surrealist works (by Eileen Agar and Ithell Colquhoun) and paintings in the Neo-Romantic tradition (by Cecil Collins and Daphne Hardy-Henrion).

The catalogue to *The Mirror and the Square* opened with an explanation of the exhibition's rationale, as follows:

> OF THE TWO symbols from which this exhibition takes its name the Mirror, held up to nature, presents the spectator with a literal image of the visible world ... The square ... is ... an abstract geometrical concept ... Between these poles, symbolised by the mirror and the square, lies the labyrinth of contemporary styles in painting and sculpture through which the enquiring spectator must find his way.

The exhibition's aspiration to show the range of art, as if on a spectrum, from the mirror (as figurative work) to the square (as abstraction) was echoed in F. H. K. Henrion's catalogue and poster designs, showing an eighteenth-century etched portrait overlaid with red and white squares (Figure 8.10). The catalogue essay suggested that all of contemporary art existed between these two poles – from social realist art (or 'the mirror') to abstraction (or 'the square') – and that by learning to understand and appreciate the formal qualities of all of the gradations on this scale, observers would have some kind of mastery over contemporary art. What was obscured and obfuscated in this narrative of British contemporary art were the ideas, ideologies and motivations behind these works. This seemingly straightforward, educative presentation masked the emerging political differences of the group behind the exhibition: the politically motivated AIA.

The striking diversity of works in the London show was picked up on by a loud and varied press response. *The Times* newspaper railed against the exhibition's didactic qualities, stating the only positive was the chance to see work by established artists, while the right-wing *Daily Telegraph* reported on the 'good but unfulfilled idea' behind the exhibition given that, the newspaper said, 'British art tends more to personal expression than to concerted movements', meaning that the attempt to map out art is 'lost in the clash of individualities', displaying the newspaper's fixed preconception of the character of British art.[88] The right-wing *Spectator* was dismissive of the exhibition's educational claim. Meanwhile the centre-left *Manchester Guardian* complained that the exhibition's 'impeccably pacifist catholicity of approach ... was bound to defeat its own ends', simply demonstrating, in the end, that some artists 'paint in this way and some in that'. The Communist *Daily Worker*, under the headline 'Approach to Lunacy', described the exhibition as a sorry example of the ailments that now prevailed in British art.[89]

The most extensive analysis of *The Mirror and the Square* came from John Berger in the *New Statesman and Nation*.[90] Berger knew the AIA from the inside, having worked and exhibited with them over recent

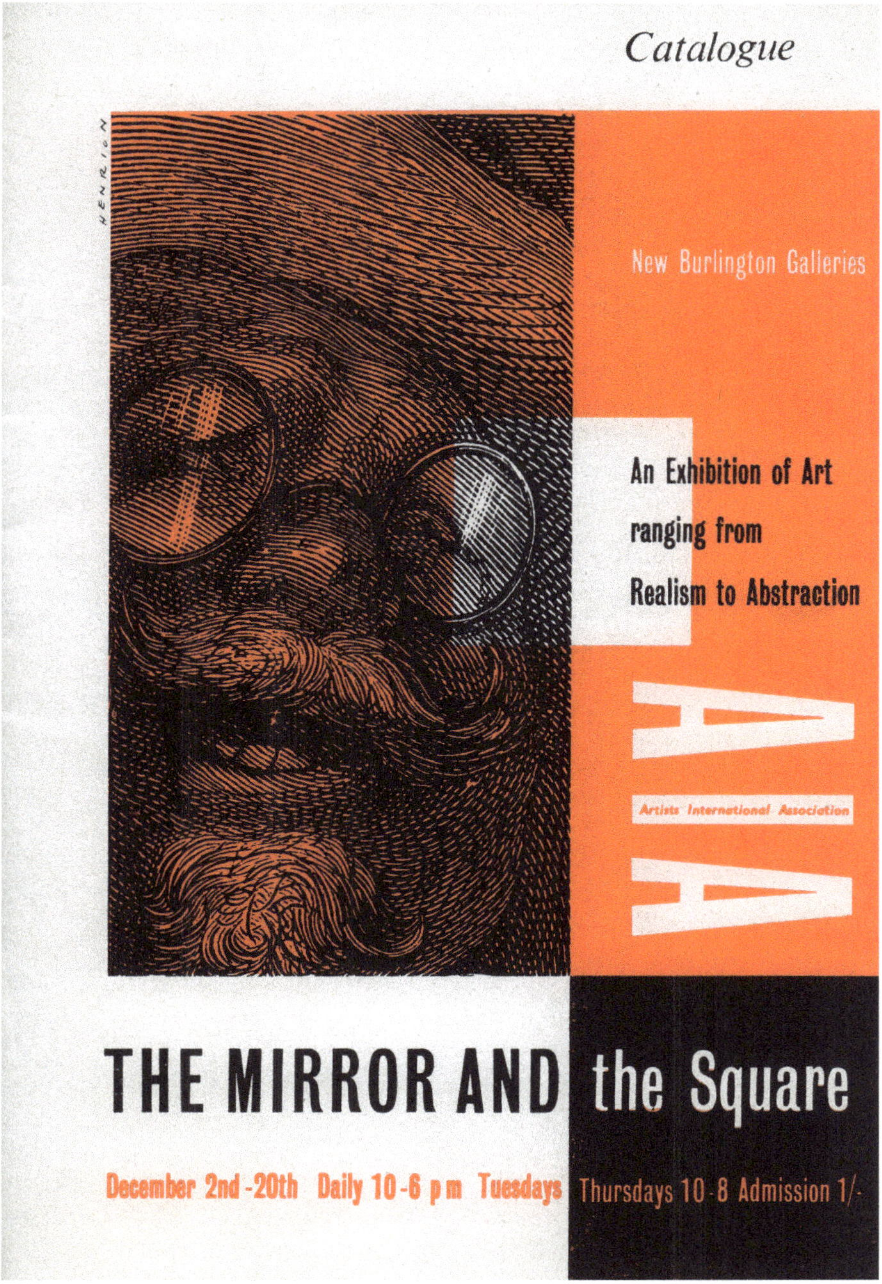

8.10 *The Mirror and the Square,* 1952, catalogue cover designed by F. H. K. Henrion. F. H. K. Henrion Archive, University of Brighton Design Archives.

years. In the *New Statesman*, Berger reported that, on the face of it, he was uninspired by *The Mirror and the Square*; however, the 'good intentions of the AIA', as he described them, made it worthwhile looking further. Rejecting the exhibition's idea that paintings could be categorised according to their superficial appearance, like the makes of cars, Berger stated: 'works of art can only be usefully sorted out by assessing their effect on the spectator. It depends on where the car takes you'. Rather than there being merely a series of styles of art, Berger suggested instead three main categories of art: 'works in the main European tradition which *deepen* the experience of the spectator'; 'eccentric "confessional" works which may *extend* his experience'; and 'decorations which, if applied in a functional context, can *embellish* experience'. Berger's focus on what the works in the exhibition could *do* to spectators to engage them through deepening, extending and embellishing experience and Berger's emphasis on the particular impetus behind the works were important. Rather than centring his comments on the work's formal qualities, as most critics had, allowing the artists to be characterised as merely eccentric outsiders on the fringes of society, Berger was making a bid for artists to be considered important mediators of inner life and of social and political ideas and ideals. This was something AIA members had tried to do through exhibitions over the past twenty years.[91] Berger's view was that by removing themselves from the cut and thrust of political debate, including in their lack of political engagement through exhibitions in which they focused on matters of form and style, the AIA were fast making themselves irrelevant, remote and removed from society.[92]

A smaller selection of eighty works from *The Mirror and the Square* was sent on a tour round Britain, travelling from the south coast at Worthing and Southsea to the Midlands at Leicester and Nottingham and to the north at Wakefield. The critical reception of the touring exhibition showed both widespread, highly conservative attitudes to contemporary art in postwar Britain and the openness, from some quarters, to being exposed to new ideas and experimental work. But what was conspicuously missing across all coverage was an awareness of or engagement with the thing that had characterised the AIA for the two decades until this point: its prominent politics, which had been vociferously at the forefront of all previous endeavours. This increasingly apolitical identity, whereby the AIA moved from being a group of political activists who saw 'the dissociation of art from everyday life' as being a major ill to being just another artists' exhibiting society, was completed that same year of 1953 when the AIA gave in to the prevailing culture that separated art from politics. The clause in the AIA's constitution that had required all members to be politically engaged was replaced by a clause ensuring the AIA members' 'intellectual freedom', something made easier by the disorganisation of the Left caused by Stalin's death earlier that year.[93] *The Mirror and the Square* was the last gasp of the AIA as a politically engaged exhibiting organisation, albeit one already in deep disarray. More importantly, it showed the precarity of exhibitions as a focus for cohesive political arguments and agendas.

Notes

1 Beverley Pick in *Display and Presentation: Exhibitions, Window and Outdoor Display* (London: C. Lockwood, 1957), p. 12. The way in which exhibition design had been transformed by war was also commented on by Paul Stiff in 'Austerity, Optimism: Modern Typography in Britain', *Typography Papers*, 8, 2009, p. 10.

2 Luckhurst, *The Story of Exhibitions*, pp. 9–11 and Bird, *Press, Parliament and People*, p. 126.

3 I have written about this at length elsewhere, most significantly in *The Festival of Britain*.

4 There is an excellent and ever-growing literature on the history and activities of the Independent Group: see David Robbins (ed.), *The Independent Group: Postwar Britain and the Aesthetics of Plenty* (Cambridge, MA: MIT Press, 1990); Anne Massey, *The Independent Group: Modernism and Mass Culture in Britain, 1945–9* (Manchester: Manchester University Press, 1995); Hal Foster and Benjamin H. D. Buchloh (eds), *The Independent Group: A Special Issue*, *October*, no. 94, Fall 2000; Isabelle Moffat, '"A horror of abstract thought": Postwar Britain and Hamilton's 1951 Growth and Form Exhibition', *October*, no. 94, 2000, pp. 89–112; Victoria Walsh, *Nigel Henderson: Parallel of Life and Art* (London: Thames & Hudson, 2001); Claire Zimmerman and Victoria Walsh, 'New Brutalist Image 1949–55: "atlas to a new world" or, "trying to look at things today"', *British Art Studies*, issue 4, 2016; Ben Highmore, *The Art of Brutalism: Rescuing Hope from Catastrophe in 1950s Britain* (New Haven, CT and London: Yale University Press, 2017); Kevin Lotery, *The Long Front of Culture: The Independent Group and Exhibition Design* (Cambridge, MA: MIT Press, 2020).

5 Stuart Hall, 'Popular Culture and the State', in Tony Bennett, Colin Mercer and Janet Woollacott (eds), *Popular Culture and Social Relations* (Milton Keynes: Open University Press, 1986), pp. 43–4.

6 Hall, 'Popular Culture and the State', p. 22.

7 Early in 1946 the Control Office for Germany had started working with MOI to plan this exhibition, TNA FO945/533.

8 The COI was established in April 1946. *COI Review*, May 1949, p. 3. *COI Review*, May 1948 carried a structure on p. 2, explaining that 'Exhibitions Division came under the Controller for "Home", designing and producing official displays and exhibition, other than trade fairs, for presentation at home and overseas, and is responsible for siting and touring of these exhibitions'. David Welch, *Protecting the People: The Central Office of Information and the Reshaping of Post-War Britain, 1946–2011* (London: British Library, 2019).

9 TNA FO945/533, May 1946. Letter of 11 June 1946 describes it being only partially complete.

10 TNA FO945/533, the MOI was replaced by the COI during the planning period.

11 TNA FO945/533, letter of 12 August 1946. Designer James Holland worked on the exhibition, with Milner Gray contributing (in a freelance capacity).

12 TNA FO945/533, letter of 22 August 1946.

13 TNA FO945/533 'Questions asked by members of the public while visiting'.

14 TNA FO945/533, letter from E. M. Tobin.

15 Quoted in Garfield, *Our Hidden Lives*, p. 377.

16 Derek Fraser, *The Evolution of the British Welfare State* (London: Macmillan, 1973), p. 199.
17 *Printed Advertising*, April 1948, p. 53. *Hospital Domestic Aids Exhibition* at the Empire Tea Bureau, Regent Street, from 14 July 1947, was opened by Bevan who declared 'good tools are essential to reduce unnecessary work in hospitals'; Warwick Digital Collections, TUC archives, 292/842/2/36, Hospitals, 1936–1959. Technicolor Halas & Batchelor cartoon, 'Charley your very good health', also released by COI in 1948, explained the NHS.
18 *The Health of the People* was illustrated in Black's *Exhibition Design* and discussed in *The Times* (7 May 1948). The exhibition was shown on a 10,000 square foot site. See also Welch, *Protecting the People*, pp. 29–30.
19 *COI Review*, no. 1, October 1948, p. 11.
20 Broadcast on 16 May, as Welch notes in *Protecting the People*, p. 30.
21 *COI Review*, no. 1, October 1948, p. 11.
22 *Britain Can Make It* had drawn heavily on this idea.
23 Wellcome Collection, special collection, script of *The Health of the Nation*, 811058i, section D.
24 Black, *Exhibition Design*, p. x; and Wellcome Collection, special collection, *The Health of the Nation*, 811058i.
25 The exhibition toured to Wolverhampton. *COI Review*, no. 1, October 1948, p. 12.
26 RIBA GolEr/405/3 'ABCA National Health Service'.
27 RIBA GolEr/405/3 'ABCA National Health Service', Letter from Ministry of Health to Ernö Goldfinger, 30 September 1946.
28 Ryan, *Daily Mail Ideal Home Exhibitions*.
29 *The Nation and the Child* was a collaboration between the Ministries of Health, Food, Education and Labour; *Advertiser's Weekly*, 25 March 1948, p. 608; Paul Reilly article in *Printed Advertising*, April 1948, p. 53.
30 *Printed Advertising*, April 1948, p. 54; Alex Mold, 'Exhibiting Good Health: Public Health Exhibitions in London, 1948–71', *Medical History*, vol. 62, no. 1, 2018, pp. 1–26.
31 'Our London Correspondence: The Miner Comes to Town', *Manchester Guardian*, 11 September 1947, p. 4. Images at RIBA 50105; RIBA63441; RIBA63442 and RIBA63443.
32 Pauline Behr later took the married name Pauline Baines: www.theguardian.com/artanddesign/2020/dec/08/pauline-baines-obituary (accessed 2 March 2022).
33 Milner Gray, 'Designer: Robin Day ARCA, FSIA', *Art and Industry*, May 1952, p. 156.
34 Gray, 'Designer: Robin Day ARCA, FSIA', p. 156.
35 Peter Moro interview with Andrew Saint, BL Sound Archives, transcript p. 39; Fair, *Peter Moro and Partners*. The *Jet Aircraft* exhibition at Charing Cross was a Moro and Day collaboration for the Ministry of Supply, 1946–7 (RIBA 106450/106452/016449/016453), the display illustrated in *Art and Industry*, May 1952, p. 157 and the poster on p. 158. Moro and Day's most noteworthy exhibition collaboration was probably the signposting scheme for the Festival of Britain at London's South Bank. 'Recent Exhibitions Designed by P. Moro and R. Day', *Architect and Building News*, 13 February 1948, pp. 138–40.
36 Fair, *Peter Moro and Partners*.
37 Peter Moro interview with Louise Brodie, BL Sound Archives, part 14 of 15.
38 Peter Moro interview with Andrew Saint, BL Sound Archives, transcript, pp. 39–40; TNA INF1/132, memo from Milner Gray to Lord Davidson, 17 April 1941.
39 'Our London Correspondence: The Miner Comes to Town', *Manchester Guardian*, 11 September 1947, p. 4.
40 'The Miner Comes to Town', *Printed Advertising*, April 1948, p. 31.
41 *Architect and Building News*, 13 February 1948, p. 140.

42 *Printed Advertising*, April 1948, p. 32.
43 LTM album Q8 24/9/48.
44 LTM album Q8 18/1/49.
45 *COI Review*, no. 1 October 1948, p. 12.
46 Peter Larkham (ed.), *The Rebuilding of British Cities: Exploring the Post-Second World War Reconstruction* (Birmingham: University of Central England School of Planning and History, 2004) points out that between 1951 and 1952 two hundred planning exhibitions were produced in Britain. See also Atkinson, '"The First Modern Townscape"?' *The Builder and the State*, jointly arranged by the Ministry of Works, Ministry of Health and the Department of Scientific and Industrial Research, was held at the Building Trades Exhibition at Olympia from November to December 1947 and Hasler worked on it. COI exhibition *Research in Housing Standards* was held in 1947 to explore application of science to the study of building materials.
47 Britain's secret atomic bomb project, codenamed Tube Alloys, had started early in the Second World War in collaboration with Canada. The Manhattan Project saw the United States and Britain co-operating on the production of nuclear weapons. By 1946, with the arrest for espionage of British spy Alan Nunn May, it was clear Britain would need to work independently on its nuclear project.
48 *The Daily Express* exhibition was organised by Chapman Pincher, the *Daily Express* Science Reporter; Sophie Forgan, 'Atoms in Wonderland', *History and Technology*, vol. 19, 2003, p. 178.
49 The Atomic Scientists Association was formed in 1946 by those who had taken part in the wartime Atomic Energy project in Britain, Canada and the United States. *Architects Journal*, 13 November 1947.
50 Mobile exhibitions discussed by Lugon in 'The Ubiquitous Exhibition'.
51 *Atom Train* toured from November 1947 to April 1948, through 25 towns from Chester to Carlisle and ending up at Paddington Station after being seen by 100,000 people, according to letter of 9 March 1948 in TNA PREM8/910. By the end it had been seen by 146,000 people in all, according to Christopher Laucht, 'Atoms for the People: The Atomic Scientists' Association, the British State and Nuclear Education in the *Atom Train* Exhibition, 1947–8', *British Journal for the History of Science*, vol. 45, no. 4, December 2012, pp. 591–608, p. 593. See also Catherine Jolivette (ed.), *British Art in the Nuclear Age* (Farnham: Ashgate, 2014).
52 Rotblat later won the Nobel Peace Prize in 1995.
53 TNA PREM8/910, letter of 12 March 1948.
54 'An Atomic Power Station "Perhaps in Ten Years"', *Manchester Guardian*, 7 November 1947.
55 'If an Atomic Bomb Hit Manchester', *Manchester Guardian*, 4 February 1947, p. 4.
56 Cassandra, 'A Dagger for Baby', *Daily Mirror*, 22 April 1948, p. 4.
57 Peter Moro interview with Louise Brodie [part 14 of 15].
58 *Atom Train*'s catalogue sold 46,000 copies, according to Laucht, 'Atoms for the People', p. 593.
59 *Atom Train* catalogue, Atomic Scientists' Association, 1947, pp. 3, 28.
60 'Atom Train: A Travelling Exhibition on Atomic Energy Designed by Peter Moro and Robin Day', *Architects Journal*, 13 November 1947.
61 *Manchester Guardian*, 20 November 1947, p. 3.
62 'A Radio-activity Exhibition', *Manchester Guardian*, 22 January 1947, p. 8.
63 Atkinson, *The Festival of Britain*.
64 Roland Barthes discusses text and image standing in a complementary relationship in 'Rhetorique de l'image', *Communications 4* (translated) and *Elements of Semiology* (London: Cape Editions, 1967). Translated as 'Rhetoric of the Image', in *Image-Music-Text* (London: Fontana, 1977), p. 41 and Umberto Eco, 'Articulations of the Cinematic Code', *Cinemantics*, vol. 1, January 1970. Peter Wollen, *Signs and Meaning in the Cinema* (London: Thames & Hudson, 1969).

65 Adrian Thomas, 'Exhibition Scripts', in Misha Black, *Exhibition Design* (London: Architectural Press, 1950), p. 118.
66 *Architects Journal*, 13 November 1947.
67 Laucht, 'Atoms for the People', p. 607.
68 Cassandra, 'A Dagger for Baby', *Daily Mirror*, 22 April 1948, p. 4.
69 Jacob Bronowski, *The Common Sense of Science* (London: Heinemann, 1951). Atkinson, *The Festival of Britain*, pp. 121–3.
70 Raphael Samuel, *Island Stories: Unravelling Britain: Theatres of Memory, Volume II* (London: Verso, 1998), p. 92.
71 Jo Littler, '"Festering Britain": The 1951 Festival of Britain, National Identity and the Representation of the Commonwealth', in A. Ramamurthy and Simon Faulkner (eds), *Visual Culture and Decolonisation in Britain* (Farnham: Ashgate, 2006).
72 'Telling the World in Pictures', *COI Review*, no. 2, May 1949, p. 13.
73 *Manchester Guardian*, 22 June 1949, p. 4.
74 British Pathe, 'King George VI opens Colonial Month', 1949.
75 *Manchester Guardian*, 22 June 1949, p. 4.
76 See listing in 'Exhibition of Colonial Art, 1949' (A35), Royal Anthropological Institute: www.therai.org.uk/archives-and-manuscripts/archive-contents/exhibition-of-colonial-art-1949-a35 (accessed 2 March 2022).
77 Milton Brown, 'An African at the Colonial Exhibition', *Daily Worker*, 29 June 1949: www.therai.org.uk/archives-and-manuscripts/archive-contents/exhibition-of-colonial-art-1949-a35 (accessed 2 March 2022).
78 Littler, '"Festering Britain"'.
79 Misha Black had resigned as AIA chair in 1944. James Boswell took over as chair from Black but he was disturbed by the growing split between members who were politically radical and those who were not so refused to stand for reelection and in 1946 he left the Communist Party; Egbert, *Social Radicalism*, p. 529.
80 Radford, *Art for a Purpose*, pp. 161–4.
81 Particularly through the chairmanship of the AIA of Richard Carline from 1951, who had been UNESCO's first Art Counsellor.
82 'Letters', *Manchester Guardian*, 17 May 1951.
83 Lynda Morris, 'The Sheffield Peace Congress', in Morris and Grunenberg (eds), *Picasso: Peace and Freedom*, p. 64, describes the role of Marjorie Abbatt in Artists for Peace.
84 See documentation in 2 Willow Rd archive and Tate collections material in Hepworth archive.
85 Report of AGM in *AIA Newsletter*, December 1950, pp. 1–2.
86 At AIA's October 1951 EGM.
87 A proposal by art critic and novelist John Berger that there should be more than two categories of art shown in the exhibition 'Realistic, Abstract and Subjective' was accepted. A proposal by Peter László Peri that each artist should send in both an abstract and a realist work was rejected; *AIA Newsletter*, issue of December 1951–January 1952, p. 1. With the organising of *The Mirror and the Square* well under way, the *AIA Newsletter* explained to members that it would pose 'for the layman the essence of the conflict which he will see visually expressed on the walls – the conflict between different, often opposed, mutually interactive aspects of experience and creation'; TGA *AIA Newsletter*, July/August 1952, p. 3.
88 *Daily Telegraph*, 5 December 1952, p. 9; *The Times*, 8 December 1952, p. 3.
89 *Manchester Guardian*, 3 December 1952, p. 3, *The Daily Worker*, December 1952 and *Spectator* responses were alluded to in *AIA Newletter*, February 1953, p. 1.
90 Berger in *The New Statesman and Nation*, 20 December 1952, p. 752.
91 Including in the AIA's manifesto, *The Full Employment of Artists* (1946) and James Boswell's *The Artist's Dilemma* (1947). AIA had formed a sub-committee to

investigate the economic position of the artist in the early 1950s led by Stephen Bone. See *AIA Bulletin*, April–May 1950, p. 4.

92 After *The Mirror and the Square* closed in London at the end of 1952, AIA member Owen Lewis reflected in the *AIA Newsletter* on the contradictory press response to the exhibition; *AIA Newsletter*, February 1953, p. 1.

93 Misha Black wrote in the *Daily Worker* of 2 April 1958, which was celebrating the founding of the AIA twenty-five years earlier, that he could 'no longer support the Communists who were our comrades in the prewar united front movement'; Egbert, *Social Radicalism*, p. 534.

Conclusion

In the 1952 article 'Exhibitions: In or Out?' designer Milner Gray reviewed the current state of exhibition design, noting, with sadness, fewer exhibitions being made and those that were being marked by 'a quite surprising dreariness' and 'emptiness'.[1] After nearly two decades of innovation in the forms and possibilities of this 'strangely rewarding' medium, Gray noted the 'threatened extinction' of story-telling exhibitions: 'the boil has burst and the bailiffs are in'.[2] Several things had precipitated this end, Gray thought: the expense of creating exhibitions and, more importantly, designers' urge to over-complicate with 'architectural fiddlesticks', which was 'getting in the way of the story'.[3] Gray's sentiments were shared by his close collaborator Misha Black, who observed exhibition designers getting so carried away with their love of design trickery for its own sake that display forms had become untethered from the specific problem to be solved.[4]

Despite this sense of foreboding from Black and Gray, the end was not up for propaganda exhibitions: Misha Black's survey *Exhibition Design* of 1950 showed such exhibitions thriving across trade, industry, agricultural and international contexts.[5] Black, who had built his early career through designing international exhibitions, from his work on the Rio Tinto stand at the 1929 Ibero-American Exposition in Seville onwards, contributing elements to the 1938 Glasgow *Empire Exhibition* and the 1939 New York World's Fair, continued to lead the design of British contributions to multiple international exhibitions after World War Two. These included the 1952 *Colombo Plan Exhibition*, where Black acted as consultant to the government of a newly independent Ceylon, designing the United Kingdom Pavilion and South East Asian Territories Pavilion. He designed the UK Pavilion at the 1953 *Rhodes Centenary Exhibition* in Bulawayo, Southern Rhodesia (still under British colonial rule), exhibits for the Tenth Triennale in Milan, a vast exhibit on Power for Progress for the British Industry

Pavilion at Brussels Expo 58 and exhibitions in Mexico and Israel as consultant to UNESCO, which from its inception in 1946 had a lively programme of exhibitions, with one of their first addressing the question of human rights.[6]

As the Cold War intensified, exhibitions became a way of the British government engaging across national borders, acting as the focus for fraught exchanges between opposing powers and platforms for international diplomacy, as many recent accounts show.[7] Biennials and triennials, with their capacity to reconfigure global geographies and to support new regional cultural collaborations, were also gaining momentum in this period. Allowing for the 'production of locality', to use anthropologist Arjun Appadurai's phrase, these international art events were becoming useful sites for creating local identities in the face of cultural centrism.[8] The Bureau International des Expositions (BIE), formed in 1931 as the intergovernmental organisation in charge of overseeing and regulating World Expos, continues to guide Britain's regular contributions to exhibitions, biennials and triennials to this day.[9]

By 1953, the end of the period of this book, exhibitions had been used for two decades in Britain as domestic propaganda, for public communication of messages by activist groups and by the government. While the culture of international expos was booming, by the mid-1950s exhibitions were far less frequently used by the British government for conveying political and economic information to home audiences. This was symptomatic of the increasing specialisation of government communication means, with print and broadcast forms identified as more useful and appropriate for conveying political and economic information, as critic Raymond Williams identified in his landmark study *Television: Technology and Cultural Form*.[10] Public interest in visiting such exhibitions was waning too. As a mode of mass domestic communication, exhibitions had already been found wanting: often unable to connect with the public in the way their makers had intended, appearing lofty, esoteric or patronising, overly verbose, excessively directive and only allowing for limited numbers of viewings on small and dispersed sites.

Some designers, such as Clifford Hatts, attributed exhibition's decline as a medium for mass domestic communication in the postwar world to the advent of mass television-watching by the British public, with television recognised as the more 'successful' communicative medium, having a 'condensed clarity' lacking in exhibitions.[11] Hatts's belief in television related to his decades of design work across the two media, building his early career working on exhibitions through Misha Black and Milner Gray's highly acclaimed practice, Design Research Unit, including on the Festival of Britain. Then, in the mid-1950s, Hatts had moved to work as a BBC TV designer, joining exhibition designer Richard Levin who had been appointed the BBC's Head of Design in 1953. In his book *Television by Design* of 1961, Levin proposed that the distinction between television and any other form of communication 'the world has ever known' was in its being linked by networks.[12]

Many others whose early careers had been built through commercial exhibition design, including Misha Black and Milner Gray, also took on design work for television, as part of the mixed ecology of their multi-disciplinary design practices. But most showed a continuing engagement with ideas of 'integration', the complementary possibilities offered by different media that had so intrigued László Moholy-Nagy in his writings of the 1930s (as discussed in the Introduction). This was an interest that continued to preoccupy designer F. H. K. Henrion, as indicated in his 1956 speech 'Design for Television', which outlined the multiple roles available to designers for enhancing television as visual medium; in his admiration for US multi-media designer Will Burtin's postwar exhibitions about everyday life for the United States Information Agency; and in Henrion's continuing experimentation with the relationships between two- and three-dimensional design.[13]

Exhibitions as the 'beehive of creativity'

Many major contributors to shaping exhibitions' use across all of the moments of this book had a common status as incomers to Britain. This is no coincidence: these exhibitions, operating outside of the established cultural institutions and mounted in Underground stations, village halls and on bombsites, mirrored the displaced status of their makers, untethered by war from networks and institutions in their places of origin. This form of exhibition was created in the crucible of exile, providing the 'beehive of creativity' that Czech-born philosopher Vilém Flusser, himself a wartime refugee to Britain, had described as surrounding the expellee, with their urge to synthesise 'new information'.[14] Exhibitions, for many of these newly arriving artists and designers, had operated initially as nodes of resistance and social connectors, offering the focus for interventions into contemporary discourse and providing a training ground for work in multi-disciplinary practice.

Ultimately exhibitions provided the bridge to new life and work, either in the United Kingdom or at another remove. In a 1967 essay, Misha Black wrote at length about his, by then, four decades of design practice, citing his experience of working on the 1951 Festival of Britain exhibitions as creative, extraordinarily formative indicative structures for collective working.[15]

It may seem ironic that the archetypally patriotic work of creating national propaganda at the heart of government was largely led by recent arrivals in Britain. Identifying what such designers contributed develops our understanding of British visual culture as characterised by diversity over a long period. This is one of the motivating ideas that has driven and shaped my writing of this book.

The fact that this kind of propaganda exhibition in Britain has remained largely invisible, written out of historical accounts, except when seen as a critical mass in the Festival of Britain, is partly a reflection of the perception of such cultural forms as of low status. The lack of remaining evidence should not deter us, however, from reconsidering these exhibitions as integral to British art, design and visual cultures in the tumultuous twenty years from 1933 to 1953 that are the focus of this book. They allow additional insights into the extended creativity of the documentary movement in Britain and into aspects of political culture 'from below' that have largely been overlooked. They provide a vivid index to concerns, debates and wider social and cultural movements of the time, through alternative means, in their use across projection, promotion, policy and activism; as manifestos, demonstrations, counter-arguments, solidarities, warfare and welfare.

Evidence of the impact of these small events remains limited and piecemeal. As I have shown, this was often because exhibitions were only one element in a wider campaign or movement, because they were intended to be short-lived and immediate, because they happened outside archiving institutions or because the public response was muted and limited. As I complete this manuscript, which has been so long in the making, I am still making contact with people who can provide new pieces of the complex and time-consuming jigsaw from which this book has been created. It is possible that the exhibitions I have long assumed were undocumented may have fulsome archives in lofts, garages or even in prominent archives that somehow I have failed to find, but I will leave it to others to pursue them.

Notes

1 Gray, 'Exhibitions In or Out', pp. 110–12.
2 Gray, 'Exhibitions In or Out', p. 112.
3 Gray, 'Exhibitions In or Out', pp. 115–16.
4 Misha Black, 'Are Exhibition Designers Too Good to Be Good?', *Display: Design and Presentation*, April 1949, pp. 18–19.
5 Black, *Exhibition Design*.
6 Christopher Pearson, *Designing UNESCO: Art, Architecture and International Politics at Mid-Century* (Farnham: Ashgate, 2010), p. 42; Fabian Röderer, 'The Power of Photographic Display', *Photography & Power: Daguerrotype Studies in the History and Theory of Photography*, no. 2(26), 2019, pp. 137–48.
7 See Atkinson et al. (eds), *Exhibitions Beyond Boundaries*; Harriet Atkinson and Verity Clarkson (eds), 'Design as an Object of Diplomacy Post-1945', *Design and Culture*, vol. 9, no. 2, 2017, pp. 117–262; Jane Pavitt and David Crowley (eds), *Cold War Modern* (London: V&A Publishing, 2010); Susan Reid on 'The Soviet Pavilion at Expo 58', in Alla Aronova and Alexander Ortenberg (eds), *A History of Russian Exposition and Festival Architecture, 1700–2014* (Abingdon: Routledge, 2018); Robert H. Haddow, *Pavilions of Plenty: Exhibiting American Culture Abroad in the*

1950s (Washington, DC: Smithsonian Institution Press, 1997); Greg Castillo, *Cold War Cold War on the Home Front: The Soft Power of Midcentury Design* (Minneapolis, MN: Minnesota University Press, 2009); Jack Masey and Lloyd Morgan, *Cold War Confrontations: US Exhibitions and their Role in the Cultural Cold War* (Zurich: Lars Muller, 2008).

8 Arjun Appadurai, 'The Production of Locality', in Richard Fardon (ed.), *Counterworks: Managing the Diversity of Knowledge* (Abingdon: Routledge, 1995), pp. 204–25; Caroline A. Jones, *The Global Work of Art: World's Fairs, Biennials, and the Aesthetics of Experience* (Chicago, IL: University of Chicago Press, 2017); Sabine B. Vogel, *Biennials – Art on a Global Scale* (Vienna: Springer, 2010); and Anthony Gardner, *Biennials, Triennials, and Documenta: The Exhibitions That Created Contemporary Art* (Oxford: Wiley-Blackwell, 2015).

9 www.biennialfoundation.org/network/biennial-map/ (accessed 12 May 2022). It is evident that the British government still sees value in making major investment in such events, as I know from a recent seminar I led for civil servants at the British Department of International Trade, currently developing Britain's pavilion for Osako Expo 2025.

10 Raymond Williams, *Television: Technology and Cultural Form* (London: Fontana, 1974), later published as *Television* (London: Routledge, 1990), pp. 12–14, and *Communications* (Harmondsworth: Penguin, 1976).

11 This was the claim of Clifford Hatts, who moved from exhibitions work to set design at the BBC; graphic designer Clifford Hatts, in an interview with the author, 2003.

12 Richard Levin, *Television by Design* (London: Bodley Head, 1961).

13 F. H. K. Henrion, 'Design for Television', *Journal of the Royal Society of Arts*, vol. 104, no. 4976, 27 April 1956, pp. 439–55; F. H. K. Henrion, 'A Tribute to Will Burtin', *Typographic 1: Journal of the Society of Typographic Designers*, Autumn 1972, p. 2.

14 Flusser, *The Freedom of the Migrant*, pp. 3, 81, 86.

15 M. Middleton, *Group Practice in Design* (London: Architectural Press, 1967), pp. 118–19.

Bibliography

Archival sources and private papers

British Film Institute
British Library, St Pancras
National Sound Archives
Henry Moore Institute, Leeds
Betty Rea collection
Imperial War Museum
D series of Home Front photographs
Ministry of Information folder
Oral History collections
Leo Baeck Institution
Library of Congress, Washington
London Metropolitan Archives
LCC Architects' Department collection
London Transport Museum Archives, Acton
Photographic collections
Marx Memorial Library
Mass Observation Archive www.amdigital.co.uk/collection/mass-observation-online
Museum of Domestic Design & Architecture, Middlesex
Charles Hasler collection
MOI Digital https://moidigital.ac.uk
Home Intelligence Reports
National Archives, Kew
Board of Trade BT54
Foreign Office FO945
Home Office HO334
Ministry of Information INF1
Prime Minister's Office Records PREM
National Archives, Maryland
Office of War Information

People's History Museum
Labour History Archive and Study Centre
RIBA Library, London
Ernö Goldfinger papers
Godfrey Samuel papers
Ove Arup papers
Royal Society of the Arts, London
Hans Schleger papers
Tate Archive
London Group collection
Artists Association collection
Felicia Browne collection
Klaus Hinrichsen collection
Julian Trevelyan collection
Transport for London Archives
Tyne & Wear Archives
University of Brighton Design Archives
F. H. K. Henrion collection
H. A. Rothholz collection
Brighton School of Art collection
Victoria & Albert Museum Archives of Art & Design
Misha Black collection
Warwick University Modern Records Centre Archives
Trades Union Congress papers
Wellcome Collection
Working Class Movement Library

Primary sources

Journals and periodicals

Action
Advertisers Weekly
Architect & Building News
Architects Journal
Architectural Association Journal
Architectural Review
Art and Industry
The Australian Quarterly
Bulletin of the British Anti-War Movement
COI Review
Commercial Art
Daily Mirror
Daily Worker
Design for Today
The Designer
Display
Forward
Imperial Institute Annual Report
International African Opinion
Jewish Chronicle
Journal of the National Smoke Abatement Society

Journal of the Royal Institute of British Architects
Journal of the Royal Society of Arts
Left Review
The Listener
The Manchester Guardian
Modern Publicity Commercial Art Annual
The New Leader
News Chronicle
The New Statesman and Nation
Picture Post
Printed Advertising
Shelf Appeal
Smokeless Air
The Spectator
The Star
The Studio
The Times
Weekend Review
Woman Today
Wood

Recorded sources

Hatts, Clifford, interview with the author, London, 2 May 2003
Hollis, Richard, interview with F. H. K. Henrion in 1986 Imperial War Museum Catalogue 9592/ 1986–11–12: www.iwm.org.uk/collections/item/object/80009378
Levin, Richard, interviewed by Norman Swallow and Alan Lawson on 3 September 1991: https://historyproject.org.uk/interview/richard-levin
Moro, Peter, interview with Louise Brodie, BL National Sound Archives
Thornycroft, Priscilla (Oral History, IWM catalogue number 31966): www.iwm.org.uk/collections/item/object/80031580

Published articles

Bayer, Herbert, 'Aspects of Design of Exhibitions and Museums', *Curator: The Museum Journal*, vol. 14, no. 3, 1961, pp. 257–88
Bayer, Herbert, 'Fundamentals of Exhibition Design', *Production Manager*, vol. 6, no. 2, 1937
Black, Misha, 'Are Exhibition Designers Too Good to Be Good?', *Display: Design and Presentation*, April 1949, pp. 18–19
Black, Misha, 'The British Exhibition', *Graphis*, March–April 1946, pp. 212–13
Kallmann, G. S., 'The Wartime Exhibition', *Architectural Review*, October 1943, pp. 95–106
Medley, Robert, 'Hitler's Art in Munich', *Axis*, Early Winter 1937
Nash, Paul, 'Finding a Living in Art To-day', *The Journal of Careers*, vol. 12, no. 130, May 1933, pp. 5–8
Neurath, Marie, 'An Isotype Exhibition on Housing', *Journal of RIBA*, vol. 54, 1947, pp. 600–3
Neurath, Otto, 'Health Education through Isotype', *The Lancet*, 25 August 1945, pp. 236–43
Read, Herbert, 'Picasso's Guernica', *London Bulletin*, no. 6, October 1938, p. 6 [reprinted in G. Schiff (ed.), *Picasso in Perspective* (Hoboken, NJ: Prentice Hall, 1976)]

Stargardt, A. W., 'Allies inside Germany: The German Resistance Movement against Nazi-Fascism', *The Australian Quarterly*, vol. 16, no. 3, September 1944, pp. 23–9

Tallents, Stephen, 'The Birth of British Documentary', *Journal of the University Film Association*, vol. 20, no. 1, 1968, pp. 15–21

Wheeler, Monroe, 'A Note on the Exhibition', *The Bulletin of the Museum of Modern Art*, vol. 9, nos. 5–6, June 1942, pp. 18–20

Published books

Ademola, Ade, 'Colour Bar Notoriety in Britain', in Nancy Cunard (ed.), *Negro: Anthology Made by Nancy Cunard* (London: Wishart and Co., 1934)

Art & Industry: Report of the Committee Appointed by the Board of Trade Under the Chairmanship of Lord Gorell on the Production and Exhibition of Articles of Good Design and Every-Day Use (London: HMSO, 1932)

Bernays, Edward L., *Propaganda: The Public Mind in the Making* (Port Washington, NY: Kennikat Press, 1928)

Bilbo, Jack, *Jack Bilbo: An Autobiography (The first forty years of the complete and intimate life-story of an Artist, Author, Sculptor, Art Dealer, Philosopher, Psychologist, Traveller and a Modernist Fighter for Humanity)* (self-published, 1948)

Bird, Francis, *Press, Parliament and People* (London: William Heinemann, 1946)

Black, Misha, 'Design for Ceremony and Exhibitions', in Herbert Read (ed.), *The Practice of Design* (London: Lund Humphries, 1946)

Black, Misha, *Exhibition Design* (London: Architectural Press, 1950)

Black, Misha, 'Propaganda in Three Dimensions' [1949], in Avril Blake (ed.), *The Black Papers on Design: Selected Writings of the Late Sir Misha Black* (Oxford: Pergamon Press, 1983)

Boswell, James, *The Artist's Dilemma* (London: The Bodley Head, 1947)

Brockway, Fenner, *Inside the Left* (London: Allen & Unwin, 1942)

Bronowski, Jacob, *The Common Sense of Science* (London: Heinemann, 1951)

Clark, Kenneth, *Another Part of the Wood: A Self Portrait* (London: Hamish Hamilton, 1985)

Connolly, Cyril, *Enemies of Promise* (London: Routledge Kegan Paul, 1938)

Evans, Myfanwy (ed.), *The Painter's Object* (London: Gerald Howe, 1937)

Games, Abram, *Over My Shoulder* (London: Studio, 1960)

Garfield, Simon, *Our Hidden Lives: The Everyday Diaries of a Forgotten Britain 1945–1948* (London: Random House, 2004)

Gray, Milner, 'A Century of Commercial Design 1851–1951', in *Designers in Britain*, vol. 3 (London: Allan Wingate, 1951)

Hardy, Forsyth, *John Grierson: On Documentary* (London: Collins, 1946)

Havinden, Ashley, *Advertising and the Artist* (London: Studio Publications, 1956)

Higham, Charles, *Looking Forward: Mass Education through Publicity* (London: Nisbet, 1920)

Hinrichsen, Klaus, 'Visual Art behind the Wire', in David Cesarani and Tony Kushner (eds), *The Internment of Aliens in Twentieth Century Britain* (London: Frank Cass, 1993)

Huxley, Aldous, *Brave New World* (London: Chatto & Windus, 1932)

Kiesler, Frederick, *Contemporary Art Applied to the Store and Its Display* (London: Pitman, 1930)

Lambert, Richard S., *Propaganda* (London: Thomas Nelson & Sons, 1938)

Levin, Richard, *Television by Design* (London: Bodley Head, 1961)

Luckhurst, Kenneth, *The Story of Exhibitions* (London: Studio, 1951)
Martin, Kingsley, *Editor: A Second Volume of Autobiography, 1931–45* (London: Hutchinson, 1968)
Mayor Gallery, *Unit One: Spirit of the 30s*, May–June 1984, exh. cat.
Mercer, F. A. and Grace Lovat Fraser (eds), *Modern Publicity in War* (London and New York: Studio Publications, 1941)
Middleton, M., *Group Practice in Design* (London: Architectural Press, 1967)
Moholy-Nagy, László, *Painting Photography Film* (London: Lund Humphries, 1986)
Moholy-Nagy, László, *Vision in Motion* (Chicago, IL: Hillison and Etten, 1947)
Moholy-Nagy, László, *Von Material zu Architektur* (Munich: Albert Langen Verlag, 1929) [published in English as *The New Vision*, 1932]
Moholy-Nagy, Sibyl, *Experiment in Totality* (New York: Harper and Brothers, 1950)
Naipaul, V. S., *The Enigma of Arrival* (New York: Alfred A. Knopf, 1987)
Nash, Paul, *Room and Book* (London: Soncino Press, 1932)
Neurath, Otto, *From Hieroglyphics to Isotype* (London: Hyphen Press, 2010)
Orwell, George, *The Road to Wigan Pier* (London: Left Book Club, 1937)
Penrose, Roland, *Picasso: His Life and Work* (London: Gollancz, 1958)
Pick, Beverley, *Display and Presentation: Exhibitions, Window and Outdoor Display* (London: C. Lockwood, 1957)
Ponsonby, Arthur, *Falsehood in Wartime* (London: Allen & Unwin, 1928)
Read, Herbert, *Art and Industry: The Principles of Industrial Design* (London: Faber & Faber, 1934)
Read, Herbert (ed.), *Five on Revolutionary Art* (London: Wishart, 1935)
Read, Herbert (ed.), *Surrealism* (London: Faber & Faber, 1936)
Robeson, Paul, *Paul Robeson Speaks* (Secaucus, NJ: Citadel, 1978)
Sartre, Jean-Paul, *The Age of Reason* (London: Penguin, 1967)
Schmidt, José, *Lehre und Arbeit am Bauhaus 1919–32* (Dusseldorf: Marzona, 1984)
Summerson, John, 'The MARS Group and the Thirties', in John Bold and Edward Chaney (eds), *English Architecture: Public and Private* (London: The Hambledon Press, 1993)
Tallents, Stephen, *The Projection of England* (London: Faber & Faber, 1932)
Thoene, Peter, *Modern German Art* (London: Pelican, 1938)
Thomas, Adrian, 'Exhibition Scripts', in Misha Black, *Exhibition Design* (London: Architectural Press, 1950)
Trevelyan, Julian, *Indigo Days: The Art and Memoirs of Julian Trevelyan* (London: MacGibbon & Kee, 1957)
Uhlman, Fred, *The Making of an Englishman* (London: Gollancz, 1960)
Unit 1: The Modern Movement in English Architecture, Painting and Sculpture (London: Cassell, 1934)
Vorobyyov, K. I., 'Museums in Industrial Enterprises', translated by Caroline Rees in Arseny Zhilyaev (ed.), *Avant-Garde Museology* (Minneapolis, MN: University of Minnesota Press, 2015)
The War as Seen by Children catalogue published by Refugee Children's Evacuation Fund, with a foreword by J. G. Siebert (London, Geo. Marshall & Co., n.d.)
War Pictures at the National Gallery, London (London: National Gallery, 1944)
Williams, Eric, *The Negro in the Caribbean* (Washington: The Associates in Negro Folk Education, 1942)
Woolf, Virginia, *Three Guineas* (New York: Harcourt, Brace & Co., 1938)

Secondary sources

Acland, Charles R. and Haidee Wasson, *Useful Cinema* (Durham, NC: Duke University Press, 2012)

Albera, Francois and Maria Tortajada, *Cinema beyond Film* (Amsterdam: Amsterdam University Press, 2010)

Anthony, Scott, *Public Relations and the Making of Modern Britain* (Manchester: Manchester University Press, 2012)

Anthony, Scott and James G. Mansell, *The Projection of Britain: A History of the GPO Film Unit* (London: Palgrave Macmillan, 2011)

Appadurai, Arjun, 'The Production of Locality', in Richard Fardon (ed.), *Counter-works: Managing the Diversity of Knowledge* (Abingdon: Routledge, 1995)

Aral, Jenna Lundin, 'Information as Spectacle: Second World War Exhibitions by the Ministry of Information', *Journal of Design History*, vol. 31, issue 1, February 2018

Atkinson, Harriet, *The Festival of Britain: A Land and Its People* (London: I.B. Tauris, 2012)

Atkinson, Harriet, '"The First Modern Townscape"? The Festival of Britain, Townscape and New Picturesque', in John Pendlebury et al. (eds), *Alternative Visions of Post-War Reconstruction: Creating the Modern Townscape* (Abingdon: Routledge, 2014)

Atkinson, Harriet, '"Lines of Becoming" Misha Black and Entanglements through Exhibition Design', *Journal of Design History*, vol. 34, issue 1, March 2021

Atkinson, Harriet and Verity Clarkson (eds), 'Design as an Object of Diplomacy Post-1945', *Design and Culture*, vol. 9, no. 2, 2017

Atkinson, Harriet, 'Visual storytelling in the Ministry of Information's wartime exhibitions', in Beatriz Lopez, James Smith and Guy Ward (eds), *British Writing, Propaganda and Cultural Diplomacy in the Second World War and Beyond* (London: Bloomsbury Academic, 2024).

Atkinson, Harriet, Verity Clarkson and Sarah Lichtman (eds), *Exhibitions beyond Boundaries* (London: Bloomsbury Academic, 2022)

Aulich, Jim, *War Posters: Weapons of Mass Communication* (London: Thames & Hudson, 2011)

Aynsley, Jeremy, *Designing Modern Germany* (London: Reaktion, 2009)

Aynsley, Jeremy, *Graphic Design in Germany 1890–1945* (London: Thames & Hudson, 2000)

Aynsley, Jeremy, 'Pressa Cologne, 1928: Exhibitions and Publication Design in the Weimar Period', *Design Issues*, vol. 10, no. 3, Autumn 1994

Azoulay, Ariella, *The Civil Contract of Photography* (New York: Zone Books, 2008)

Baker, Michael H. C., *London Transport in the 1930s* (Hersham: Ian Allan, 2007)

Bakhtin, Mikhail, *Rabelais and His World* (Bloomington, IN: Indiana University Press, 1984)

Bal, Mieke, *Double Exposures: The Subject of Cultural Analysis* (London: Routledge, 1996)

Barron, Stephanie and Sabine Eckmann (eds), *Exiles & Emigrés: The Flight of European Artists from Hitler* (New York: Harry N. Abrams, 1997)

Barthes, Roland, *Image-Music-Text* (London: Fontana, 1977)

Barthes, Roland, *Mythologies* (London: Jonathan Cape, 1972)

Baudrillard, Jean, *Simulacra and Simulation* (London: Semiotext(e), 1981)

Benjamin, Walter, *One-Way Street and Other Writings* (London: Harcourt Brace Jovanovich, 1978)
Benjamin, Walter, 'The Work of Art in the Age of Mechanical Reproduction', in *Illuminations* (London: Bodley Head, 2015)
Bennett, Tony, 'The Exhibitionary Complex', *New Formations*, vol. 4, Spring 1988
Berger, John, *Understanding a Photograph* (London: Penguin, 2013)
Berman, Marshall, *All That Is Solid Melts into Air* (London: Verso, 1983)
Berridge, Kate, *Madame Tussaud: A Life in Wax* (New York: W. Morrow, 2006)
Black, Jonathan, 'For the People's Good: Hans Schleger (1898–1976), Poster Design and British National Identity, 1935–60', *Visual Culture in Britain*, vol. 13, no. 2, 2012
Black, Naomi, *Virginia Woolf as Feminist* (Ithaca, NY: Cornell University Press, 2004)
Blake, Avril, *Misha Black* (London: Design Council, 1984)
Blake, Jody, 'The Truth about the Colonies, 1931: Art Indigene in Service of the Revolution', *Oxford Art Journal*, vol. 25, no. 1, 2002
Bohm-Duchen, Monica, *Art in Exile in Great Britain 1933–45* (London: Belmont Press, 1986)
Bohm-Duchen, Monica (ed.), *Insiders Outsiders: Refugees from Nazi Europe and Their Contribution to British Visual Culture* (London: Lund Humphries, 2019)
Borchardt-Hume, Achim (ed.), *Albers and Moholy-Nagy: From the Bauhaus to the New World* (London: Yale University Press, 2006)
Bornstein, George, *Material Modernism: The Politics of the Page* (Cambridge: Cambridge University Press, 2001)
Borthwick, Alistair, *The Empire Exhibition Fifty Years On* (Edinburgh: Mainstream Publishing, 1988)
Bourdieu, Pierre, 'Forms of Capital', in J. E. Richardson (ed.), *Handbook of Theory of Research for the Sociology of Education* (Westport, CT: Greenwood Press, 1986)
Bowness, Sophie and Clive Phillpot, *Britain at the Venice Biennale 1895–1995* (London: The British Council, 1995)
Bradshaw, David, 'British Writers and Anti-Fascism in the 1930s', Part I, *Woolf Studies Annual*, vol. 3, 1997; Part II, *Woolf Studies Annual*, vol. 4, 1998
Breakell, Sue, '"The Exercise of a Peculiar Art-Skill": Kenneth Clark's Design Advocacy and the Council of Industrial Design', *Visual Culture in Britain*, vol. 16, no. 1, 2015
Breakell, Sue, 'Exhibiting "The Taste of Everyday Things": Kenneth Clark and CEMA's Wartime Exhibitions of Design', in Liz Farrelly and Joanna Weddell (eds), *Design Objects and the Museum* (London: Bloomsbury, 2016)
Brimblecombe, Peter, *The Big Smoke: A History of Air Pollution since Medieval Times* (London and New York: Routledge, 1988)
Brinson, Charmian and Richard Dove, *A Matter of Intelligence: M15 and the Surveillance of Anti-Nazi Refugees, 1933–50* (Manchester: Manchester University Press, 2015)
Brinson, Charmian and Richard Dove (eds), *Politics by Other Means: The Free German League of Culture in London, 1939–1946* (London: Vallentine Mitchell, 2010)
Brinson, Charmian and Richard Dove (eds), *Working for the War Effort: German-speaking Exiles in the Performing Arts in Britain after 1933* (Elstree: Valentine Mitchell, 2021)

Britton, S., '"Come and see the empire by the all red route!": Anti-imperialism and Exhibitions in Inter-war Britain', *History Workshop Journal*, vol. 69, no. 1, 2010

Brown, David, 'Refugee Artists in Great Britain', *Art and Artists*, April 1984

Buchanan, Tom, 'The Lost Art of Felicia Browne', *History Workshop Journal*, issue 54, 2002

Buchloh, Benjamin H. D., 'From Faktura to Factography', *October*, no. 30, Autumn 1984

Buckley, Cheryl, *Designing Modern Britain* (London: Reaktion, 2007)

Burke, C., E. Kindel and S. Walker, *Isotype Design and Contexts, 1925–1971* (London: Hyphen Press, 2014)

Bush, Barbara, *Representing Empire* (Abingdon: Routledge, 2006)

Butler, Judith, *Notes Toward a Performative Theory of Assembly* (Cambridge, MA: Harvard University Press, 2015)

Campbell, Louise, 'The MARS Group, 1933–1939', *RIBA Transactions*, issue no. 8, vol. 4, no. 2, 1985

Carline, Richard, 'John Heartfield in England', in *John Heartfield 1891–1968: Photomontages* (London: Arts Council of Great Britain, 1969)

Carullo, Valeria, *Moholy-Nagy in Britain, 1935–7* (London: Lund Humphries, 2019)

Castillo, Greg, *Cold War on the Home Front: The Soft Power of Midcentury Design* (Minneapolis, MN: Minnesota University Press, 2009)

Cesarani, David and Tony Kushner (eds), *The Internment of Aliens in Twentieth Century Britain* (London: Frank Cass, 1993)

Chantrell, Glynnis (ed.), *The Oxford Dictionary of Word Histories* (Oxford: Oxford University Press, 2002)

Chapman, James, *The British at War: Cinema, State and Propaganda, 1939–45* (London: I.B. Tauris, 2011)

Chomsky, Noam, *Manufacturing Consent: The Political Economy of the Mass Media* (New York: Pantheon, 1988)

Clark, Toby, *Art and Propaganda in the Twentieth Century: The Political Image in Mass Culture* (London: Weidenfeld & Nicolson, 1997)

Clarke, Gill, *Conflicting Views: Pacifist Artists* (Bristol: Sansom & Co., 2018)

Coe, P. and M. Reading, *Lubetkin and Tecton: Architecture and Social Commitment* (London: Arts Council, 1981)

Cohen, Jean-Louis, *Architecture in Uniform: Designing and Building for the Second World War* (Paris: Canadian Centre for Architecture, 2011)

Cohen, Scott, 'The Empire from the Street: Virginia Woolf, Wembley, and Imperial Monuments', *Modern Fiction Studies*, vol. 50, no. 1, Spring 2004

Colomina, Beatriz, *Privacy and Publicity: Modern Architecture as Mass Media* (Cambridge, MA and London: MIT Press, 1996)

Colpus, Eve, *Female Philanthropy in the Interwar World* (London: Bloomsbury, 2018)

Condell, Caitlin and Emily M. Orr (eds), *E. McKnight Kauffer, The Artist in Advertising* (New York: Rizzoli Electa, 2020)

Corthorn, Paul, 'Labour, the Left and the Stalinist Purges of the Late 1930s', *The Historical Journal*, vol. 48, no. 1, March 2005

Cowling, Elizabeth *Visiting Picasso: The Notebooks and Letters of Roland Penrose* (London: Thames & Hudson, 2006)

Cox, Howard and Simon Mowatt, 'Horatio Bottomley and the Rise of *John Bull* Magazine', *Media History*, vol. 25, no. 1, 2019

Crampsey, Bob, *The Empire Exhibition of 1938: The Last Durbar* (Edinburgh: Mainstream, 1988)

Crary, Jonathan, *Suspensions of Perception: Attention, Spectacle and Modern Culture* (Cambridge, MA: MIT Press, 1999)

Danchev, Alex (ed.), *100 Artists' Manifestos* (London: Penguin, 2011)

Darling, Elizabeth, '"Enriching and enlarging the whole sphere of human activities": The Work of the Voluntary Sector in Housing Reform in Inter-War Britain', in Christopher Lawrence and Anna-K. Mayer (eds), *Regenerating England: Science, Medicine and Culture in Inter-War Britain* (Leiden: Brill, 2000)

Darling, Elizabeth, 'Institutionalizing English Modernism 1924–33: From the Vers Group to MARS', *Architectural History*, vol. 55, 2012

Darling, Elizabeth, *Re-forming Britain: Narratives of Modernity before Reconstruction* (London: Routledge, 2007)

Darling, Elizabeth, '"The star in the profession she invented for herself": A Brief Biography of Elizabeth Denby, Housing Consultant', *Planning Perspectives*, vol. 20, July 2005

Darling, Elizabeth, '"To induce humanitarian sentiments in prurient Londoners": The Propaganda Activities of London's Voluntary Housing Associations in the Inter-War Period', *The London Journal*, vol. 27, no. 1, 2002

Dean, David, *The Architect as Stand Designer: Building Exhibitions 1895–1983* (London: Scolar, 1985)

Deepwell, Katy, 'Anti-fascist Activities amongst Women Artists in the 1930s in Britain', in Concha Lomba Serrano and Alberto Castán Chocarro (eds), *Las mujeres en el sistema artístico (1804–1939)/Women in the Art Scene (1804–1939)* (Madrid: Universidad de Madrid, 2022)

Deepwell, Katy, *Women Artists between the Wars: 'A fair field and no favour'* (Manchester: Manchester University Press, 2010)

Demos, T. J., *The Migrant Image: The Art and Politics of Documentary during Global Crisis* (Durham, NC: Duke University Press, 2013)

de Smet, Catherine, *Le Corbusier Penseur du Musée* (Paris : Flammarion, 2019)

Dickson, Rachel, Sarah MacDougall and Ulrike Smalley, '"Astounding and Encouraging": High and Low Art Produced in Internment on the Isle of Man during the Second World War', in Gilly Carr and Harold Mytum (eds), *Cultural Heritage and Prisoners of War: Creativity behind Barbed War* (Abingdon: Routledge, 2012)

Driver, Felix and David Gilbert, 'Capital and Empire: Geographies of Imperial London', *GeoJournal*, vol. 51, nos. 1/2: *European Capital Cities*, 2000

Druick, Zoe and Deane Williams (eds), *The Grierson Effect: Tracing Documentary's International Movement* (London: Palgrave Macmillan, 2014)

Duberman, Martin B., *Paul Robeson: A Biography* (New York: Ballantine Books, 1989)

Dunnett, James and Gavin Stamp, *Ernö Goldfinger: Works 1* (London: Architectural Association, 1983)

Durkheim, E., 'From Mechanical to Organic Solidarity', *Sociology: Introductory Readings*, vol. 2, no. 1, 2010

Eco, Umberto, 'Articulations of the Cinematic Code', *Cinemantics*, vol. 1, January 1970

Egbert, Donald Drew, *Social Radicalism in the Arts: Western Europe; A Cultural History from the French Revolution to 1968* (London: Duckworth, 1970)

Eliot, Simon and Marc Wiggam (eds), *Allied Communication to the Public During the Second World War* (London: Bloomsbury Academic, 2019)

Ellul, Jacques, *Propaganda: The Formation of Men's Attitudes* (New York: Vintage, 1973)

Elwall, Robert, *Ernö Goldfinger* (London: Academy Editions, 1996)

England, Jane, *Jack Bilbo & the Moderns*, exh. cat. (London: England & Co. Gallery, 1990)

Erten, Erdem, 'Shaping "The Second Half Century": The Architectural Review, 1947–1971' (PhD thesis, Massachusetts Institute of Technology, 2004)

Fair, Alistair, *Peter Moro and Partners* (Liverpool: Liverpool University Press, 2021)

Farmer, Richard, *Cinemas and Cinemagoing in Wartime Britain, 1939–45: The Utility Dream Palace* (Manchester: Manchester University Press, 2016)

Farmer, Richard, *The Food Companions: Cinema and Consumption in Wartime Britain, 1939–45* (Manchester: Manchester University Press, 2011)

Flusser, Vilém, *The Freedom of the Migrant: Objections to Nationalism*, ed. Anke K. Finger, trans. Kenneth Kronenberg (Urbana, IL: University of Illinois Press, 2003)

Ford, Sue Grayson, *Brave New Visions: The Émigrés Who Transformed the British Art World* (London: Sotheby's, 2019)

Forgan, Sophie, 'Atoms in Wonderland', *History and Technology*, vol. 19, 2003

Foss, Brian, *War Paint: Art, War, State and Identity in Britain 1939–1945* (New Haven, CT and London: Yale University Press, 2007)

Foster, Alicia, *Radical Women: Jessica Dismorr and Her Contemporaries* (London: Lund Humphries, 2019)

Foster, Hal and Benjamin H. D. Buchloh (eds), *The Independent Group: A Special Issue*, *October*, no. 94, Fall 2000

Fox, Jo, 'Careless Talk: Tensions within British Domestic Propaganda during the Second World War', *Journal of British Studies*, vol. 51, no. 4, October 2012

Fox, Jo, *Film Propaganda in Britain and Nazi Germany: World War II* (Oxford: Berg, 2006)

Fraser, Derek, *The Evolution of the British Welfare State* (London: Macmillan, 1973)

Frosh, Paul, *The Image Factory: Consumer Culture, Photography and the Visual Content Industry* (Oxford and New York: Berg, 2003)

Frowein, Cordula, 'German Artists in War-Time Britain', *Third Text*, vol. 15, Summer 1991

Fyrth, Jim, *The Signal Was Spain: The Aid Spain Movement 1936–1939* (London: Lawrence and Wishart, 1986)

Gardiner, Juliet, *The Thirties: An Intimate History* (London: HarperPress, 2010)

Gardner, Anthony, *Biennials, Triennials, and Documenta: The Exhibitions That Created Contemporary Art* (Oxford: Wiley-Blackwell, 2015)

Gilbert, Martin, *The Roots of Appeasement* (London: Weidenfeld and Nicolson, 1966)

Golan, Romy, *Muralnomad: The Paradox of Wall Painting, Europe 1927–1957* (London: Yale University Press, 2009)

Golan, Romy, 'The World Fair: A Transmedial Theatre', in Jordan Mendelson et al. (eds), *Encounters with the 30s* (Madrid: Fabrica, 2012)

Gold, John, *The Experience of Modernism* (London: Routledge, 1997)

Gold, John, 'New Architecture and the Search for Modernity', in Robert Freestone and Marco Amati (eds), *Exhibitions and the Development of Modern Planning Culture* (Farnham: Ashgate, 2014)

Gopal, Priyamvada, *Insurgent Empire: Anticolonial Resistance and British Dissent* (London: Verso, 2019)

Gottlieb, Julie, 'Varieties of Feminist Responses to Fascism', in Nigel Copsey (ed.), *Varieties of Anti-Fascism: Britain in the Inter-War Period* (Basingstoke: Palgrave Macmillan, 2010)

Gramsci, Antonio, *Selections from the Prison Notebooks* (London: Lawrence & Wishart, 1971)

Greenhalgh, Paul, *Ephemeral Vistas: The Expositions Universelles, Great Exhibitions and World's Fairs, 1851–1939* (Manchester: Manchester University Press, 1988)

Grosvenor, Ian and Sian Roberts, 'Art, Anti-fascism, and the Evolution of a "Propaganda of the Imagination": The Artists International Association 1933–1945', in Frederik Herman, Sjaak Braster and Maria del Mar del Pozo Andres (eds), *Exhibiting the Past: Public Histories of Education* (Berlin: De Gruyter, 2022)

Guttsman, Willi, 'The Influence and Failure of Weimar Radicalism in Emigré Art in Britain', *Third Text*, vol. 5, no. 15, 1991

Guy, Kate, Hajra Williams and Claire Wintle (eds), *Histories of Exhibition Design in the Museum: Makers, Process and Practice* (London: Routledge, 2023)

Haddow, Robert H., *Pavilions of Plenty: Exhibiting American Culture Abroad in the 1950s* (Washington, DC: Smithsonian Institution Press, 1997)

Hall, Stuart, 'Popular Culture and the State', in Tony Bennett, Colin Mercer and Janet Woollacott (eds), *Popular Culture and Social Relations* (Milton Keynes: Open University Press, 1986)

Hall, Stuart, 'The Social Eye of Picture Post' [1973], *Typography Papers*, 8, 2009

Halpern, Orit, *Beautiful Data: A History of Vision and Reason since 1945* (Durham, NC and London: Duke University Press, 2014)

Hannigan, Dave, *Barbed Wire University: The Untold Story of the Interned Jewish Intellectuals Who Turned an Island Prison into the Most Remarkable School in the World* (Guildford: Lyons Press, 2021)

Harker, Ben, '"On Different Levels Ourselves went Forward": Pageantry, Class Politics and Narrative Form in Virginia Woolf's Late Writing', *ELH*, vol. 78, no. 2, Summer 2011

Harrison, Charles, *English Art and Modernism* (New Haven, CT: Yale University Press, 1994)

Haskell, Frances, 'Botticelli, Fascism and Burlington House – The "Italian Exhibition" of 1930', *The Burlington Magazine*, vol. 141, no. 1157, August 1999

Havinden, Michael et al. (eds), *Ashley Havinden: Advertising and the Artist* (Edinburgh: National Galleries of Scotland, 2003)

Hayum, Andree, 'Mussolini Exports the Renaissance: The Burlington House Exhibition of 1930 Revisited', *The Art Bulletin*, vol. 101, no. 2, 2019

Hediger, Vinzenz and Patrick Vonderau (eds), *Films That Work: Industrial Film and the Productivity of Media* (Amsterdam: Amsterdam University Press, 2014)

Hein, George E., *Learning in the Museum* (London: Routledge, 1998)

Henning, Michelle, 'The Pig in the Bath: New Materialisms and Cultural Studies', *Radical Philosophy*, vol. 145, September/October 2007

Highmore, Ben, *The Art of Brutalism: Rescuing Hope from Catastrophe in 1950s Britain* (New Haven, CT and London: Yale University Press, 2017)

Huhtamo, Erkki, 'Messages on the Wall: An Archaeology of Public Media Displays', in *Urban Screens Reader* (Amsterdam: Institute of Network Cultures, 2009)

Huyssen, Andreas, *Twilight Memories: Marking Time in a Culture of Amnesia* (New York and London: Routledge, 1995)

Irving, Henry, 'The Ministry of Information on the British Home Front', in Simon Eliot and Marc Wiggam (eds), *Allied Communication to the Public during the Second World War* (London: Bloomsbury Academic, 2019)

Jackson, Angela, *British Women and the Spanish Civil War* (Abingdon: Routledge, 2002)

Jackson, Iain and Jessica Holland, *The Architecture of Edwin Maxwell Fry and Jane Drew* (Farnham: Ashgate, 2014)

Jolivette, Catherine (ed.), *British Art in the Nuclear Age* (Farnham: Ashgate, 2014)

Jolles, Adam, *The Curatorial Avant-Garde: Surrealism and Exhibition Practice in France, 1925–1941* (University Park, PA: Pennsylvania State University Press, 2013)

Jolles, Adam, 'Stalin's Talking Museums', *Oxford Art Journal*, vol. 28, no. 3, 2005

Jones, Caroline A., *The Global Work of Art: World's Fairs, Biennials, and the Aesthetics of Experience* (Chicago, IL: University of Chicago Press, 2017)

Kaal, Harm and Casper Kirkels, 'Public Order Acts and Their Effects on Street Politics in 1930s Europe: A Case Study of Britain and Netherlands', *Société française d'histoire urbaine*, vol. 2, no. 55, 2019

Kelly, Jessica, *No More Giants: J. M. Richards, Modernism and The Architectural Review* (Manchester: Manchester University Press, 2022)

Kettenacker, Lothar, 'The Anglo-Soviet Alliance and the Problem of Germany, 1941–5', *Journal of Contemporary History*, vol. 17, no. 3, July 1982

Kinchin, P. and J. Kinchin, *Glasgow's Great Exhibitions* (Bicester: White Cockade, 1988)

Kinross, Robin, 'Design in Central-European London', *Typography Papers*, 8, 2009

Kinross, Robin, *Modern Typography: An Essay in Critical History* (London: Hyphen Press, 1992)

Klonk, Charlotte, *Spaces of Experience: Art Gallery Interiors from 1800 to 2000* (New Haven, CT: Yale University Press, 2009)

Kort Zegers, Peter and Douglas Druick, *Windows on the War, Soviet TASS Posters at Home and Abroad 1941–1945* (New Haven, CT: Yale University Press, 2011)

Krauss, Rosalind, 'Photography's Discursive Spaces: Landscape/View', *Art Journal*, vol. 42, no. 4, *The Crisis in the Discipline*, Winter 1982

Kuijt, M., 'Exposing the Colonial Exhibition: Dutch Anti-Colonial Activism in a Transnational Context', *Reinvention: An International Journal of Undergraduate Research*, vol. 12, issue 2, 2019

Lambert, Susan, *Paul Nash as Designer* (London: Victoria & Albert Museum, 1975)

Larkham, Peter, 'Exhibiting Planning in Wartime Britain', in Robert Freestone and Marco Amati (eds), *Exhibitions and the Development of Modern Planning Culture* (Farnham: Ashgate, 2014)

Larkham, Peter (ed.), *The Rebuilding of British Cities: Exploring the Post-Second World War Reconstruction*. (Birmingham: University of Central England School of Planning and History, 2004)

Laucht, Christopher, 'Atoms for the People: The Atomic Scientists' Association, the British State and Nuclear Education in the *Atom Train* Exhibition, 1947–8', *British Journal for the History of Science*, vol. 45, no. 4, December 2012

Leahy, Helen Rees, *Museum Bodies: The Politics and Practices of Visiting and Viewing* (London: Routledge, 2016)

Lee, Hermione, *Virginia Woolf* (London: Vintage, 1996)

Lee, Hermione and Joanna P. Gardner-Huggett, 'Margaret Gardiner: Collecting as Activism', *The British Art Journal*, vol. 6, no. 2, Autumn 2005

Lefebvre, Henri, *The Production of Space* (Oxford: Blackwell, 1984)

Leslie, Esther, *Hollywood Flatlands: Animation, Critical Theory and the Avant-Garde* (London: Verso, 2002)

Lilly, Amy M., 'Three Guineas, Two Exhibits: Woolf's Politics of Display', *Woolf Studies Annual*, no. 9, part 1, 2003

Lincove, David A., 'The British Library of Information in New York: A Tool of British Foreign Policy, 1919–1942', *Libraries & the Cultural Record*, vol. 46, no. 2, 2011

Linehan, Thomas, 'Communist Culture and Anti-Fascism', in Nigel Copsey (ed.), *Varieties of Anti-Fascism: Britain in the Inter-war Period* (London: Palgrave Macmillan, 2010)

Lipke, W., *David Bomberg: A Critical Study of His Life and Work* (London: Evelyn Adams & Mackay, 1967)

Littler, Jo, '"Festering Britain": The 1951 Festival of Britain, National Identity and the Representation of the Commonwealth', in A. Ramamurthy and Simon Faulkner (eds), *Visual Culture and Decolonisation in Britain* (Farnham: Ashgate, 2006)

Lotery, Kevin, *The Long Front of Culture: The Independent Group and Exhibition Design* (Cambridge, MA: MIT Press, 2020)

Lugon, Olivier, 'Dynamic Paths of Thought: Exhibition Design, Photography and Circulation in the Work of Herbert Bayer', in Francois Albera and Maria Tortajada (eds), *Cinema Beyond Film: Media Epistemology in the Modern Era* (Amsterdam: Amsterdam University Press, 2010)

Lugon, Olivier, 'Photography and Exhibition in Germany around 1930', in *Object: Photo, Modern Photographs 1909–1949: The Thomas Walther Collection at the Museum of Modern Art* (New York: Museum of Modern Art, 2014)

Lugon, Olivier, 'The Ubiquitous Exhibition: Magazines, Museums and the Reproducible Exhibition after World War II', in Thierry Gervais (ed.), *The 'Public' Life of Photographs* (London: MIT Press, 2016)

Luke, Megan, *Kurt Schwitters: Space, Image, Exile* (Chicago, IL: University of Chicago Press, 2014)

Lyon, Janet, *Manifestoes: Provocations of the Modern* (Ithaca, NY and London: Cornell University Press, 1999)

Maasri, Zeina, *Cosmopolitan Radicalism: The Visual Politics of Beirut's Global Sixties* (Cambridge: Cambridge University Press, 2020)

MacCarthy, Fiona, *All Things Bright and Beautiful: Design in Britain, 1830 to Today* (London: Allen and Unwin, 1972)

Mackay, Robert, *Half the Battle: Civilian Morale in Britain during the Second World War* (Manchester: Manchester University Press, 2002)

Mackenzie, S. P., 'The Coming of ABCA, 1941–1942', in *Politics and Military Morale: Current Affairs and Citizenship Education in the British Army 1914–1950* (Oxford: Oxford University Press, 2011)

Martin, Courtney J., 'Exiles, Émigrés and Cosmopolitans: London's Postwar Art World', in Okwui Enwezor, Katy Siegel and Ulrich Wilmes (eds), *Postwar: Art between the Pacific and the Atlantic* (Munich: Prestel, 2016)

Masey, Jack and Lloyd Morgan, *Cold War Confrontations: US Exhibitions and Their Role in the Cultural Cold War* (Zurich: Lars Muller, 2008)

Massey, Anne, *The Independent Group: Modernism and Mass Culture in Britain, 1945–9* (Manchester: Manchester University Press, 1995)

Massey, Doreen, 'A Global Sense of Place', *Marxism Today*, June 1991

Matera, Marc, *Black London: The Imperial Metropolis and Decolonization in the Twentieth Century* (Berkeley, CA: University of California Press, 2015)

McAleer, John and John MacKenzie, *Exhibiting the Empire: Cultures of Display and the British Empire* (Manchester: Manchester University Press, 2015)

McKellar, Erin, 'Designing the Child's World: Ernö Goldfinger and the Role of the Architect, 1933–1946', *Journal of Design History*, vol. 33, no. 1, 2020

McKellar, Erin, 'Living, Working, Playing: Ernö Goldfinger's Planning Exhibitions, 1943–46', in Gaia Caramellino and Stephanie Dadour (eds), *The Housing Project: Discourse, Ideals, Models and Politics in 20th Century Exhibitions* (Leuven: Leuven University Press, 2020)

McLaine, Ian, *Ministry of Morale: Home Front Morale and the Ministry of Information in World War II* (Abingdon: Routledge, 2021)

McNair, Brian, *Fake News: Falsehood, Fabrication and Fantasy in Journalism* (Abingdon: Routledge, 2021)

Meakin, Kerry, *The Professionalization of Window Display in Britain, 1919–1939* (London: Bloomsbury Academic, 2024).

Meakin, Kerry, 'Women in British Window Display during the 1920s and 1930s', *History of Retailing and Consumption*, vol. 7, no. 2, 2021

Mendelson, Jordana, 'Josep Renau and the 1937 Spanish Pavilion in Paris', in Jorge Ribalta (ed.), *Public Photographic Spaces: Exhibitions of Propaganda, from Pressa to The Family of Man 1928–55* (Barcelona: Museu d'Art Contemporani de Barcelona, 2008)

Mileaf, Janine, 'Body to Politics: Surrealist Exhibition of the Tribal and the Modern at the Anti-Imperialist Exhibition and the Galerie Charles Ratton', *Anthropology and Aesthetics*, no. 40, Autumn 2001

Mitchell, W. J. T., 'Ekphrasis and the Other', in *Picture Theory: Essays on Verbal and Visual Representation* (Chicago, IL and London: University of Chicago Press, 1994)

Moffat, Isabelle, '"A horror of abstract thought": Postwar Britain and Hamilton's 1951 Growth and Form Exhibition', *October*, no. 94, 2000

Mohan, John, 'Neglected Roots of Regionalism? The Commissioners for the Special Areas and Grants to Hospital Services in the 1930s', *Social History of Medicine*, vol. 10, issue 2, August 1997

Mold, Alex, 'Exhibiting Good Health: Public Health Exhibitions in London, 1948–71', *Medical History*, vol. 62, no. 1, 2018

Morgan, Kenneth O., *The Oxford History of Britain* (Oxford: Oxford University Press, 2001)

Morris, Lynda and Robert Radford, *The Story of the Artists International Association 1933–1953* (Oxford: MOMA Oxford, 1983)

Morris, Lynda and Christoph Grunenberg (eds), *Picasso: Peace and Freedom* (London: Tate, 2010)

Nead, Lynda, *The Tiger in the Smoke: Art and Culture in Post-War Britain* (New Haven, CT and London: Yale University Press, 2017)

Nieland, Justus, *Happiness by Design: Modernism and Media in the Eames Era* (Minneapolis, MN: University of Minnesota Press, 2020)

Noakes, Lucy, *Dying for the Nation: Death, Grief and Bereavement in Second World War Britain* (Manchester: Manchester University Press, 2020)

O'Doherty, Brian, 'Inside the White Cube', *Artforum*, vol. 14, no. 7, March 1976

O'Doherty, Brian, *Inside the White Cube* (San Francisco: Lapis, 1986)

Oldfield, Sibyl, *The Black Book* (London: Profile, 2020)

Olin, Margaret, *Touching Photographs* (Chicago, IL: University of Chicago, 2011)

Orr, Emily *Designing the Department Store: Display and Retail at the Turn of the Twentieth Century* (London: Bloomsbury, 2019)

Pachnicke, Peter and Klaus Honnef (eds), *John Heartfield* (New York: Harry N. Abrams, 1992)

Parker, R. A. C., *Chamberlain and Appeasement: British Policy and the Coming of the Second World War* (Basingstoke: Red Globe Press, 1993)

Parkin, Simon, *The Island of Extraordinary Captives: A True Story of an Artist, a Spy and a Wartime Scandal* (London: Sceptre, 2022)

Pavitt, Jane and David Crowley (eds), *Cold War Modern* (London: V&A Publishing, 2010)

Pearson, Christopher, *Designing UNESCO: Art, Architecture and International Politics at Mid-Century* (Farnham: Ashgate, 2010)

Peat, A., 'Scottish Internationalisms at the 1938 Empire Exhibition: Between Britain, Europe, and Empire', *Open Library of Humanities*, vol. 6, no. 1, 2020

Peto, James and Donna Loveday (eds), *Modern Britain 1929–1939* (London: Design Museum, 1999)

Pezzini, Barbara, '"Aid to Russia" A Wartime Modern Art Exhibition in a Modernist Setting', *The British Art Journal*, vol. 5, no. 3, 2004

Pohlmann, Ulrich, '"Not Autonomous Art but a Political Weapon": Photography Exhibitions as a Means for Aestheticising Politics and Economy in National Socialism', in *Public Photographic Spaces: Exhibitions of Propaganda from Pressa to the Family of Man, 1928–55* (Barcelona: MACBA, 2009)

Pollen, Annebella, *Art without Frontiers: The Story of the British Council, Visual Arts, and a Changing World* (London: British Council, 2023)

Powers, Alan, *Abbatt Toys: Modern Toys for Modern Children* (London: Design for Today, 2020)

Powers, Alan, 'A Popular Modernism? Timber Architecture in Britain 1936–39', *Architectural Theory Review*, vol. 25, nos. 1–2, 2021

Preston, Paul, *The Spanish Civil War* (London: Harper Perennial, 2006)

Price, Hollie, *Picturing Home: Domestic Life and Modernity in 1940s British Film* (Manchester: Manchester University Press, 2021)

Public Photographic Spaces: Exhibitions of Propaganda from Pressa to the Family of Man, 1928–55 (Barcelona: MACBA, 2009)

Puchner, Martin, *Poetry of the Revolution: Marx, Manifestos, and the Avant-Garde* (Princeton, NJ and Oxford: Princeton University Press, 2006)

Radford, Robert, *Art for a Purpose: The Artists' International Association 1933–1953* (Winchester: Winchester School of Art Press, 1987)

Radford, Robert, 'Edith Tudor-Hart: Photographs from the Thirties', *Camerawork*, July 1980

Rancière, Jacques, 'The Surface of Design', in the *Future of the Image* (London: Verso, 2007)

Reid, Susan, 'The Soviet Pavilion at Expo 58', in Alla Aronova and Alexander Ortenberg (eds), *A History of Russian Exposition and Festival Architecture, 1700–2014* (Abingdon: Routledge, 2018)

Remy, Michael, *Towards a Dictionary of Surrealism in England* (Nancy: Groupe-Édition Marges, 1978)

Rewcastle, Martin and Nicholas Serota, 'Art and the Social Purpose of the Whitechapel Art Gallery', in *Art for Society: Contemporary British Art with a Social or Political Purpose* (London: Whitechapel Art Gallery, 1978)

Robbins, David (ed.), *The Independent Group: Postwar Britain and the Aesthetics of Plenty* (Cambridge, MA: MIT Press, 1990)

Roberts, S., 'Education, Art and Exile: Cultural Activists and Exhibitions of Refugee Children's Art in the UK during the Second World War', *Paedagogica Historica*, vol. 53, no. 3, 2017

Röderer, Fabian, 'The Power of Photographic Display', *Photography & Power: Daguerrotype Studies in the History and Theory of Photography*, no. 2(26), 2019

Rose, Gillian, *Doing Family Photography: The Domestic, the Public and the Politics of Sentiment* (Abingdon: Routledge, 2010)

Rose, Peter, '"It Must Be Done Now": The Arts and Crafts Exhibition at Burlington House, 1916', *The Journal of the Decorative Arts Society 1850–the Present*, no. 17, 1993

Samuel, Raphael, *Island Stories: Unravelling Britain: Theatres of Memory, Volume II* (London: Verso, 1998).

Satterthwaite, Tim, *Modernist Magazines and the Social Ideal* (London: Bloomsbury, 2020)

Saunders, Nicholas J., *Trench Art: Materialities and Memories of War* (Abingdon: Routledge, 2020)

Schnapp, Jeffrey T., 'Epic Demonstrations: Fascist Modernity and the 1932 Exhibition of the Fascist Revolution', in Richard J. Golsan (ed.), *Fascism, Aesthetics, and Culture* (Hanover, NH: University Press of New England, 1992)

Schnapp, Jeffrey T., 'Fascism's Museum in Motion', *Journal of Architectural Education*, vol. 45, no. 2, February 1992

Scholz, Sally J., 'Political Solidarity and Violent Resistance', *Journal of Social Philosophy*, vol. 38, issue 1, March 2007

Scott, James C., *Domination and the Arts of Resistance* (New Haven, CT and London: Yale University Press, 1990)

Sekula, Allan, 'On the Invention of Photographic Meaning', in Victor Burgin (ed.), *Thinking Photography Communications and Culture* (London: Palgrave, 1982)

Shanks, Michael, *Experiencing the Past: On the Character of Archaeology* (London: Routledge, 1992)

Shasore, Neal, *Designs on Democracy: Architecture and the Public in Interwar London* (Oxford: Oxford University Press, 2022)

Shaughnessy, Adrian, *FHK Henrion: The Complete Designer* (London: Unit Editions, 2013)

Shaw, Elizabeth, 'Radical Émigré Artists and Their Return to DDR', *Third Text*, vol. 5, no. 15, 1991

Snowman, Daniel, *The Hitler Émigrés: The Cultural Impact on Britain of Refugees from Nazism* (Vintage Digital, 2010)

Staniszewski, Mary Anne, *The Power of Display: A History of Exhibition Installations at the Museum of Modern Art* (Cambridge, MA: MIT Press, 1998)

Stemp, Robin, 'Nan Youngman: Part 1', *The Artist*, November 1987

Stephenson, Andrew, '"Strategies of Situation": British Modernism and the Slump c.1929–1934', *Oxford Art Journal*, vol. 14, no. 2, 1991

Stiff, Paul, 'Austerity, Optimism: Modern Typography in Britain', *Typography Papers*, 8, 2009

Struk, Janina, *Photographing the Holocaust: Interpretations of the Evidence* (London: I.B. Tauris, 2004)

Suga, Yasuko, 'Modernism, Commercialism and Display Design in Britain', *Journal of Design History*, vol. 19, no. 2, 2006

Suga, Yasuko, 'The Reimann School and Studios', in *Designs on Britain: An Exhibition by Jewish Museum London* (London: Jewish Museum London, 2017)

Suga, Yasuko, 'State Patronage of Design? The Elitism/Commercialism Battle in the General Post Office's Graphic Production', *Journal of Design History*, vol. 13, no. 1, 2000

Sugg Ryan, Deborah, 'The Daily Mail Ideal Home Exhibition, 1908–39', in Maggie Andrews and Mary M. Talbot (eds), *All the World and Her Husband: Women in Twentieth-Century Consumer Culture* (London: Cassell, 2000)

Sugg Ryan, Deborah, *Daily Mail Ideal Home Exhibitions* (London: Hazar Publishing, 1997)

Sugg Ryan, Deborah, 'Staging the Imperial City: The Pageant of London, 1911', in Felix Driver and David Gilbert (eds), *Imperial Cities: Landscape, Display and Identity* (Manchester: Manchester University Press, 2003)

Tagliaventi, Alessia, 'Photography at MoMA: Four Landmark Exhibitions', in *Photoshow: Landmark Exhibitions That Defined the History of Photography* (London: Thames & Hudson, 2014)

Taylor, John, 'Picturing the Past', *Ten.8*, no. 11, 1983

Taylor, Philip M., *The Projection of Britain: British Overseas Publicity and Propaganda 1919–1939* (Cambridge: Cambridge University Press, 1981)

Teitelbaum, Matthew, *Montage and Modern Life 1919–1942* (Cambridge, MA and London: MIT Press, 1992)

Thirties: British Art and Design before the War (London: Arts Council of Great Britain, 1979)

Thomas, Zoe, *Women Art Workers and the Arts and Crafts Movement* (Manchester: Manchester University Press, 2020)

Thorsheim, Peter, *Inventing Pollution: Coal, Smoke and Culture in Britain since 1800* (Athens, OH: Ohio University Press, 2006)

Turner, Fred, *The Democratic Surround: Multimedia & American Liberalism from World War II to the Psychedelic Sixties* (Chicago, IL and London: University of Chicago Press, 2013)

Turner, Fred, '*The Family of Man* and the Politics of Attention in Cold War America', *Public Culture*, vol. 24, no. 1, 2012

Tymkiw, Michael, *Nazi Exhibition Design and Modernism* (Minneapolis, MN: University of Minnesota Press, 2018)

Udovicki-Selb, Danilo, 'Le Corbusier and the Paris Exhibition of 1937: The Temps Nouveaux Pavilion', *Journal of the Society of Architectural Historians*, vol. 56, no. 1, 1997

Vinzent, Jutta, 'Muteness as Utterance of a Forced Reality Jack Bilbo's Modern Art Gallery (1941–1948)', in Shulamith Behr (ed.), *Arts in Exile in Britain 1933–1945: Politics and Cultural Identity* (Leiden: Brill, 2005)

Vogel, Sabine B., *Biennials – Art on a Global Scale* (Vienna: Springer, 2010)

Walkowitz, Rebecca L., *Cosmopolitan Style: Modernism beyond the Nation* (New York: Columbia University Press, 2006)

Walker, Sue, 'Effective Antimicrobial Resistance Communication: The Role of Information Design', *Palgrave Communications*, vol. 5, 2019

Walsh, Victoria, *Nigel Henderson: Parallel of Life and Art* (London: Thames & Hudson, 2001)

Ward, Janet, *Weimar Surfaces: Urban Visual Culture in 1920s Germany* (Berkeley, CA: University of California Press, 2001)

Wasensteiner, Lucy, *The Twentieth Century German Art Exhibition: Answering Degenerate Art in 1930s London* (Abingdon: Routledge, 2018)

Weber, Anne-Katrin, *Television before TV: New Media and Exhibition Culture in Europe and the USA, 1928–1939* (Amsterdam: Amsterdam University Press, 2022)

Webster, Wendy, *Mixing It: Diversity in World War Two Britain* (Oxford: Oxford University Press, 2018)

Welch, David, *Persuading the People: British Propaganda in World War II* (London: British Library, 2016)

Welch, David, *Protecting the People: The Central Office of Information and the Reshaping of Post-War Britain, 1946–2011* (London: British Library, 2019)

Williams, Raymond, *Culture and Society* (London: Chatto & Windus, 1958)

Williams, Raymond, *Marxism and Literature* (Oxford: Oxford University Press, 1977)

Williams, Raymond, *Television: Technology and Cultural Form* (London: Fontana, 1974)

Wolff, Janet, *The Aesthetics of Uncertainty* (New York: Columbia University Press, 2008)

Wollaeger, Mark, *Modernism, Media and Propaganda: British Narrative from 1900 to 1945* (Princeton, NJ: Princeton University Press, 2006)

Wollen, Peter, *Signs and Meaning in the Cinema* (London: Thames & Hudson, 1969)

Woodham, Jonathan M., *A Dictionary of Modern Design* (Oxford: Oxford University Press, 2004)

Woodham, Jonathan, 'Images of Africa and Design at the British Empire Exhibitions between the Wars', *Journal of Design History*, vol. 2, no. 1, 1989

Yusaf, Shundana, *Broadcasting Buildings: Architecture on the Wireless, 1927–1945* (Cambridge, MA: MIT Press, 2014)

Zanot, Francesco, 'The Film und Foto Exhibition of 1929', *Photoshow: Landmark Exhibitions that Defined the History of Photography* (London: Thames & Hudson, 2014)

Zervigón, Andrés Mario, 'The Peripatetic Viewer at Heartfield's Film und Foto Exhibition Room', *October*, no. 150, Fall 2014

Zimmerman, Claire and Victoria Walsh, 'New Brutalist Image 1949–55: "atlas to a new world" or, "trying to look at things today"', *British Art Studies*, issue 4, 2016

Index

Note: Page numbers in *italic* indicate illustrations

EU authorised representative for GPSR:
Easy Access System Europe, Mustamäe tee 50,
10621 Tallinn, Estonia
gpsr.requests@easproject.com

www.ingramcontent.com/pod-product-compliance
Ingram Content Group UK Ltd.
Pitfield, Milton Keynes, MK11 3LW, UK
UKHW061858100926
8556IPUK00009B/922
* 9 7 8 1 5 2 6 1 5 7 4 1 6 *